BIOLOGICAL NITROGEN FIXATION, SUSTAINABLE AGRICULTURE AND THE ENVIRONMENT

Current Plant Science and Biotechnology in Agriculture

VOLUME 41

Scientific Editor
R.J. Summerfield, *The University of Reading, Department of Agriculture, P.O. Box 236, Reading RG6 2AT, Berkshire, UK*

Scientific Advisory Board
J. Hamblin, *Research Director, Export Grains Centre Ltd., WA, Australia*
H.-J. Jacobsen, *Universität Hannover, Hannover, Germany*

Aims and Scope
The book series is intended for readers ranging from advanced students to senior research scientists and corporate directors interested in acquiring in-depth, state-of-the-art knowledge about research findings and techniques related to all aspects of agricultural biotechnology. Although the previous volumes in the series dealt with plant science and biotechnology, the aim is now to also include volumes dealing with animals science, food science and microbiology. While the subject matter will relate more particularly to agricultural applications, timely topics in basic science and biotechnology will also be explored. Some volumes will report progress in rapidly advancing disciplines through proceedings of symposia and workshops while others will detail fundamental information of an enduring nature that will be referenced repeatedly.

The titles published in this series are listed at the end of this volume.

Biological Nitrogen Fixation, Sustainable Agriculture and the Environment

Proceedings of the 14th International Nitrogen Fixation Congress

Edited by

YI-PING WANG

*College of Life Sciences,
Peking University,
Beijing, China*

MIN LIN

*Institute of Biotechnology,
Chinese Academy of Agricultural Sciences,
Beijing, China*

ZHE-XIAN TIAN

*College of Life Sciences,
Peking University,
Beijing, China*

CLAUDINE ELMERICH

*Institut des Sciences du Végétal,
CNRS-UPR2355, Gif-sur-Yvette and Institute Pasteur, Paris, France*

and

WILLIAM E. NEWTON

*Virginia Polytechnic Institute and
State University,
Blacksburg, VA, USA*

Library of Congress Cataloging-in-Publication Data

ISBN-10 1-4020-3569-1 (HB)
ISBN-10 1-4020-3570-5 (e-book)
ISBN-13 978-1-4020-3569-1 (HB)
ISBN-13 978-1-4020-3570-5 (e-book)

Published by Springer,
P.O. Box 17, 3300 AA Dordrecht, The Netherlands.

www.springeronline.com

Printed on acid-free paper

All Rights Reserved
© 2005 Springer
No part of this work may be reproduced, stored in a retrieval system, or transmitted
in any form or by any means, electronic, mechanical, photocopying, microfilming, recording
or otherwise, without written permission from the Publisher, with the exception
of any material supplied specifically for the purpose of being entered
and executed on a computer system, for exclusive use by the purchaser of the work.

Printed in the Netherlands.

14TH INTERNATIONAL CONGRESS ON NITROGEN FIXATION

was organized by

NATIONAL LABORATORY OF PROTEIN ENGINEERING
AND PLANT GENETIC ENGINEERING,
COLLEGE OF LIFE SCIENCES, PEKING UNIVERSITY
BIOTECHNOLOGY RESEARCH INSTITUTE, CHINESE ACADEMY OF
AGRICULTURAL SCIENCES

and supported by

Peking University
Ministry of Education, P. R. China
National High Technology Research and Development Program of China
(863 Program)
The National Key Fundamental Research Program of China (973 Program)
National Natural Science Foundation of China
UNESCO
National Laboratory of Protein Engineering and Plant Genetic Engineering
Chinese Society of Biotechnology
Chinese Society for Microbiology
Sichuan Habio Bioengineering Co., Ltd
Shenfeng Biotech Engineering Co., Ltd of Angang Industry

14TH INTERNATIONAL CONGRESS ON NITROGEN FIXATION

was organized by the following committees

National Organizers

National Laboratory of Protein Engineering and Plant Genetic Engineering,
College of Life Sciences, Peking University
Biotechnology Research Institute, Chinese Academy of Agricultural Sciences

National Organizing Committee

Y.P. Wang, College of Life Sciences, Peking University, Beijing, China
M. Lin, Institute of Biotechnology, CAAS, Beijing, China
Y.X. Jing, Institute of Botany, CAS, Beijing, China

National Program Committee

S.J. Shen (S.C. Shen), Shanghai
J.L. Li, Beijing
Q.R. Cai (K.R. Tsai), Xiamen
S.F. Chen, Beijing
W.X. Chen, Beijing
C. Ge, Beijing
J.F. Huang, Beijing
Y.X. Jing, Beijing
F.D. Li, Wuhan
J. Li, Beijing
J.D. Li, Beijing
Z.T. Li, Beijing
M. Lin, Beijing
Q.S. Ma, Nanning
Y.S. Qiu, Guangzhou
S.H. Shen, Beijing
Y.X. Zhu, Beijing

J.L. Tang, Nanning
Z.X. Tian, Beijing
Y.P. Wang, Beijing
B. Wu, Nanning
X.T. Wu, Fuzhou
L. Xu, Shanghai
J. Q. Xu, Changchun
S.S. Yang, Beijing
G.Q. Yu, Shanghai
Y.X. Yu, Chongqing
C.G. Zhang, Shenyang
X.P. Zhang, Chendu
Z.Z. Zhang, Shenyang
J.D. Zhao, Beijing
J.C. Zhou, Wuhan
J.B. Zhu, Shanghai

14ᵀᴴ INTERNATIONAL CONGRESS ON NITROGEN FIXATION

International Steering Committee

W.E. Newton, USA
R. Palacios, Mexico
C. Elmerich, France
H. Hennecke, Switzerland
B.E. Smith, UK

F.O. Pedrosa, Brazil
A. Pühler, Germany
I. Tikhonovich, Russia
T. Finan, Canada
Y.P. Wang, China

International Program Advisory Committee

N. Boonkerd, Thailand
W. Broughton, Switzerland
M. Buck, UK
E.C. Cocking, UK
F. Dakora, South Africa
H. Das, India
F. de Bruijn, France
R. Dixon, UK
Z.M. Dong, Canada
D. Emerich, USA
R. Fani, Italy
P. Gresshoff, Australia
R. Haselkorn, USA
R. Holm, USA
C. Kennedy, USA
A. Kolb, France
J.K. Ladha, Philippines
P. Ludden, USA

K.A. Malik, Pakistan
E. Martinez-Romero, Mexico
K. Minamisawa, Japan
M. O'Brian, USA
M. Rahman, Bangladesh
B. Rolfe, Australia
C. Ronson, New Zealand
T. Ruiz-Argueso, Spain
S.J. Shen (S.C. Shen), China
H.P. Spaink, The Netherlands
G. Stacey, USA
S. Tabata, Japan
Nguyen Thanh Hein, Vietnam
R. Thorneley, UK
C. Vance, USA
D. Werner, Germany
F. O'Gara, Ireland
K. Pawlowski, Germany

Greetings from Prof. San Chiun Shen

"Progress in Nitrogen Fixation Research relies on the integration of knowledge from different fields, plus scientific collaboration and a free exchange of views between scientists. Accumulation of research data over many years is the guarantee of innovation. To be at the top in science does not mean to be at the front of the stage. The best work will always be recognized, and will lead us all along the path of truth and understanding. This is the driving force of science."

National Laboratory of Plant Molecular Genetics, Institute of Plant Physiology and Ecology, Shanghai Institutes of Biological Sciences, Chinese Academy of Sciences, Shanghai, P. R. China

生物固氮研究的进展有赖于不同学科的整合，科学家间的交流与合作，创新源于多年工作的积累，批专不言个自成蹊，砥砺科学精神。

沈善炯 二〇四年五月

TABLE OF CONTENTS

PREFACE..xxiii

DEDICATION..1

In Memory of Professor Jia-Xi Lu
Xintao Wu..3

A Life Affair with Rhizobia – In Memory of Professor Chen Hua-Kui,
Academician of the Chinese Academy of Sciences
Fu-Di Li...7

KEYNOTE ADDRESS..11

Sustainable Agriculture/Forestry and Biological Nitrogen Fixation
*Dietrich Werner, Suresh K. Mahna, Vertica Mahobia, B.N. Prasad, Jose
M. Barea, Heidemarie Thierfelder, Peter Müller and Pablo Vinuesa*.......................13

**SECTION I. NITROGENASE: THE ORIGIN, THE ENZYMES,
AND THE CHEMISTRY**..19

On the Origin and Evolution of *nif* Genes
Renato Fani..21

Recent Progress in the Biochemistry of Mo-Nitrogenase
William E. Newton..25

MgATP Binding and Hydrolysis in Nitrogenase Catalysis
Sanchayita Sen and John W. Peters...29

Structural Modeling and Chemical Simulation of the FeMoco Active Center
of Nitrogenase
Shaowu Du, Qiutian Liu, Beisheng Kang, Changren Chen and Rong Cao..............35

A Non-Metal System for Nitrogen Fixation Utilizing Buckminsterfullerene
Yoshiaki Nishibayashi, Sakae Uemura, Shin-ichi Takekuma and Zen-ichi Yoshida.......39

Chemical Simulation of the Molybdenum Coordination Mode in the FeMo
Cofactor by a Series of Molybdenum- or Tungsten-Polycarboxylate Complexes
with Sulfide Bridges
Dong-Mei Li, Ji-Qing Xu, Yong-Heng Xing, Ying-Jie Lin and Xiao-Hua Zhou..........41

Activation of Homocitrate Ligand of FeMo-Cofactor as Mimicked by the Coordination of Homocitrato and its Homologues
Z.H. Zhou, S. Ya Hou, Z. Xing Cao, Y. Fu Deng, H. Lin Wan, and K. Rui Tsai............43

Poster Summaries………………………………………………………………….46

SECTION II.   GENETICS AND REGULATION IN HETEROTROPHS AND PHOTOSYNTHETIC BACTERIA……………………………………………51

Genetic Regulation of Nitrogen Fixation: Integration of Multiple Signals
Richard Little, Isabel Martinez-Argudo, Neil Shearer, Philip Johnson and Ray Dixon………………………………………………………………………….53

The Structural Basis for *nif* Gene Activation
S. R. Wigneshweraraj, P. C. Burrows, P. Bordes, J. Schumacher, M. Rappas, R.D. Finn, W.V. Cannon, X. Zhang and M. Buck…………………………………59

Heterocyst Differentiation and Nitrogen Fixation in *Anabaena*
Robert Haselkorn………………………………………………………………….65

Signal Transduction in Heterocyst Differentiation of the Cyanobacterium *Anabaena* sp. PCC 7120
Yinhong Zhao, Yunming Shi, Guohua Yang, Ying Zhang, Xu Huang and Jindong Zhao………………………………………………………………………...69

P_{II} Signal Transduction in Cyanobacteria: Novel Mechanisms of Global Nitrogen Control
Karl Forchhammer, Annette Heinrich, Nicole Kloft, Mani Maheswaran and Ulrike Ruppert…………………………………………………………………….73

Molecular Signal of Nitrogen Starvation, Insight from the Cyanobacterium *Anabaena* PCC 7120
Cheng-Cai Zhang, Sophie Laurent and Sylvie Bédu……………………………...77

Electron Transport to Nitrogenase in *Rhodospirillum rubrum*
Tomas Edgren and Stephan Nordlund……………………………………………79

Regulation of Nitrogen Fixation in the Photosynthetic Bacterium *Rhodospirillum rubrum*:   Roles of GlnD and P_{II} Proteins
Yaoping Zhang, Edward L. Pohlmann and Gray P. Roberts……………………..83

Functional Analysis of the NifA GAF Domain of *Azospirillum brasilense*:
Effect of Tyr-Phe Mutations in NifA and Interaction of the Mutated Proteins
with GlnB using the Yeast Two-Hybrid System
Sanfeng Chen, Li Liu, Xiaoyu Zhou, Claudine Elmerich and JiLun Li................87

Effects of 2-Oxoglutarate and the P_{II} Homologues $nifI_1$ and $nifI_2$ on Nitrogenase Activity
in Cell Extracts of *Methanococcus maripaludis*
J.A. Dodsworth and J.A. Leigh 89

Mechanism of Regulation of Nitrogen Fixation by NifLA in *Pseudomonas
Stutzeri* A1501
*Zhihong Xie, Jie Pan, Shuzheng Ping, Ming Chen, Guoying Wang, Yi Yang,
Min Lin and Claudine Elmerich*................91

Rhizobial FixL/FixJ System is a Paradigmn of the Two-Component
Signal Transducing Systems: A Two-Cylinder Reciprocating Engine Model
Of Phosphorylation Reactions
Hiro Nakamura................93

Poster Summaries................95

SECTION III. FROM GENOMES TO FUNCTION................109

A Report on the Genome of *Herbaspirillum seropedicae* Strain Z78
Fabio O. Pedrosa and Genopar Consortium................111

Functional Analysis of Genes of Unknown Functions in *Sinorhizobium
meliloti* 1021
*J. Cheng, J. Fowler, A. Cowie, R. Zaheer, C. Patten, C. Sibley, P. Chain,
Z. Yuan, C. Baron, T. Charles, B. McCarry, P. Summers, J. Xu, E. Weretilnyk,
G.B. Golding and T.M. Finan*................115

What Can Bacterial Genome Research Teach Us about Iron Uptake
in *Sinorhizobium meliloti*?
A. Pühler, A. Becker, J. Buhrmester, T.-C. Chao, S. Rüberg and S. Weidner................119

Structural and Functional Genome Analysis of *Lotus japonicus* and
Mesorhizobium loti
S. Sato, T. Kaneko, Y. Nakamura, E. Asamizu, T. Kato and S. Tabata................123

Analyzing a *Sinorhizobium meliloti* 1021 ORFeome in a Functional Genomics Platform
Brenda K. Schroeder, Brent L. House, Michael W. Mortimer, Scott C. Maloney, Casey A. Taylor, Kristel L. Ward, Hope T. Ziemkiewicz, Scott Clark, John J. Bovitz, Hao Jin, Svetlana Yurgel and Michael L. Kahn....................127

Functional Genomic Analysis of the SDR Family in *Sinorhizobium meliloti*
Sirin Adham, Asha Jacob, David Capstick, Trevor Charles...............129

Proteome Analysis on Bacteroid Differentiation of *Bradyrhizobium Japonicum* USDA 110
Shigeyuki Tajima, Le T-P Hoa, Rie Hamaguchi and Mika Nomura.................131

Global Changes in Gene Expression of *S. meliloti* 1021 Nodule Bacteria in *nif*A and *nif*H Mutant Background
Zhe-Xian Tian, Hua-Song Zou, Jian Li, Yuan-Tao Zhang, Guan-Qiao Yu, Silvia Rüberg, Anke Becker and Yi-Ping Wang.................133

Toward Deciphering the Genome of *Frankia alni* Strain ACN14a.
P. Normand., S. Felix, N. Alloiso, J. Marechal, C. Lavire, A.M. Berry, B.C. Mullin, J., Tomkins, N. Choisne, N. Demange, Y.C. Truong Cong, A. Coulloux, D. Vallenet, S. Cruveiller, C. Médigue.................137

Poster Summaries.................139

SECTION IV. ESTABLISHMENT OF SYMBIOSIS AND NODULE FUNCTION.................145

Cell Cycle and Symbiosis
Adam Kondorosi, José Maria Vinardell, Toshiki Uchiumi, Peter Mergaert, Eva Kondorosi.................147

Expression Pattern of DMI Genes in *Medicago* Nodules
Rossana Mirabella, Marijke Hartog, Carolien Franken, René Geurts and Ton Bisseling.................153

Actinorhizal Symbioses
Katharina Pawlowski, Anna Zdyb, Bettina Hause, Cornelia Göbe, Ivo Feussner, Kirill Demchenko.................157

Signaling for Nodulation in a Water-Tolerant Legume
Marcelle Holsters, Ward Capoen, Jeroen den Herder and Sofie Goormachtig.........161

Genetic and Molecular Analysis of Nod Factor Signalling in *Medicago truncatula*
F. Debellé, C. Bres, J. Lévy, B. Ben Amor, J.F. Arrighi, F. Maillet, J.M. Ane, C. Rosenberg, J. Dénarié, S. Shaw, G. Oldroyd, S. Long, R. Penmetsa, D. Cook, R. Geurts, T. Bisseling, G. Duc and C. Gough......................................165

Activation and Perception of Calcium Oscillations during *nod* Factor Signalling
Cynthia Gleason, Raka Mitra, Péter Kaló, Christine Galera, Clare Gough, Jean Dénarié, Sharon R. Long and Giles E.D. Oldroyd......................................169

Functional Genomics of the Regulation of Nodule Number in Legumes
Peter M. Gresshoff, Gustavo Gualtieri, Titeki Laniya, Arief Indrasumunar, Akira Miyahara, Sureeporn Nontachaiyapoom, Tim Wells, Bandana Biswas, Pick Kuen Chan, Paul Scott, M. Kinkema, M. Djordjevic, Dana Hoffmann, Lisette Pregelj, Diana M. Buzas, Dong Xi Li, Artem Men, Qunyi Jiang, Cheol-Ho Hwang and Bernard J. Carroll......................................173

"Activator" and "Inhibitor" Leading to Generation and Stabilization of Symbiotic Organ Development in Legume
Masayoshi Kawaguchi......................................179

Regulatory Mechanisms of SYMRK Kinase Activity
Satoko Yoshida and Martin Parniske......................................183

A Method to Enable Continuous Liquid Introduction into an Apoplast of Soybean-Plantlet Leaf and its Application to Elucidating the Supernodulation Trait of Nod1-3
Yasuhiro Arima, Hiroko Yamaya and Tadashi Yokoyama......................................187

Amino Acid Cycling by *Rhizobium leguminosarum* in Pea Nodules
James White, Alex Bourdes, Arthur Hosie, Seonag Kinghorn, Philip Poole......................................189

Root Nodule Extensins in Infection Thread Development
Elizabeth A. Rathbun and Nicholas J. Brewin......................................193

CASTOR and *POLLUX*, the Twin Genes that are Responsible for Endosymbioses in *Lotus japonicus*
Haruko Imaizumi-Anraku, Naoya Takeda, Martin Parniske, Makoto Hayashi, Shinji Kawasaki......................................195

The Expression of the *Rhizobium tropici guaB* Gene is Required in the Early Stages of Bean Nodulation
Mónica Collavino, Daniel H. Grasso, Pablo M. Riccillo and O. Mario Aguilar........199

Does the Rhizobial Type III Secretion System of Rhizobium Species
NGR234 Affect the Defence Response in Plant Leaves?
O. Schumpp, W.J. Deakin, C. Stahelin and W.J. Broughton..............................201

Mathematical Modeling for Nodule Size Distribution
Jun-ichi Ikeda, Ayako Fukunaga, Yuko Suga and Kaneaki Hori.........................203

Early Events in the Nodulation of *Casuarina glauca* by *Frankia*
*L. Laplaze, S. Svistoonoff, M. Obertello, B. Peret, F. Auguy, M.O. Sy,
V. Hocher, D. Autran, M. Nicole, C. Franche, D. Bogusz*..................................205

Gene Expression in the Root Nodules of *Elaeagnus umbellata*
*Chung-Sun An, Ho-Bang Kim, Sang-Ho Lee, Jung Jang Hyun, Chang-Jae
Oh, Hyoungseok Lee*..207

Diffusible Signal Factors in Nodulation of *Discaria trinervis* by *Frankia*
Luciano A. Gabbarini and Luis G. Wall..209

Cloning and Expression Studies of *fur*F and *cap* F in *Frankia* Strain R43
João Vieira, Pedro Moradas-Ferreira and Fernando Tavares...........................211

Characterization of Mutants from *Rhizobium* sp. NGR234 with Defective
Expolysaccharide Synthesis
*C. Staehelin, W. D'Haeze, M.Y. Gao, R.W. Carlson, B.J. Pellock, G.C. Walker,
W.R. Streit and W.J. Broughton*...213

Isolation of Two New *Sinorhizobium meliloti* Transcriptional Regulators
Required for Nodulation
*Li Luo, Shi-Yi Yao, Anke Becker, Silvia Rüberg, Guan-Qiao Yu, Jia-Bi Zhu and
Hai-Ping Cheng*..215

Rhizobial Control of Host-Specificity
*N.M. Boukli, W.J. Deakin, K. Kambara, H. Kobayashi, C. Marie, X. Perret,
A. Le Quéeré, B. Reuhs, M. Saad, O. Schumpp, P. Skorpil, C. Staehelin,
W. Streit and W.J. Broughton*...217

The *Sinorhizobium melilot nifA* Exerts its Signaling Effect for Nodulation in
the Host Plants
Huasong Zou, Guanqiao Yu and Jiabi Zhu..219

Evolutionary Stability of *Rhizobium* Mutualism Depends on Legume Host Sanctions
R. Ford Denison and E. Toby Kiers...221

Poster Summaries...225

SECTION V. TOWARDS SUSTAINABLE AGRICULTURE AND PROTECTION OF THE ENVIRONMENT...251

Ecological Significance of Lumichrome and Riboflavin as Signals in the Rhizosphere of Plants
Felix D. Dakora, Sheku Kanu and Viviene N. Maitru..253

Rhizospheric Plant-Microbe Interactions for Sustainable Agriculture
Kauser A. Malik, F.Y. Hafeez, M.S. Mirza, S. Hameed, G. Rasul and R. Bilal..257

Genetic Programs for Development of Nodules and Arbuscular Mycorrhiza in Legumes: Solid Facts and Unsolved Problems
A.Y. Borisov, V.E. Tsyganov, A.O. Ovtsyna, N.A. Provorov, and I.A. Tikhonovich...261

Microbial Inoculants: A Challenge to Tune the Microbial Metabolites and Signals for Plant Responsiveness in the Field
R. Remans, C. Snoeck, E. Luyten, S. Dobbelaere, E. Somers, A. Croonenborghs, J. Michiels and J. Vanderleyden..265

Are Legumes Doing Their Job? The Effect of Herbicides on N_2 Fixation in Southern Australian Agricultural Systems
Elizabeth A. Drew, Vadakattu VSR Gupta, David K. Roget...............................269

Non-Symbiotic Bacterial Diazotrophs in Crop-Farming Systems: Can Their Potential for Plant Growth Promotion be Better Exploited?
Ivan R. Kennedy, A.T.M.A. Choudhury, Mihály L. Keeskés, Rodney J. Roughley and Nguyen Thanh Hien...271

Effect of Legume Nodule Hydrogen Uptake Status on Hydrogen-Oxidizing Rhizobacteria and the Rotation of Crops
Zhongmin Dong..273

Phosphorous Use Efficiency for Symbiotic Nitrogen Fixation in Common Bean (*Phaseolus vulgaris*) and its Consequence on Soil P Dynamic
Jean-Jacques Drevon, Nora Alkama, Mathew Blair, Aurelio Garcia, German Hernandez, Philippe Hinsinger, Benoit Jaillard, Aline Lopez, Paula Rodino..........277

Mutational Analysis to Study the Role of Genetic Factors in Pea Adaptation to Stresses during Development its Symbioses with Rhizobium and Mycorrhizal Fungi
Viktor E. Tsyganov, Alexander I. Zhernakov, Anna V. Khodorenko, Pavel Y. Kisutin, Andrei A. Belimov, Vera I. Safronova, Tatyana S. Naumkina, Alexey Y. Borisov, Peter Lindlbad, Karl-Josef Dietz, Igor A. Tikhonovich.....................279

Effects of Waterlogging on Nitrogen Fixation of a Supernodulating Soybean Genotype, Sakukei 4
T. Matsunami, G.H. Jung, Y. Oki, W.H. Zhang and M. Kokubun.........................283

Effect of N Application Rate on Nitrogen Fixation and Transfer from Vetch to Barley in Mixed Stand
Hyowon Lee, Wongo Kim, Hyungsu Park and Sugon Kim................................285

Analysis of the Promotive Effects of Deep Placement of Slow Release Fertilizers on Growth and Seed Yield of Soybean by ^{15}N Dilution Method
Tewari Kaushal, Masaru Onda, Sayuri Ito, Akihiko Yamazaki, Hiroyuki Fujikake, Norikuni Ohtake, Kuni Sueyoshi, Yoshihiko Takahashi and Takuji Ohayama..287

Role of Rhizobial Endophytes as Nitrogen Fixers in Promoting Plant Growth and Productivity of Indian Cultivated Upland Rice (*Orysa sativa L.*) Plants
R.K. Singh, R.P.N. Mishra and H.K. Jaiswal...289

Growth, Photosynthesis, Nodule Nitrogen and Carbon Fixation in Desi and Kabuli Genotypes of Chickpea (*Cicer Arietinum L.*) Under Salt Stress
Garg Neera and Ranju Singla...293

Nodule Co-Occupancy of *Agrobacterium* and *Bradyrhizobium* with Potential Benefit to Legume Host
Sohail Hameed, Fathia Mubeen, Kauser A. Malik and Fauzia Hafeez...................295

Salt Stress Adaptation in Alfalfa Bacteroids: Importance of Protein Betaine
Karine Mandon, Alexandre Boscari, Jean Charles Trinchant, Laurence Dupont, Geneviève Alloing, Didier Héouart, and Daniel Le Rudulier....................297

Isolation of Genes for Salt Tolerance from *Sinorhizobium* LT11
W. Payakapong, P. Tittabutr, N. Teaumroong, D. Borthakur and N. Boonkerd..301

Regulation of Heme and Iron Metabolism in *Bradyrhizobium japonicum*
Jianhua Yang, Yali Friedman and Mark R. O'Brian......................................303

Rhizobial Genes Essential for Salt Tolerance
Ju Quan Jiang, Wei Wei, Li Shi Xie, Lei Wang and Su Sheng Yang......................307

Effect of Low-pH and Aluminum on the Nodulation Signal Transduction in *Medicago sativa*
Min Yang, Xiaofeng Li, Yongxiong Yu and Minghua Gu..................................309

Further Investigation of the Roles of Poly-3-Hydroxybutyrate (PHB) and Glycogen in *Sinorhizobium meliloti Medicago* sp. Symbiosis
Chunxia Wang, Marsha Saldanha, Xiaoyan Sheng, Kris Shelswell, Trevor C. Charles and Bruno W. Sorbral...311

Poster Summaries...313

SECTION VI. MOLECULAR ECOLOGY AND THE DISCOVERY OF NEW NITROGEN FIXERS..337

Diazotrophic Endophytes in Rice: Colonization and Nitrogen Fixation of *Herbaspirillum* and *Clostridium* Species
Kiwamu Minamisawa, Mu You, Tadashi Abe, Bin Ye, Asami Saito, Makoto Kawahara and Tadashi Sato..339

New Perspectives of Grass-Associated Nitrogen Fixation
B. Reinhold-Hurek, F. Battistoni, A. Krause, A. Ramakumar, F. Friedrich, A. Sarkar, M. Böhm, S. Gemmer, L. Miché, L. Zhang and T. Hurek......................345

Electron Donation and Alternative Nitrogenases in Cyanobacterial Dinitrogen Fixation
Hermann Bothe and Gudrun Boison..349

Role of Unicellular Diazotrophs in the Oceanic Nitrogen Fixation
Joseph P. Montoya..355

New Nitrogen-Fixing Microorganisms from the Oceans: Biological Aspects
and Global Implications
Jonathan P. Zehr, Barbara Methe and Rachel Foster......................................361

Biodiversity and Phylogeny of Rhizobial Germplasm in China
Wen Xin Chen, En Tao Wang and Wen Feng Chen..367

The Species Paradigm in Bacteriology: From a Cross-Disciplinary Species
Concept to a New Prokaryotic Species Definition (with Emphasis on
Rhizobia)
Kristina Lindström, Zewdu Terefework, Petri Auvinen, Lars Paulin, Stefan............373
Weidner, Anke Becker, Silvia Rüberg and Helge Gyllenberg

Direct Amplification of Rhizobial *nod*C Sequences from Soil Total DNA
and Comparison to *nod*C Diversity of Root Nodule Isolates
Sarita Sarita, Parveen K. Sharma, Ursula B. Priefer and Juergen Prell..............377

Recent Studies on the *Rhizobium*-Cereal Association
Frank B. Dazzo, Youssef G. Yanni, Rizk Rizk, M. Zidan, Abu-Bakr M.
Gomaa, Andrea Squartini, Yu-Xiang Jing, Feng Chi and Shi-Hua Shen................379

Ascending Migration of Endophytic Rhizobia from Roots to Leaves inside
Rice Plants
Feng Chi, Shi-Hua Shen, Hai-Ping Cheng, Yu-Xiang Jing and F.B. Dazzo............381

The Ecology of *Azorhizobium caulinodans* ORS571 Colonizing on Root
Surfaces of *Arabidopsis thaliana*
Taichiro Iki, Toshihiro Aono, Oyaizu Hiroshi...383

Evolutionary Implication of Nitrogenase-Like Proteins in the Plant Kingdom
and Prospects for *nif* Gene Transfer in Model Eukaryotes
Q. Cheng, J. Yang, A. Day, M. Dowson-Day and R. Dixon................................387

Novel *Mimosa*-Nodulating Strains of *Burkholderia* from South America
Euan K. James, Wen-Ming Chen, Sergio M. De Faria, Jean L. Simões-Araujo,
Roseangela Straliotto, Rosa M. Pitard, Jui-Hsing Chou, Yi-Ju Chou,
Edmundo Barrios, Alan R. Prescott, Janet I. Sprent, J. Peter W. Young..............391

Investigation Approach on Associative Diazotrophs in Plant Microecosystem with Multi-Field of View
Song Wei..395

Poster Summaries..398

AUTHOR INDEX..431

PREFACE

The 14th International Nitrogen Fixation Congress was held in Beijing, China from October 27th through November 1st, 2004. This volume constitutes the proceedings of the Congress and represents a compilation of the presentations by scientists from more than 30 countries around the World who came to Beijing to discuss the progress made since the last Congress and to exchange ideas and information. This year marked the 30th anniversary of the first Congress held in Pullman, Washington, USA, in 1974. Since then, this series of Congresses has met five times in North America (three in the United States and once each in Canada and Mexico), once in South America (Brazil), four times in Western Europe (once each in Spain, The Netherlands, Germany and France), once in Eastern Europe (Russia), and once in Australia; and now for the first time in Asia. China was a most appropriate choice because China is a big country with the largest population in the World, about 1.3 billion people, which is about 22% of the World's population. It is traditionally an agricultural country, even though China has only 7% of the available farming land. This situation explains why agriculture and its productivity are major issues for the Chinese people, its government and the scientists in the field.

There are many serious problems facing us in the World today: the global population is growing rapidly and most of the people in poverty are in developing countries; the increase in the grain-production rate on a global scale continues to decrease (from 3% in the 1960's to 2% in the 1980's and 1990's to an estimated 1.8% in 2010); and too much chemical fertilizer is being applied and causing even-more serious environmental problems. These continuing problems clearly indicate that the World needs to develop effective forms of sustainable agriculture that result in protection of our environment. These issues have led to increased attention, especially in Asia, Africa, and Latin America, to research on biological nitrogen fixation. As a result, China, for example, has launched highly effective research programs and provided significant support for all areas of nitrogen fixation. Therefore, to host this Congress was a very special occasion for the Chinese scientists who work in this area.

Spectacular progress has occurred in this field over the last 3 years and we were fortunate to have the major players in attendance at the Congress. The content of the final program involved extensive consultation with the members of the International Steering Committee and the International Program Advisory Committee. We thank them all for their very valuable input. Our program covers the most recent theories on how nitrogenase may have evolved, how the enzyme catalyzes substrate reduction and how it is regulated in response to molecular oxygen, fixed nitrogen, and light. It then considers the current advances in genomics, transcriptomics, and proteomics. Besides information on several rhizobium genomes whose sequencing was achieved some time ago, new information on the genome of endophytic bacteria, such as *Herbaspirillum* and *Azoarcus* as well as that of *Frankia*, was discussed at the Congress. The machinery involved in establishing a nitrogen-fixing symbiotic relationship and its requirements for effective functioning is extremely complex. Recent information on the common pool of genes required for nodule formation in legumes, efficient mychorrization, and actinorhizal symbiosis as well progress in the signal-transduction pathways leading to the organogenesis of the nodule was presented.

A large part of this volume reflects the concern of having a productive agriculture that will respect the environment. In particular, the reader will find information on the physiological and molecular mechanisms that enable adaptation to high saline soils, selection of the legumes genotypes with high productivity or with the super-nodulation phenotype, together with data on bioremediation of harmful chemicals and preparation and formulation of bacterial inoculants. Molecular ecology is another subject of importance, since most of soil, rhizosphere and endophytic nitrogen-fixing organisms remain to be discovered. However, a very large number of new nitrogen-fixing species have been discovered recently, and an up-to-date description of the taxonomy of rhizobia as well as molecular tools to study bacteria-plant interactions are described in this volume.

We want to express our sincere thanks to all those who worked hard and long to help us in the organization of this Congress. We are especially grateful to those people who stayed with us when we were confronted by the unexpected SARS outbreak in Beijing in 2003, which cost us a one-year postponement of the Congress. We also thank all our scientific colleagues, in particular Profs. Shanjiong Shen, Jilun Li, and Qirui Cai for their understanding and support. We want to emphasize our gratitude for the financial support that we received from Peking University, National Laboratory of Protein Engineering and Plant Genetic Engineering, The National Key Fundamental Research Program of China (973 program), The National High Technology Research and Development Program of China (863 program), the National Natural Science Foundation of China, UNESCO, Sichuan Habio Bioengineering Co., Ltd and Shenfeng Biotech Engineering Co., Ltd of Angang Industry.

Finally, we all thank Vicki Newton for her extraordinarily efficient expertise during the editing of this proceedings volume. We also thank Beiyan Nan for his assistance with computing and related problems. Most importantly, we want to thank our families and friends for their moral support during the last three years and all the participants for making this Congress such an enjoyable and interesting experience for us. We hope you enjoyed it as much as we did.

Yi-Ping Wang
Min Lin
Zhe-Xian Tian
Claudine Elmerich
William E. Newton

Beijing, November, 2004

14TH INTERNATIONAL CONGRESS ON NITROGEN FIXATION

DEDICATION

The editors dedicate this volume to

Jia-Xi Lu
and
Hua-Kui Chen

in acknowledgement of their contributions
to nitrogen fixation research in China

Professor Lu Jiaxi (1915-2001)

IN MEMORY OF PROFESSOR LU JIAXI

Wu Xintao
State Key Laboratory of Structural Chemistry, Fujian Institute of Research on the Structure of Matter, Chinese Academy of Sciences, Fuzhou, Fujian 350002, P. R. China.

Professor Lu was born of a scholarly family in 1915. In 1934, he graduated with a major in chemistry and then worked at Xiamen University as a lecturer for three years. In 1937, after passing a highly competitive examination, he received a special national postgraduate fellowship and then went to study physical chemistry at the Department of Chemistry, University College, London, U.K. After two years under the supervision of Professor S. Sugden, a research fellow of the Royal Society, he obtained his Ph.D. degree at the very young age of 24. Recommended by Professor S. Sugden in the autumn of 1939, he was admitted to the California Institute of Technology (Caltech) in the USA and became engaged in the study of structural chemistry under the guidance of Linus Carl Pauling (1901–1994), the laureate of the 1954 Nobel Prize in chemistry and also the 1962 Nobel Peace Prize. This study started a very important period in Professor Lu's career as he re-oriented his own research toward structural chemistry.

During his stay in the USA, Professor Lu took part in many significant research activities involving molecular and crystal structures. In 1944, he was awarded a prize for R&D achievement by the US National Defense Research Committee for his meritorious service in the area of combustion and explosion research.

In the winter of 1945, he returned to his homeland despite the aftermath of World War II and was appointed as Professor of Chemistry and concurrently Dean of the faculty at Xiamen University. In 1955 and 1956, he was elected a Chinese Academy of Sciences founding member and First-class Professor, becoming one of the youngest CAS members and senior higher-learning lecturers in the country.

Professor Lu was an erudite scholar, an inspiring teacher, and an eloquent lecturer. He was also well versed in communication as an instructive tutor. His teaching style was always to explain the underlying concepts in simple terms, such that his teaching was attractive, enchanting, and full of charm, wit, and humor so that it was an enduring and memorable experience for his students. According to him, "One would not be a good teacher if he cannot bring up several students who excel himself."

Professor Lu explored science ceaselessly and innovatively. During his stay in England (1937–1939) as a young post-graduate in preparation of his doctorate dissertation under the tutelage of Professor S. Sugden, his major was in radioactive chemistry, particularly on artificial radioactivity. He was the first chemist to successfully prepare highly concentrated radioactive halogens for quantitative research and later became one of China's pioneering nuclear chemists.

At Caltech, he participated in a series of research projects to determine the structures of urea-hydrogen peroxide addition compounds, S-N and As-S cluster compounds, and polysulfide molecules. The multi-centered bonding characteristics in the clusters of S, N, and other non-transitional elements evoked great interest for him and later, as part of his domestic work, he extended the concept to the chemical modeling of the biologically active center of nitrogenase and transition-metal cluster compounds. In developing new methods for structural analysis, he devised a Lu's Diagram, the *Lp* factor's reciprocal in Weissenberg's photography in the equi-inclination. This method was widely applied in the international community of structural analysts before the emergence of computerized technology.

Around 1973, Professor Lu proposed his *nido*-cubane-like "string-bag" model for the biologically active center of nitrogenase, the FeMo-cofactor (FeMoco). This model involves an incomplete cubane-like [$MoFe_3S_3$] cluster core. In 1992, two US research teams jointly established the structure crystallographically showing that the FeMoco contains a double incomplete Mo-Fe-S cubane cluster, indicating that Professor Lu's incomplete cubane-like "string-bag" model does exhibit the basic features of the FeMoco.

The Fujian Institute, with Professor Lu as its founding father, succeeded in synthesizing and characterizing hundreds of cluster compounds of various types. Among the most important of these are the trinuclear molybdenum clusters and chain-type Mo-Fe-S clusters that won the 1986 CAS Awards for Science and Technology Progress and the 1989 CAS Awards for Natural Sciences. At this time, he proposed the *"Unit Construction"* concept for rational syntheses of cluster compounds and the *quasi-aromaticity* concept for trinuclear incomplete cubane-type transition-metal clusters. Due to his outstanding contributions to the field, he was granted both the first-class prize from the 1991 CAS Awards for Natural Sciences and the second-class prize from the 1993 National Awards for Natural Sciences.

Professor Lu was also a pioneer and explorer in the field of high-tech crystalline materials. Under his leadership, the Fujian Institute pinpointed opto-electric and non-linear optic crystal materials and laser devices as its main R&D directions. As a result, a series of new high-tech crystal materials were developed by Chinese scientists. The BBO and LBO crystals were renowned as the best ultra-violet frequency-doubling materials by the international community. These inventions were recognized with a special prize from the CAS Awards for Science and Technology Progress, a first-class prize from the National Awards for Inventions, and a chemistry prize granted by the Third World Academy of Sciences.

Professor Lu devoted his whole life to science, education, directing R&D undertakings, and academic administrations. Thus, it was only rational and natural that he assumed the presidency of the Chinese Academy of Sciences in the 1980s. In 1999, he was honored with a prize for science and technology achievements by Hong Kong's Ho Leung & Ho Lee Foundation, becoming one of only two mainland Chinese chemists to have won such a special honor.

Professor Lu was the pioneer and founder of structural chemistry and cluster chemistry in China. He trained hundreds of students and researchers and was an outstanding educationist in China. He had not only made great contributions to science but also played an important role in defining scientific policy in China. His noble personality, outstanding achievements, and indelible contributions to Chinese science and technology will never be forgotten, and he will live forever in the hearts of the Chinese scientists and people.

Professor Chen Hua-Kui (1914－2002)

A LIFE AFFAIR WITH RHIZOBIA – IN MEMORY OF PROFESSOR CHEN HUA-KUI, ACADEMICIAN OF THE CHINESE ACADEMY OF SCIENCES

Li Fu-Di
Huazhong Agricultural University, Shizishan Street, Hongshan District, Wuhan, Hubei 430070, PR China.

Chen Hua-Kui（1914－2002）was born in Beijing. He was educated at Peking University, from where he graduated with a bachelor's degree in 1936. One year later, he went to England to do research on rhizobia for a doctorate at Rothamsted Experimental Station. He was supervised by Dr. Henry Gerard Thornton, F.R.S. and was enrolled at the University of London, where he received PhD degree in 1939. Dr. Chen came back to China in 1940. He served briefly at the Agricultural Institute of Qinghua University, Kunming, and later he worked (1941－early1946) at the Central Institute of Agricultural Experimentation, Zhongqing. His work focused on biological nitrogen fixation and green manure. After the World War II, he served as Professor at Peking University and was appointed founding head of the Department of Soil Science, which was the first one of this discipline in China. He was invited as professor to Wuhan University late in 1947, where he founded the Department of Agricultural Chemistry and was appointed its head. After the reformation of the education system of China in 1952, Dr. Chen became Professor at Huazhong Agricultural College (now University) and head of the Department of Soil and Agricultural Chemistry. He was nominated president of the college for 1979－1984. Professor Chen was elected Academician of the Chinese Academy of Sciences in 1980.

Professor Chen was a pioneer in the study of soil microbiology in China. His research interest covered a wide spectrum of topics, including material cycling in paddy soil and nitrogen fixation. He devoted all 60 years of his scientific career to the study of rhizobia, producing a large number of important findings and publications. Here, only part of his work on rhizobia is highlighted.

Symbiosis in legumes was an active area of research at Rothamsted Experimental Station when Chen started his creative research. He was the first to confirm that root hair curling with bacteria-free culture filtrates involved a plant auxin, which led to his first publication (Chen (1938) Nature, 142, 753). The fact that strains of nodule bacteria differed in their ability to benefit host legumes was realized as early as the 1890s. However, before the 1930s, there were few publications in which the anatomy and cytology of nodules produced by different strains had been compared. Chen and Thornton studied the relationship between structure and function of nodules produced on soybean, pea and clover by effective and ineffective strains. They found that the amount of nitrogen fixed by a nodulated plant was a function of (i) the number of nodules, (ii) the volume of bacteroid-containing tissue in each, and (iii) the time for which this tissue persisted (Chen and Thornton (1940) Proc. Roy. Soc. Lond. Ser.B, 129, 208). This concept has been widely used in studies both of the relative effectiveness in nitrogen fixation of various combinations of host-plant and rhizobial strains and of the structural basis for genetic defects in the N_2-fixing capacity of nodules (see Bergersen (1974 in Quispel "The Biology of Nitrogen Fixation", p. 474).

On returning from overseas, Chen continued research on rhizobia and on other aspects of soil microbiology. He isolated for the first time in pure culture the microsymbiont of *Astragalus sinicus* (Chen and Shu (1944) Soil Sci., 58, 291). This plant is an important winter-growing green manure crop, widely grown in the rice fields of southern China and the Yangtze River Valley. Initially, a tentative name *Rhizobium astragali* was assigned according to the concept of 'cross inoculation'. It was reassigned by his former student, Chen Wen-Xin, as *Rhizobium huakuii* (Chen et al. (1991) IJSB, 41,275), and later was transferred into the genus *Mesorhoizobium* (Jarvis et al. (1997) IJSB, 47, 895). A series of publications on its biology, ecology, physiology and genetics have been produced. The research has also led to the application of inoculants in huge areas of China. The work on differentiation and viability of nodule bacteria in host cells was granted a first award by the China State Ministry of Agriculture. It clarified the long controversy in opinion on the viability and reproduction of bacteroids by showing that the highly morphologically differentiated cells in indeterminate nodules of *Astragalus sinicus*, clover and alfalfa were not able to replicate, whereas most of the cells showing no apparent morphological differentiation in determinate nodules of soybean were able to replicate (Cao et al. (1984) Scientia Sinica, 27, 593).

Chen paid great attention to the application of rhizobia to agriculture and its ecology. Since the middle of the 1960s, there has been a dramatic expansion in the use of *A. sinicus* as a green manure for rice fields in China. In a field on which *A. sinicus* has never been sown, there are no or very few compatible rhizibia. The effective inoculation of rhizobia becomes the key for successful growth of *A. sinicus* in such soils. According to Chen's suggestion, a workshop for manufacture of inoculants was set up on the campus in 1965. Peat-based inoculants were distributed to provinces of central and southern China. A total of 8 million hectares of fields growing *A. sinicus* were inoculated in Hubei Province alone between 1966 and 1975. Repeated experiments and extensive surveys showed that inoculation of efficient rhizobia was a factor for good yield of *A. sinicus*. Most of the inoculated fields produced more than 3000kg fresh shoots per hectare, whereas uninoculated fields never attained this level of productivity, with more than half of them below 1500kg per hectare. The use of nitrogen fertilizer was greatly reduced and yields of rice were increased. However, two problems arose in practice. Inoculation did not always improve the growth of *A. sinicus* in fields where it had been grown repeatedly for several years, as is the case of inoculation to soybeans. Some effective strains lost the ability to nodulate the host plant after being cultured for many generations on media, particularly when it contained gentian violet, which was sometimes added to prevent contamination by spore-forming bacteria.

Competition for nodulation between inoculants and indigenous rhizobia has been known for a long time, but it has not been solved despite a great deal of research done on it. Chen and his group attached importance to the ecological side, and considered that the distribution of rhizobial cells in soil might be a key factor affecting competitive nodulation in the soil. A concept called the "location effect" was proposed and then confirmed by a series of experiments using immunofluorescent techniques. It was suggested that for the inoculated rhizobia to overwhelm the indigenous population and to form most nodules, the inoculants have to be distributed throughout the plough layer, so as to have every chance to infect the root hairs of the host plant as the roots develop. This concept is useful for the improvement of inoculation techniques.

In the early 1970s, when *Rhizobium* plasmids were still unknown, Chen's group presumed that the high frequency of the loss of nodulation ability was probably attributed to its genetic determinant not being located on the chromosome. After the "cultural revolution", Chen recommended me to go England to work with Professor Sir John Beringer at Rothamsted. Since then, the genetics of rhizobia has been the main priority of our research with work conducted on both biodiversity and molecular genetics.

A large number of rhizobial isolates from nodules on *Astragalus sinicus* were collected by Chen's group. In one survey, two thousand isolates were obtained from 52 ecologically different sites in seven southern provinces of China. From them, the genetic diversity of 204 strains was investigated by using REP-PCR fingerprinting, RFLP analysis of PCR-amplified 16S rDNA genes, and 16S-23S rDNA intergenic spacers. Most strains (78.4%) belonged to different 16S rDNA genotypes from that of the *Mesorhizobium huakuii* type strain. When data for both 16S and 23S rDNA PCR-RFLP of twelve representative strains were analyzed, seven different genotypes were detected, and no strain shared a genotype with the *M. huakuii* type strain, but they were clustered within the *Mesorhizobium* branch (Zhang et al. (1999) Syst. Appl. Microbiol., 22, 312).

These rhizobia harbour 1-5 plasmids, among which there is a sym plasmid (Zou et al. (1997) Curr. Microbiol., 35, 215). *M. huakuii* 7653R was used to identify the *nod* gene organization by using *M. meliloti* 2011 *nod* mutants. The common nod genes, *nod*D, *nod*A, *nod*B, and *nod*C, were identified by heterologous hybridization and sequence analysis and a genetic map of the nodulation region was constructed (Zhang et al. (2000) AEM, 66, 2988). The organization of common *nod* genes on plasmids is different from that of other rhizobial species. *Nod*A is separated from *nod*BC by a distance of approximately 22 kb, and a 107-bp truncated *nod*B-like sequence is found downstream of *nod*A. The two operons are divergently transcribed, which is different from the *nod* gene organization of *R. etli, M. lot*i, and *Mesorhizobium* sp. strain N33 in which *nod*A, B, and C genes are transcribed in the same direction. To determine whether the separation of *nod*A from *nod*BC was unique to 7653R or was also present in other *A. sinicus* isolates, twenty-four representative strains, which represented different genotypes and geographic origins, were studied. Using primers to amplify the 2.0kb *nod*DBC region, no polymorphism was found in the size of PCR products, which were the same as those obtained from strain 7653R. Sequence analysis of *nod*A and 16S rDNA for seven strains that represent seven genotypes showed them to be identical to strain 7653R. These data indicate that the separation of *nod*A from *nod*BC is a general feature of *A. sinicus* rhizobia despite high genetic diversity. This finding suggests that the *nod* gene organization of these rhizobia is conserved in contrast to their chromosomal diversity.

Professor Chen was a famous educationist, esteemed mentor and admired teacher. He founded four departments (Soil Science, Agricultural Chemistry, Soil and Fertilizer, and Microbiology) successively in different universities. He also set up the Ministry Key Laboratory of Agricultural Microbiology affiliated with Huazhong Agricultural University. Chen was the first to enroll graduates majoring in soil microbiology in China early in the 1950s, and PhD candidates were enrolled in the 1980s, when the degree system was set up in China. Microbiology teachers of other agricultural universities nation-wide came to his department for refresher courses. His book, "Soil Microbiology", which was published in 1957, was the first microbiology textbook to appear in China. He and Professor Fan Qing-Sheng of Nanjing Agricultural University, as co-editors-in-chief, compiled the book "Microbiology" in 1959 for undergraduates of agricultural universities, which replaced the then-used textbook adopted from the ex-USSR. Chen had disciples everywhere in the country; among them are quite a number of distinguished scientists, including three academicians of the CAS and CAE. The quality and eminence of the scientists that he trained will be his greatest memorial.

Professor Chen was an indefatigable and passionate worker until his demise at the age of 88. He was always willing to help others and to advise young researchers. He enjoyed a high reputation as a man of science. During his long and distinguished career, he was appointed to many scientific advisory bodies. He was Vice-Chairman of Chinese Society for Microbiology, Vice-Chairman of Chinese Society of Agronomy, Chairman of Chinese Society of Soil and Fertilizer, member of Academic Degree Committee of State Council, and member of Scientific Committee of State Ministry of Agriculture. He was a selected representative of 3rd, 5th, and 6th PRC National People's Congress. Professor Chen has passed away, but he will always be remembered by his students, colleagues and friends.

KEYNOTE ADDRESS

SUSTAINABLE AGRICULTURE/FORESTRY AND BIOLOGICAL NITROGEN FIXATION

Dietrich Werner[1], Suresh K. Mahna[2], Vertica Mahobia[2], B. N. Prasad[3], Jose M. Barea[4], Heidemarie Thierfelder[1], Peter Müller[1], and Pablo Vinuesa[5]
[1]Fachbereich Biologie der Philipps Universität Marburg, Germany. [2]Botany Department, University of Ajmer, India. [3]Biotechnology Department, Tribhuvan University, Kathmandu, Nepal. [4]Estacion Experimental del Zaidin, Granada, Spain. [5]CIFN, UNAM, Cuernavaca, Mexico.

The biological nitrogen cycle is closely linked to the carbon cycle and both are affected by human activities. The pools and annual turnover rates of the global carbon cycle (in Gigatonnes C) reveal that the major pools, the atmosphere (750 Gt), the biomass of land plants (590 Gt), and the usable fossil energy sources (1150 Gt), are all of the same order of magnitude. Significantly larger are the pools of organic soil carbon (1560 Gt), the soil carbonates (1740 Gt), and the carbon in the oceans (38,000 Gt). The lithosphere carbon pool (more than 6×10^7 Gt) is several orders of magnitude larger.

With respect to the first mentioned pools, the annual turnover rates are quite significant with 120 Gt gross from photosynthesis and 60 Gt transferred to organic soil carbon. The fossil-energy use is around 6.5 Gt per year, about 67% of which go to increase the atmospheric CO_2 content and about 33% to the oceans. It is astonishing that the comparatively small amount of 4 Gt C per year is 50 % of the global change challenge by green house gases (Werner, 1998).

In contrast to the carbon cycle, we are absolutely certain that global nitrogen fixation does not affect the nitrogen pools in either the atmosphere or the lithosphere. This situation is quite different when we look at the nitrogen fertilizer use per ha and year (Kawashima et al., 2000). Western Europe, China, Bangladesh and Japan use more than 100 kg ha^{-1} year^{-1} in cereal production, a rate that significantly affects the nitrogen balance of soils and surface waters. Fertilization with ammonia also leads to an increase in soil acidity, which is a problem by itself, with high-acid soils prevalent in large areas of South America, the eastern North America, in Central Africa, and in South-East Asia.

Progress in understanding the molecular basis of the interactions of microsymbionts and host plants for nitrogen fixation is covered in all contributions of this conference and in these proceedings. Here, only some recent results on compartmentation, the most complex part of endosymbiosis are mentioned (Werner, 2004). In the genome of *Bradyrhizobium japonicum* with 9.1 Mbp, the symbiosis island comprises only about 5% of the genome. But very important genes, which are responsible for the central process of compartmentation of bacteroids within stable symbiosome membranes, such as the signal-peptidase genes *sipS, sipF and sipX*, are located outside this region (Bairl and Müller, 1998; Müller, 2004). Also, new species have been detected in this genus (Vinuesa et al. 2004; Vinuesa and Silva, 2004).

Furthermore, results known for a long time await further explanation. We always assume that the legume-rhizobia symbiosis is based on the carbon-nitrogen exchange in the established symbiosis, which is certainly the case. But this mutualistic exchange is not available during the first ten days of the interaction. Why then are the first steps of the interaction successful? One explanation is that we have here a parasitic phase of the rhizobia, living from the plants carbon sources. But, outside the plant in the rhizosphere, we also have a large supply of organic carbon sources. So, the rhizobia could stay outside the plants and still participate in the carbon sources from host plants. But in the root hairs of soybeans compared to the root hairs of wheat, there are other nutrients in much higher concentrations, e.g., calcium, iron, cobalt and molybdenum, which are essential elements for rhizobia (Table 1). So the rhizobia could, in the first stages of infection, find a source of these elements in their host with the carbon-nitrogen exchange following later.

Table 1. Calcium, iron and cobalt accumulation in root hairs of soybean (*Glycine max*). (Werner et al., Z. Naturforsch. 40c, 912, 1985)

Element	Ppm (dry matter)			
	Soybean root hair	Soybean root	Wheat root hair	Wheat root
K	11740 ± 2450	12840 ± 2640	4670 ± 1010	4780 ± 990
S	530 ± 165	560 ± 170	180 ± 55	190 ± 60
Fe	414 ± 138	31 ± 5	120 ± 35	44 ± 26
Co	7.9 ± 3.8	0.88 ± 0.4	2.6 ± 0.8	1.3 ± 1.1
Ca	2000 ± 460	287 ± 70	246 ± 60	280 ± 70
Mo	3.1 ± 0.5	5.4 ± 0.7	0.6 ± 0.12	0.5 ± 0.3

Soybeans are the dominant grain legume worldwide, with a production of about 176 million t, which is more than all other legume grain crops combined (Table 2). Besides beans, soybeans are also the legume crop with the longest history; they were cultivated 3000 years ago in the Zhou dynasty in China. Today, more than 20.000 landraces have been developed, but only around 400 are in use (Carter et al., 2004). The yield increase in soybeans production is impressive, especially in Brazil. In the last 35 years, it has increased from around 200 kg per ha in 1968 to 2800 kg per ha in 2003 (Hungria et al., 2005).

Table 2. World production
(FAO Yearbook, 2001; D. Werner, in D. Werner and W.E. Newton (eds.) Nitrogen Fixation: Agriculture, Forestry, Ecology and the Environment, Kluwer, 2005)

Crop	Million tons
Soybeans	176.6
Groundnuts	35.1
Dry beans (*Phaseolus* and *Vigna*)	16.8
Dry peas	10.5
Chickpeas	6.1
Dry faba beans	3.7
Lentils	3.1
Green beans	4.7
Green peas	7.1

Equally obvious is the geographical increase in the soybean-producing counties in the USA. There were only a few dots on the map in 1927, and now there is an almost continuous cultivation area from the Carolinas over the Midwest States to the Dakotas (Pueppke, 2005). Increases in soybean seed yield in field experiments, using *B. japonicum* HN 32 as an inoculant, are less spectacular, but nevertheless significant. For 17 locations in three provinces of China, the yield increase was 10-13% over the uninoculated control (Table 3).

Table 3. Increases in soybean seed yield in field experiments
using *B. japonicum* HN32 as inoculant.
(J.E. Ruiz Sainz et al., in D. Werner and W.E. Newton (eds.) Nitrogen Fixation: Agriculture, Forestry, Ecology and the Environment, Kluwer, 2005)

Year	Province in China	Number of locations used	Acreage used in the field experiments	Yield increase (%) over the uninoculated control
1991	Heilongjiang	5	25.665	10.1
1993	Heilongjiang	9	18.678	11.8
1995	Heilongjiang	1	2.666	11.7
1995	Sichuan	1	2.000	13.0
1996	Guangxi	1	11.200	11.7

When we go from agriculture to forestry, we have to look to the actinorhizal trees and shrubs. There, we still have the situation that the reported field data for nitrogen fixation vary widely, even within one species. For *Ceanothus velutinus*, fixation rates vary between 4 and 100 kg N per ha per year; for *Alnus rubra*, the rates are beween 85 and 320 kg N per ha per year (Table 4). These variations are much larger than those reported for grain and fodder legumes in agriculturally used fields. The more homogenous soil and water conditions, together with the more homogenous seed population and crop protection, are the main reasons for this different situation.

Table 4. Rates of symbiotic N_2-fixation of some actinorhizal trees and shrubs at field level.
(R.O. Russo, in D. Werner and W.E. Newton (eds.) Nitrogen Fixation: Agriculture, Forestry, Ecology and the Environment, Kluwer, 2005)

Actinorhizal species	Kg N . ha^{-1} . yr^{-1}
Alnus acuminata (A. jorullensis)	279
Alnus glutinosa	40-53
Alnus rubra	85-320
Casuarina equisetifolia	12-110
Coriaria arborea	192
Ceanothus velutinus	4-100

When we move further from cultivated land to other ecosystems, we have to realize that, even in legumes with around 17,000 species, only a small percentage has been studied in terms of ecology, symbiotic interactions, physiology or molecular biology. In the subfamily Caesalpinioideae, nodulation is rare and nodule structures are usually primitive. In the Mimosoideae, nodulation is common but we know important exceptions and, in the Papilionoideae, nodulation is very common with some exceptions (Sprent, 2005).

The effect of plant breeding on the symbiotic interactions with nitrogen-fixing microorganisms is especially obvious with *Azospira oryzae* and rice. The maximum dilution factor that allows recovery of bacterial endophytes is 4 or 5 orders of magnitude larger for wild rice and traditional rice compared to modern rice cultivars (Hurek and Reinhold-Hurek, 2005). Progress in the future for the whole field covered by this paper will probably continue more along the lines of slow, but successful, improvements of currently used crops and cultivars than on any promise from the area of genetic engineering. However, on the other hand, a much better understanding of the mechanisms and the functions of these systems will certainly come from molecular biology.

Some open questions raised during the last years are:

- Can signalling compounds in the rhizosphere limit nodulation and nitrogen fixation?
- How can improved symbiotic characters be best integrated into breeding programmes for legumes?
- How to utilize "new" nitrogen fixing crops, e.g., *Sesbania*?
- What about ground water quality and nitrogen fixation?
- How best to use agroforestry, legume trees and actinorhiza trees (CIFOR, ICRAF, IUFRO)?

- What about soil conservation and nitrogen-fixing trees and deep rooting annual legumes?
- What are the possible contributions by transgenic plants?
- What about the relationship with other symbioses (*Frankia*, arbuscular mycorrhiza)?
- How to maximize competitiveness and inoculum efficiency?

We thank the EU for the Project INCO-DEV ICA4 –CT-2001-10057

References
Bairl A and Mueller P (1998) Mol. Gen. Genet. 259, 161-171.
Carter TE (2004) in Biological Resources and Migration (ed. D Werner) Springer Berlin, Heidelberg, New York.
Hurek T and Reinhold-Hurek B (2005) in Nitrogen Fixation: Agriculture, Forestry, Ecology and the Environment (eds. D Werner and WE Newton) Kluwer Dordrecht, in press.
Kawashima H et al. (2000) in Biological Resource Management-Connecting Science and Policy (eds. E Balazs et al.) Springer Berlin, Heidelberg, NewYork.
Mueller P (2004) Mol. Genet. Genomics 271, 359-366.
Pueppke St (2005) in Nitrogen Fixation: Agriculture, Forestry, Ecology and the Environment (eds D Werner and WE Newton) Kluwer, Dordrecht, in press.
Russo RO (2005) in Nitrogen Fixation: Agriculture, Forestry, Ecology and the Environment (eds D Werner and WE Newton) Kluwer, Dordrecht, in press.
Ruiz-Sainz JE (2005) in Nitrogen Fixation: Agriculture, Forestry, Ecology and the Environment (eds D Werner and WE Newton) Kluwer, Dordrecht, in press.
Sprent J (2005) in Nitrogen Fixation: Agriculture, Forestry, Ecology and the Environment (eds D Werner and WE Newton) Kluwer, Dordrecht, in press.
Vinuesa P et al. (2004) IJSEM , online Sept. 24 th.
Vinuesa P and Silva C (2004) in Biological Resources and Migration (ed. D Werner) Springer, Berlin, Heidelberg, New York.
Werner D et al. (1985) Z. Naturforschung 40c, 912-913.
Werner D (1998) Natuerliche CO_2 Senken. Report, Marburg.
Werner D (2004) in Plant Surface Microbiology (eds. A Varma et al.) Springer, Berlin, Heidelberg, New York.
Werner D (2005) in Nitrogen Fixation: Agriculture, Forestry, Ecology and the Environment (eds D Werner and W E Newton) Kluwer, Dordrecht, in press.

SECTION I.

NITROGENASE: THE ORIGIN, THE ENZYMES, AND THE CHEMISTRY

ON THE ORIGIN AND EVOLUTION OF *NIF* GENES

Renato Fani
Dip.to di Biologia Animale e Genetica, Via Romana 17-19,
I-50125 Firenze, Italy.

The development up of metabolic pathways represents a crucial step in molecular and cellular evolution. If Oparin's idea on the origin of life is correct, the exhaustion of the prebiotic supply of amino acids, bases, and other compounds must have imposed an increasing selective pressure, which favored those primordial heterotrophic cells that became capable of synthesizing molecules whose concentration was decreasing in the environment. Thus, the emergence of biosynthetic pathways allowed primitive organisms to become less dependent on exogenous sources of organic compounds. How did these metabolic abilities originate? This is still an open question, but several theories have been proposed to account for the establishment of anabolic routes. The "patchwork" hypothesis (Jensen, 1976) is supported both by comparisons of genes and genomes and by "directed evolution experiments". According to this hypothesis, metabolic pathways may have been assembled through the recruitment of primitive enzymes that could react with a wide range of chemically related substrates. Such relatively slow, non-specific enzymes may have enabled primitive cells that contained small genomes to overcome their limited coding capabilities. Either single or multiple duplication events of the ancestral genes, followed by evolutionary divergence of the copies produced, led to the appearance of enzymes with narrowed and refined substrate specificity. Thus, in the course of molecular evolution, duplication of DNA regions may have played a key role in both increasing the gene loading and shaping metabolic pathways. It is quite possible that most of the main metabolic pathways were built up in the DNA world before the appearance of the Last Common Ancestor (LCA).

It is still under debate as to whether nitrogen fixation, the biological conversion of atmospheric N_2 to ammonia, was one of the metabolic abilities of the LCA. Very little is known about the origin and evolution of *nif* genes, as well as about the molecular mechanisms, such as paralogous DNA duplication and/or domain shuffling, which may have been involved in shaping the nitrogen-fixation process. To understand both the time frame and the complex genetic events that have marked the history of nitrogen fixation and to try to reconstruct the evolutionary history of *nif* genes, a detailed analysis of their products has been undertaken.

1. A cascade of gene and operon duplications interconnecting nitrogen fixation and bacterial photosynthesis: *nifDKEN* and *bchnNBYZ*

Detailed analysis of *nif*-gene products revealed that *nifDK* and *nifEN*, which encode the α and β subunits of nitrogenase and the two subunits of the NifNE complex involved in FeMo-cofactor biosynthesis, respectively, belong to the same paralogous gene family (Fani et al., 2000). This analysis also allowed their possible evolutionary history to be traced. According to the proposed model, the four genes are the result of two successive duplication events (Figure 1). The first in-tandem duplication event involved an ancestral gene that encoded a slow enzyme with low substrate specificity that was able to catalyze several reactions. This enzyme might have been either a nitrogenase or a detoxyase, depending on the composition of the early atmosphere (either neutral or reducing, respectively). The first duplication step led to a bicistronic operon that, in turn, underwent a paralogous operon duplication event that gave rise to the ancestors of the present-day *nifDK* and *nifEN* operons. According to this model, these duplication events occurred in the early stages of cellular evolution before the appearance of the LCA. The two operons were then maintained in archaeal and bacterial diazotrophs.

The remarkable sequence similarity found for the so-called "Chlorophyll iron protein" subunits and the nitrogenase proteins (Burke et al., 1993) suggests an evolutionary link between nitrogen fixation and bacteriochlorophyll biosynthesis. Two of the key enzymes involved in the latter pathway, the protochlorophyllide reductase and the chlorin reductase, are complexes, each of which is composed of three different subunits encoded by *bchLNB* and *bchXYZ*, respectively. A previous analysis revealed that the six *bch* genes form three paralogous pairs (*bchL-X, bchN-Y, bchB-Z*) and that BchL exhibited a high degree of sequence similarity to the NifH protein. The comparison of the amino acid sequences of all the available NifDK, NifEN, BchNB, and BchYZ proteins revealed that they share a degree of sequence similarity that was sufficiently high to suggest that the corresponding genes belong to a large and complex paralogous gene family, i.e., they have a common ancestry, so strongly supporting the evolutionary link between nitrogen fixation and bacterial photosynthesis. Even though the pathway leading to the extant *nifDKEN* and *bchNBYZ* genes is still unclear, there is a tentative plausible scenario, which relies on the fact that, in some photosynthetic bacteria, the two triads of genes (*bchLNB* and *bchXYZ*) are organized in tricistronic operons resembling the *nifHDK* organization. Assuming that these nine genes form three triads of paralogous genes (*nifH-bchL-bchX*; *nifD-bchN-bchY* and *nifK-bchB-bchZ*), it can be surmised that an ancestral *nifHDK* operon duplicated to give rise to *bchLNB*. Then, the *bchLNB* operon duplicated and produced *bchXYZ*. Because no photosynthetic archaeon has been discovered up to now, the operon duplication events might have occurred in the bacterial lineages after the divergence of Bacteria and Archaea from the LCA.

2. A paralogous gene family interconnecting nitrogen fixation with both lysine and leucine biosynthesis; *nifV, leuA, lys20*

The idea that the ability to fix N_2 is an ancient phenotype and is interconnected with other basic metabolic routes also finds support from the analysis of homocitrate synthase (HCase), the product of *nifV* gene. This enzyme catalyzes the condensation between acetyl CoA and α-ketoglutarate (α-KG) to give homocitrate (the organic moiety of the nitrogenase Fe-Mo cofactor); it shares a high degree of sequence similarity both with α-isopropylmalate synthase (α-IPMase), encoded by *leuA*, and with the product of *lys20*, another HCase. The Lys20 protein and α-IPMase represent the first enzymes in the AAA pathway for lysine biosynthesis and for leucine biosynthesis.

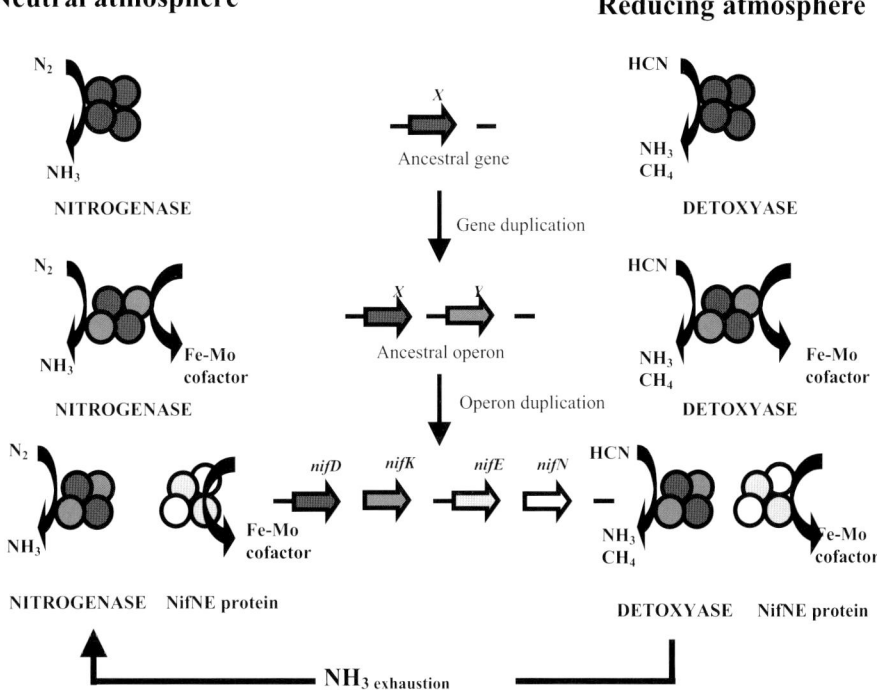

Figure 1. Origin and evolution of *nifD*, *nifK*, *nifE* and *nifN* genes (adapted from Fani et al., 2000).

The *nifV*, *lys20*, and *leuA* genes are the outcome of duplications of an ancestral gene, which most probably coded for an enzyme that was able to react with both a-KG and α-ketovalerate (α-KV), the substrate of the extant αIPMase, in agreement with the

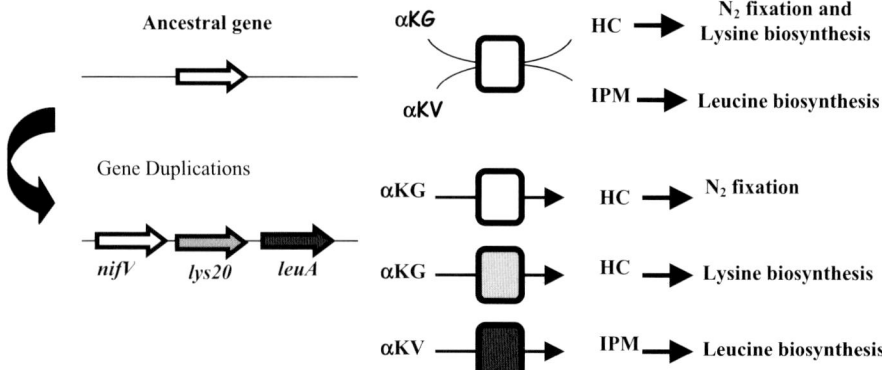

Figure 2. A possible pathway for the evolution of *nifV*, *leuA*, and *lys20*.

"patchwork" hypothesis (Figure 2). HC was then channeled into both lysine biosynthesis and N_2 fixation, whereas α-IPM was directed towards leucine biosynthesis. Different duplication events led to the narrowing of substrate specificity.

3. Conclusions

The development of nitrogen fixation represented an ancient metabolic innovation that was not only crucial for the extant life, but was critical during early cellular evolution as the prebiotic supply of fixed-nitrogen sources decreased. The ancestral *nif* pathway might have originated and developed early in cellular evolution with the entire process carried out by a limited number of *nif* genes encoding multifunctional, unspecific enzymes that could react with a wide range of related substrates. These primordial enzymes interconnected nitrogen fixation with other metabolic routes, such as bacterial photosynthesis and biosynthesis of leucine and lysine. Paralogous gene and operon duplications, gene recruitment and elongation events, and an extensive horizontal transfer of *nif* genes played crucial roles in shaping the entire pathway that was complete before the appearance of the LCA. It is likely that, in the LCA cell population, the *nif* genes coded for mono-functional enzymes and were organized in operons. After the divergence of Bacteria and Archaea from the LCA, the *nif* pathway was retained in some prokaryotic phylogenetic lineages, lost in most of them, and horizontally transferred to others (Fani et al., 2000; Fani, 2004) giving rise to the extant scenario in which the *nif* genes appear to be randomly distributed in the prokaryotic world.

Burke DH (1993) Proc. Natl. Acad. Sci. USA 90, 7134-7138.
Fani R et al. (2000) J. Mol. Evol. 51, 1-11.
Fani R (2004) in Microbial Evolution: Gene establishment, survival and exchange. Miller RV and Day MJ (Eds), ASM, Washington DC, 67-81.
Jensen RA (1976) Ann. Rev. Microbiology 30, 409-25.

RECENT PROGRESS IN THE BIOCHEMISTRY OF Mo-NITROGENASE

William E. Newton
Department of Biochemistry, Virginia Polytechnic Institute & State University, Blacksburg, VA 24061, USA.

1. Introduction.
Mo-nitrogenase consists of two components; the Fe protein with two MgATP-binding sites and a single [4Fe-4S] cluster and the MoFe protein with two pairs of two different metalloclusters, FeMoco and the P-cluster. The Fe protein is a specific reductant for the MoFe protein, which contains the site(s) of substrate reduction. Three-dimensional structures of the components, various complexes of the two proteins, and variant MoFe proteins are all known. The Fe protein structures are the subject of the following contribution (Sen and Peters, this volume). Mo-nitrogenases from a variety of genera exhibit a high level of primary sequence identity and, except for *C. pasteurianum*, they form catalytically active heterologous enzymes.

Mo-nitrogenase catalyzes the biological nitrogen-fixation reaction and also the

$$N_2 + 8\,H^+ + 8\,e^- + 16\,MgATP \rightarrow 2\,NH_3 + H_2 + 16\,MgADP + 16\,P_i$$

reduction of other small-molecule "alternative" substrates, all of which have the same requirements as N_2 reduction. Most notable is acetylene, which is reduced to ethylene. The iso-electronic carbon monoxide (CO) is not a substrate but is a potent inhibitor of all nitrogenase-catalyzed substrate reductions, except for the reduction of H^+.

2. The MoFe protein-Fe Protein Complex.
During turnover, the Fe protein, as the specific electron donor to the MoFe protein, first associates with the MoFe protein, which contains the substrate-reduction site. It is likely that electrons are initially transferred from the [4Fe-4S] cluster of the Fe protein to the MoFe protein's P cluster; they then traverse the 15Å distance to the FeMoco, where substrate is bound and reduced. It is still unclear how, when, and where the eight electrons necessary for the reduction of each N_2 are stored within the MoFe protein and how the required protons are delivered to this site.

Because the Fe protein is similar to "nucleotide switch" proteins, like ATPases and GTPases, similar trapping techniques, using AlF_4^- together with ADP, have been applied to produce a stable nitrogenase complex. The 3Å-resolution crystal structure of the resulting 2:1 Fe protein-MoFe protein complex, which may approximate the transition state for inter-component electron transfer, confirmed predictions of earlier modeling studies. The complexed Fe protein, with all subunits having an associated Mg^{2+}-ADP-AlF_4^- moiety, showed a significant conformational change, which involved a 13° rotation of the subunits, resulting in a more compact structure. In contrast, the MoFe protein was little changed from its uncomplexed structure (1). The two proteins dock along the Fe protein's two-fold symmetry axis, which bisects its single [4Fe-4S] cluster, and the pseudosymmetric αβ-interface of the MoFe protein. The more-compact Fe protein buries its [4Fe-4S] cluster in the protein-protein interface about 14Å from the P-cluster. The resulting arrangement supports the proposed electron-transfer pathway.

Two other structures of a 2:1 Fe protein/MoFe protein complex have been solved. The first structure involved a variant (L127Δ) Fe protein, which has Leu-127 deleted from the switch-II region to mimic nucleotide binding, and is, therefore, in the equivalent of a permanently MgATP-bound state. As such, the L127Δ Fe protein forms a tightly bound inactive 2:1 complex with the MoFe protein (2), which resembles that of the AlF_4^--stabilized complex as far as the MoFe-protein component and the protein-protein interfaces are concerned. However, a more open conformation is adopted by the complexed L127Δ Fe protein, which resembles the conformation of the uncomplexed nucleotide-free, rather than the complexed nucleotide-bound, native Fe protein.

The native Fe protein in the second 2:1 complex structure, which resulted from chemical cross-linking the two native component proteins through Glu-112 and βLys-400, adopted a structure that is even more open than when it is uncomplexed (3). In contrast, the structure of the MoFe-protein component is again effectively unchanged. What is changed, however, is the relative orientation of the two components and the presence of a completely different interface area. It's possible that the cross-linked complex represents an "initial encounter" state that then proceeds through a series of conformational changes on the Fe protein to reorient the Fe protein in the "electron-transfer competent" state, which is likely represented by the AlF_4^--stabilized complex.

3. The MoFe Protein – Structure and Reactivity.
A major result of the structural investigations was the determination of the exact composition and distribution of the FeMoco and P-cluster clusters within the MoFe protein (4,5). One pair of prosthetic groups resides within each αβ-subunit pair and is separated by about 70Å from the other pair. The FeMoco is buried within the α-subunit and is covalently bound to only two amino-acid residues (αCys-275 and αHis-442) and has no close involvement with the β-subunit. In contrast, the P cluster is located at the interface of the α- and β-subunits with each subunit providing an equal number of ligating Cys residues. This arrangement was the basis of the MoFe protein often being

treated as a dimer of dimers with each αβ-dimer functioning independently of the other, however long-range interactions between Fe-protein binding sites do occur (6).

3.1. Is there a central atom in FeMoco and, if so, what is its nature?

FeMoco is (or contains) the substrate-binding and -reducing site(s) and consists of two sub-clusters, one [Mo-Fe$_3$-S$_3$] and one [Fe$_4$-S$_3$], that are bridged by three non-protein-based sulfides (7). Recently, a very high resolution structure (8) has provided evidence for a single light atom, suggested (but not proven) to be nitrogen, possibly as nitride (N^{3-}), within the central cavity of the FeMoco and equidistant to all six of the central Fe atoms. Both its exact nature and its function are still uncertain; does it serve either a mechanistic or a structural role or both?

Spectroscopic ($^{14/15}$N ESEEM and ENDOR) probes of this question were based on the hypothesis that, if central atom has a mechanistic role (e.g., it was part of a recently fixed N$_2$), it should exchange during turn over under ^{15}N$_2$. However, neither the ^{14}N ESEEM/ENDOR spectra changed nor were any new ^{15}N species detected after such turnover (9). Unfortunately, this negative result can be interpreted as: there is no central atom; if it is there, it is not nitrogen; or, if it is nitrogenous, it does not exchange with N$_2$. We have also started to address this difficult question using ^{57}Fe nuclear resonance vibrational spectroscopy (NRVS). To simplify the analysis, we have compared the spectra of ^{57}Fe-labeled samples of holo-MoFe protein, apo-MoFe protein (with no FeMoco), and isolated FeMoco to identify FeMoco-specific modes. The FeMoco spectrum is dominated by a band at 190 cm^{-1}, which is also present in the holo-protein spectrum. The spectra of simple Fe-S compounds have no similar 190 cm^{-1} band, which is too high in frequency to be a Fe-S-Fe bending mode. However, the spectrum of [Fe$_6$N(CO)$_{15}$]$^{3-}$ (10) does exhibit such a band with significant Fe motion. Preliminary simulation of the FeMoco spectrum, using a simple 20-atom Fe$_8$S$_9$NY$_2$ model (Y= end capping ligands), also shows the 190 cm^{-1} band, which is lost when the central N atom is removed from the simulation. We believe these data are the first supporting evidence for the presence of a centrally located light atom in FeMoco (11).

3.2. Is there any evidence for where substrates bind? Fe or Mo?

Theoretical studies and model-compound chemistry have produced plausible models that support N$_2$ binding to either Mo or Fe (12). Direct studies of the reactivity of Mo-nitrogenase also remain ambiguous in answering this question. Several recent probes have focused on the relationship of catalyzed N$_2$ reduction and C$_2$H$_2$ reduction. Using an altered MoFe protein, with Ser replacing Gly at position 69 in the α-subunits (αSer-69), C$_2$H$_2$, which normally acts as a non-competitive inhibitor of N$_2$ reduction, was now a competitive inhibitor (13). These data were rationalized by invoking two C$_2$H$_2$–reducing sites for native MoFe protein; a high affinity site (K_m = 0.7%) that binds C$_2$H$_2$ only; and a low-affinity site (K_m = 14%) that also binds N$_2$. In the αSer-69 MoFe protein, the high affinity site has been eliminated by the substitution, so is unavailable to C$_2$H$_2$, which now has to compete directly with N$_2$. Because αGly-69 is near a 4Fe-4S face of FeMoco and at the end of a short helix that connects FeMoco to the P-cluster, this 4Fe-4S face was proposed as binding both N_2 and C_2H_2. Earlier data showed two (at least) C_2H_2-binding sites on the native MoFe protein (14) but with quite different K_m values (ca. 0.1% and ca. 1% C_2H_2), plus an additional substrate-inhibition site populated at ca. 20% C_2H_2.

We have further probed this question using C_2D_2 (15), which is reduced to both *cis*-$C_2D_2H_2$ (>90%) and *trans*-$C_2D_2H_2$ by native MoFe protein, and found that the stereoselectivity of protonation of C_2D_2 is affected by: (a) C_2D_2 concentration; (b) electron flux; (c) the presence of N_2; and (d) the pH. The results again suggest two acetylene-binding sites; a low-affinity site (K_m = 1%) that produces *trans*-$C_2D_2H_2$, and a high-affinity site (K_m = 0.1%) that produces only *cis*-$C_2D_2H_2$. Because added N_2 only inhibits *cis*-$C_2D_2H_2$ production, it likely binds to the high-affinity site. Further, when we repeated these measurements at pH 8, the substrate-inhibition phenomenon disappeared and instead, a third acetylene-binding site with a K_m >10% was activated. These results contrast with those cited above. First, the magnitudes of the kinetic constant, K_m, are different. Their higher-affinity site may reflect the average of our individual K_m values. Second, their lower-affinity site could be the "third" site that we observe at pH 8, possibly formed from the substrate-inhibition site. If so, this site would be invisible with native nitrogenase at pH 7 because of substrate inhibition. Third, our results favor N_2 binding at a higher-affinity C_2H_2-binding site rather than a lower-affinity site. Otherwise, how could you explain the partial (and competitive) inhibition by N_2 of acetylene reduction if N_2 binds only at the lower-affinity C_2H_2-binding site? These differences cast doubt on the physiological relevance of the αSer-69 MoFe protein results and their support of the 4Fe-4S face hypothesis as the site(s) of substrate reduction.

I thank the NIH (DK-37255) for support of our work.

4. References.
1. Schindelin H et al. (1997) Nature 387, 370-376.
2. Chiu H-J et al. (2001) Biochemistry 40, 641-650.
3. Schmid B et al. (2002) Biochemistry 41, 15557-15565.
4. Howard JB and Rees DC (1996) Chem. Rev. 96, 2965-2982.
5. Kim C and Rees DC (1992) Nature 360, 553-560.
6. Maritano S et al. (2001) J. Biol. Inorg. Chem. 6, 590-600.
7. Chan MK et al. (1993) Science 260, 792-794.
8. Einsle O et al. (2002) Science 297, 1696-1700.
9. Lee H-I et al. (2003) J. Am. Chem. Soc. 125, 5604-5604.
10. Della Pergola R et al. (1996) J. Chem. Soc. Dalton Trans., 747-754.
11. Xiao, Fisher, George, Smith, Newton, Sturhahn, Alp, Yoda, Cramer, unpublished.
12. Seefeldt L et al. Biochemistry (2004) 43, 1401-1409.
13. Christiansen J et al. (2000) J. Biol. Chem. 275, 11459-11464.
14. Davis L et al. (1979) Biochemistry 18, 4860-69; J. Bacteriol. (1980) 141, 1230-38.
15. Han J and Newton WE (2004) Biochemistry 43, 2947-2956.

MgATP BINDING AND HYDROLYSIS IN NITROGENASE CATALYSIS

Sanchayita Sen and John W. Peters
Department of Chemistry and Biochemistry and the Thermal Biology Institute, Montana State University, Bozeman, MT.

Nucleotide binding and hydrolysis is a fascinating aspect of the nitrogenase-catalyzed conversion of N_2 to ammonia. During catalysis, the iron protein and molybdenum-iron protein associate and dissociate in a manner that hydrolyzes two molecules of MgATP and transfer of at least one electron to the MoFe protein. Multiple cycles of iron protein association and dissociation, MgATP hydrolysis, and electron transfer are required for the complete reduction of a single molecule of nitrogen to ammonia (1, 2). There are a number of aspects of nitrogenase structure/function that are interesting areas of fundamental research. Nitrogenase can be considered an ideal model system for the study of the complex metal cluster mediated catalysis, electron transfer, complex metal cluster assembly, protein-protein interactions, and nucleotide-dependent signal transduction. In addition, the involvement of MgATP in nitrogenase catalysis is similar to the role of nucleotides in a large class of nucleotide binding proteins that couple nucleotide binding and hydrolysis to protein conformational changes transduced within a macromolecular assembly. Members of the class include G proteins, Ras p21, RecA, elongation factor Tu, myosin, and transducin, making the role of MgATP binding and hydrolysis one of the most fascinating aspects of nitrogenase research. Over the past decade, several structures of nitrogenase have been determined representing defined states presumed to mimic conformations along the pathway of catalysis. The following contribution summarizes our recent studies aimed at gaining insights into the MgATP bound or "on state" of the nitrogenase Fe protein in the context of previous structural work concerning nucleotide interactions and nitrogenase complex formation. These structures provide the basis for proposing a hypothesis concerning the set of dynamic conformational changes or conformational states that define the role of MgATP binding and hydrolysis in nitrogenase catalysis.

1. Approach
To a large extent, our approach at gaining insights into the role of MgATP binding and hydrolysis involves the use of x-ray crystallography. This approach offers the detailed

structural analysis of static states of macromolecules and researchers must be creative to gain useful and interpretable information about dynamic processes. With respect to the enzyme nitrogenase, a wealth of information concerning the biochemical properties of nitrogenase has been compiled over many years. These results have been exploited in determining the structures of nitrogenase in defined states that presumably approximate conformations that occur along the pathway of catalysis.

2. Nitrogenase Complex Structures

The first structure of the as isolated state of the nitrogenase Fe protein reported by the Rees group in 1992 was a significant advancement and provided the groundwork for numerous mutagenesis studies probing the role of MgATP binding and hydrolysis in nitrogenase catalysis (3). It was however not until 1998, when the structure of the *A. vindlandii* nitrogenase complex was first revealed, that many of the implications concerning nucleotide interactions, protein-protein interactions, and protein conformational change started to be fully recognized (4). This work involved applying an observation that had been recognized for several of the nucleotide signal transduction proteins that tetrafluoroaluminate together with MgGDP/MgADP act as a transition state analog (5-7). In the case of nitrogenase incubation of the nitrogenase components (Fe protein and MoFe protein) with either MgATP and tetrafluoraluminate under turnover conditions of simply tetrafluoraluminate and MgADP under nonturnover conditions a trapped complex between the two nitrogenase components is stabilized (8, 9). Some of the key observations that resulted from the determination of the structure of this complex include: (i) the structural affirmation and detailed description of an electron transfer pathway from the [4Fe-4S] cluster of the Fe protein to the FeMo-cofactors of the MoFe protein via the MoFe protein P cluster as intermediates; (ii) the detailed ion-pairing, hydrogen bonding and hydrophobic interactions that occur between the Fe protein and MoFe protein within the complex; (iii) the details of nucleotide interaction at the sites for MgATP binding and hydrolysis within the Fe protein; and (iv) protein conformational change within the Fe protein. Aspects concerning the latter of these observations were probably the most unexpected in the structure. It was observed that each Fe protein underwent a significant conformational change involving the rigid body reorientation of each subunit of the Fe protein homodimer. This conformational change has been described as a ~13° rotation of each monomer toward one another resulting in ore compact structure. This resulted in a translation of the [4Fe-4S] cluster of the Fe protein such that it was allowed to approach the electron acceptor, the P cluster. Although conformational change had been predicted for the Fe protein from previous spectroscopic and biochemical studies the mode and manner of such conformational changes had not been predicted (10, 11).

3. Nucleotide Bound Conformations of the Fe Protein

In general, nucleotide-dependent signal transduction involves the modulation of the nucleotide binding protein between an "on state" and an "off state". Each state is defined by either the interaction of the protein with nucleoside diphosphate "off state" or nucleoside triphosphate "on state" with transition between the two states resulting from either nucleotide displacement (MgADP/MgGDP for MgATP/MgGTP) or nucleotide hydrolysis (MgATP/MgGTP to MgADP/MgGDP and inorganic phosphate). The "on state" usually signals the formation of some macromolecular complex and the transition to the "off state" or nucleoside triphosphate hydrolysis results in dissociation of the macromolecular complex. For nitrogenase, the binding of MgATP by the Fe protein results in the formation of a macromolecular complex with the MoFe protein in which intermolecular electron transfer occurs. In terms of structures, the description of the "on state" (MgATP bound) and "off state" MgADP bound in addition to the aforementioned complex structures serve to provide a more complete understanding of the protein conformational change that occurs during nitrogenase catalysis. Our research over the last several years has been targeted at structurally characterizing these states.

The first structure in this regard that was determined in the lab was the structure of the MgADP-bound "off state" of the Fe protein from *A. vinlandii* (12). The structure revealed that, in the presence of bound MgADP, the overall conformation of the Fe protein is very similar to the conformation observed for the Fe protein in the absence of nucleotides or the partial occupancy nucleotide observed in the first structure. There were a number of more subtle differences in the conformation of various regions of the protein involved in nucleotide binding and the structure of the MgADP-bound "off state" state, together with the structure of the nitrogenase complexes, solidified our understanding of communication between the nucleotide-binding site and the site of protein-protein interaction (Fe protein–MoFe protein complex formation) via the Switch I region and the [4Fe-4S] cluster via Switch II region.

With respect to getting an overall view of the cycle at this stage, the remaining element was to structurally define the MgATP-bound state. As mentioned, there is considerable biochemical and spectroscopic evidence which suggest that the Fe protein undergoes conformational changes during nucleotide displacement (either MgATP for MgADP or nucleoside triphosphate hydrolysis (10, 11)). We have recently determined the structure of a site-direct variant of the Fe protein that has a number of biochemical and spectroscopic features in the absence of nucleotide that resemble the native Fe protein in the presence of bound MgATP (13). The detailed description of the conformational differences between the Fe protein within the nitrogenase complex, the Fe protein in the MgADP-bound "off state" and the MgATP-bound "on state" allows the full inference of the conformational changes that occur during catalysis and provides tremendous insight into understanding the role of MgATP binding and hydrolysis in nitrogenase catalysis.

Trapping the Fe protein in a MgATP-bound state for crystallography has proven to be a difficult task. Soaking native Fe protein crystals with MgATP results in destruction of the crystals which is not a surprising result given significant conformational changes are anticipated. Another approach involves using analogs of MgATP that do not hydrolyze readily, for example, the common analogs β-γ-methylene ATP, ATPγS, or AMPPMP. These offer a sound approach at defining the nucleoside triphosphate conformation although with the caveat of these analogs having slightly differing pKa values for the ionizable groups. We anticipate that this approach will yield a structure soon and, toward these ends, we have been able to generate reproducible crystals of the Fe protein in the presence of Mg^{2+} and β-γ-methylene ATP. In advance of a structure using these approaches, we have been able to determine the structure of a mutant of the Fe protein possesses a number of the biochemical and spectroscopic features of the native Fe protein with bound MgATP, in the absence of any bound nucleotides. This site-direct variant of the Fe protein, characterized extensively by Seefeldt and coworkers (14), differs from the native Fe protein by a single amino-acid deletion in each of the dimer Fe protein Switch II regions. Interestingly, it was observed that the variant forms a tight binding complex with the MoFe protein in the absence of nucleotides and the subsequent structural description of the complex revealed a very similar structure to the aforementioned structure of the nitrogenase complex stabilized in the presence of MgADP and tetrafluoroaluminate.

The overall structure of the Fe protein variant reveals a strikingly different protein conformation than observed previously for the native state, the MgADP bound state, and the Fe protein conformation observed within the nitrogenase complex structures (Figure 1).

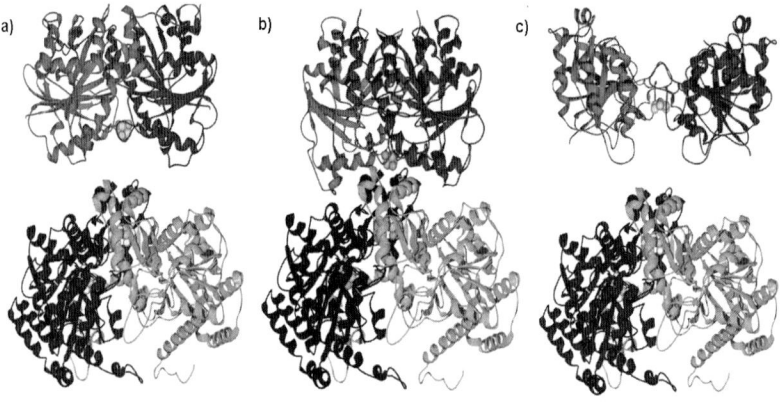

Figure 1. Fe protein MoFe protein interaction models. a) the structure of the MgADP bound Fe protein with an αβ-dimer of the MoFe protein tetramer (12) b) the structure of one-half of the nitrogenase complex (4) and c) the structure of the Leu 127deletion variant with an αβ-dimer of the MoFe protein tetramer.

The most significant structural differences that separate the Fe protein variant and the aforementioned structures of the Fe protein are largely manifested in a rigid body reorientation of the Fe protein subunits with respect to one another. Each of the Fe protein subunits is reoriented by a rigid-body rotation with respect to one another on the order of sixty-to-eighty degrees relative to the previously characterized Fe protein structures. We are moving forward with the working hypothesis that the structure of the deletion variant either resembles or provides a mimic at some level of the MgATP-bound "on state" of the Fe protein. We are therefore using this structure to rationalize a docking model (Figure 1C) to understand: (i) the conformational changes that occur during nitrogenase catalysis; and (ii) potential interactions that provide the trigger for MgATP hydrolysis and transition to the "off state". Having this structure in hand, we have the means to develop methods and approaches to probe the relationship between this variant as a structural mimic of the MgATP-bound "on state" and the actual MgATP-bound state which is more difficult to probe crystallographically.

Acknowledgements
This research is supported by the National Institutes of Health (NIH RO1 GM068553-01), the United States Department of Agriculture (2001-35318-10122), and the NASA-supported MSU Thermal Biology Institute (NAG5-8807). J.W.P. is recipient of a Camille Dreyfus Teacher/Scholar Award.

1. Drennan CL and Peters JW. (2003) Curr Opin Struct Biol 13, 220-226.
2. Rees DC and Howard JB. (2000) Curr Opin Chem Biol 4, 559-566.
3. Georgiadis MM et al. (1992) Science 257, 1653-1659.
4. Schindelin H et al. (1997) Nature 387, 370-376.
5. Sondek J et al. (1994) Nature 372, 276-279.
6. Fisher AJ et al. (1995) Biochemistry 34, 8960-8972.
7. Coleman DE et al. (1994) Science 265, 1405-1412.
8. Duyvis MG et al. (1996) FEBS Lett 380, 233-236.
9. Renner KA and Howard JB (1996) Biochemistry 35, 5353-5358.
10. Howard JB and Rees DC (1994) Annu Rev Biochem 6, 235-264.
11. Seefeldt LC and Dean DR (1997) Acc Chem Res 30, 260-266.
12. Jang SB et al. (2000) Biochemistry 39, 14745-14752
13. Sen S et al. (2004) Biochemistry 43, 1787-1797
14. Ryle MJ and Seefeldt LC (1996) Biochemistry 35, 4766-4775.

STRUCTURAL MODELING AND CHEMICAL SIMULATION OF THE FeMoco ACTIVE CENTER OF NITROGENASE

Du Shaowu, Liu Qiutian, Kang Beisheng, Chen Changren and Cao Rong
State Key Laboratory of Structural Chemistry, Fujian Institute of Research on the Structure of Matter, Chinese Academy of Sciences, Fujian, Fuzhou 350002, P. R. China

The mechanism of dinitrogen activation and reduction by the FeMoco in nitrogenase has been a hot subject in the fields of the molecular biochemistry and bioinorganic c chemistry. Numerous theoretical calculations and structural models have been proposed for the active FeMoco center during last two decades. Among these, structural models called Fuzhou Models (see Figure 1) suggested by Prof. Lu Jiaxi are the earliest cluster models (1). The two Fuzhou Models both contain $MoFe_3S_3$ "string-bag" units, which are indeed similar to the $MoFe_3S_3$ subunit in the FeMoco (2). Theoretical calculations showed that the nitrogen molecule was most likely to penetrate into these "string-bag" units and subsequently be activated by the synergistic action of the four metal centers in the way shown in Figure 1(a).

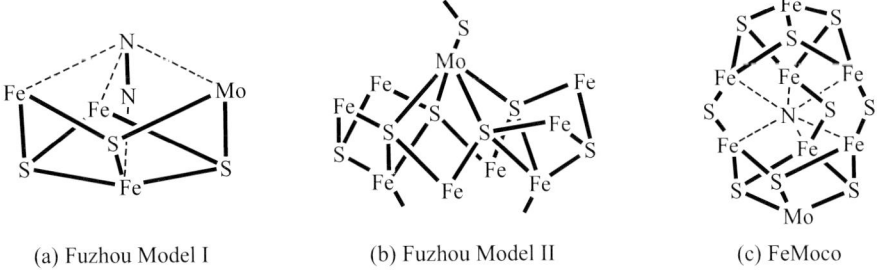

(a) Fuzhou Model I   (b) Fuzhou Model II   (c) FeMoco

Figure 1. Structural Models for the active FeMoco center.

While working on the "string-bag" model compounds for the active center of nitrogenase, we have also organized a systematic study of the synthetic chemistry and structural chemistry of many Mo/Fe/S clusters for the chemical simulation of the FeMoco. From the self-assembly reactions using monodentate thiolate ligands such as cyclohexanethiol, we were able to isolate a single cubane $MoFe_3S_4$ cluster with one additional Fe atom hanging outside the cubane (see Figure 2; ref. 3).

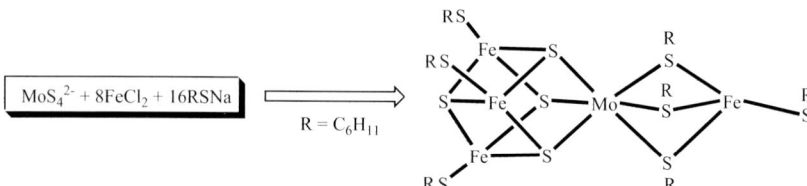

Figure 2. Preparation of a Mo/Fe/S cluster by cyclohexanethiol ligands.

In addition to the monothiolate ligands, which are extensively used in the preparation of Mo/Fe/S clusters, we also used bidentate dithiocarbamates as ligands, and we have synthesized many Mo/Fe/S clusters with a single cubane structure (Figure 3). These cubane clusters have similar structures but with different cluster core oxidation states (from +6 to +4). The R_2dtc^- ligands in these clusters are either chelated to metal atoms or bridging between Mo and Fe atoms. By comparison of the structural parameters of the monoanion cubane and the neutral cubane, the excess of electron is delocalized over the whole cluster but not localized on any particular Fe atoms (4).

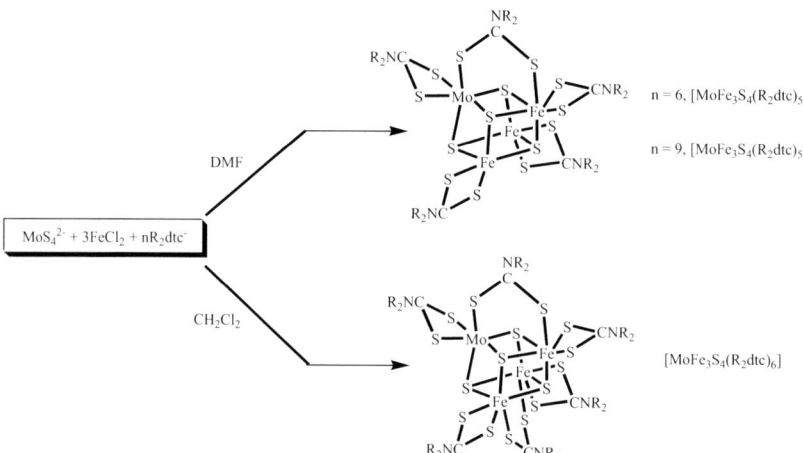

Figure 3. Syntheses of $MoFe_3S_4$ cubane clusters with bidentate R_2dtc^- ligands.

In comparison with Mo/Fe/S clusters, there are only a few known V/Fe/S clusters. We have extended our studies to the V/Fe/S chemistry also by the use of the bidentate R_2dtc^- ligands (5). Different from the Mo system, the one-pot reactions starting from VS_4^{3-}/$FeCl_2$/ R_2dtc^- failed to give V/Fe/S clusters. But when the reactions were carried out in the presence of stronger reducing reagents, such as RS^-, we were able to isolate many cubane clusters with different metal compositions (Figure 4).

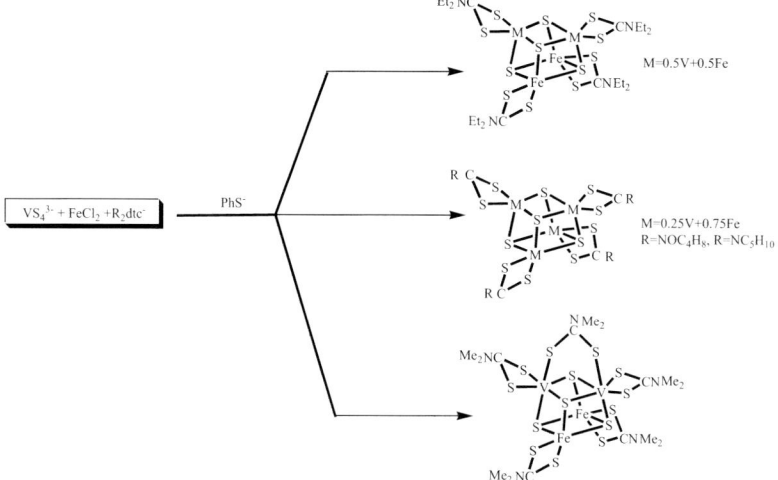

Figure 4. Preparations of V/Fe/S cubane clusters using bidentate R_2dtc^- ligands.

Besides the self-assembly reactions, we also developed a reasonable way to make the cubane clusters mentioned above by a "unit-construction approach". The idea is to combine active metal-sulfido units to build up heterometallic clusters with expected structures. For example, by coupling two linear cluster dianions $[Cl_2FeS_2MoS_2]^{2-}$ in the presence of R_2dtc^- ligands, we were successful in making a cubane cluster with two Mo and two Fe metals as depicted in Figure 5.

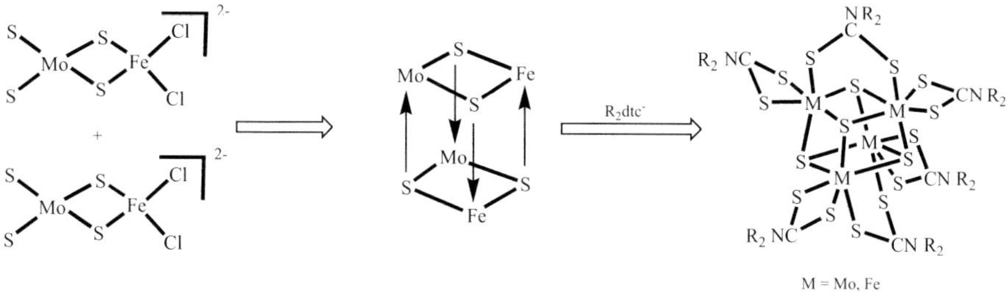

Figure 5. Synthesis of $[Mo_2Fe_2S_4(R_2dtc)_6]$ cluster by unit-construction method.

Because the unit-construction approach has been successful for the preparations of Mo/Fe/S cubane clusters, we also applied this method to make many Mo/Cu/S clusters, most of which contain two or more incomplete cubane-like cluster fragments. We have recently synthesized a double incomplete cubane-like cluster by using the butterfly dimer as a building block. Its structure consists of two $OMoCu_3S_3$ incomplete cubane-like units arranged in a "mouth-to-mouth" fashion and are linked via three thiolate ligands. This structure is very similar to that of FeMoco from the structural point of view, as shown in Figure 6 (7).

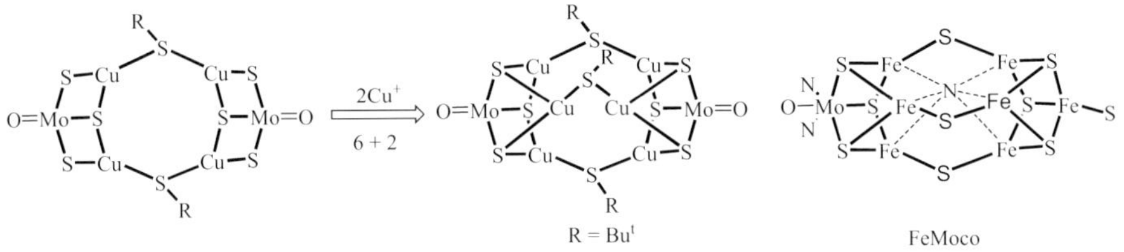

Figure 6. Synthesis of a double incomplete cubane-like Mo/Cu/S cluster by unit-construction method and the comparison of the structure with FeMoco.

1. JX Lu et al. (1992) Nitrogen Fixation and Its Research in China (ed. Hong G.-F.), 3-29. Shanghai: Springer-Verlag and Shanghai Scientific and Technical Publishers.
2. O Einsle et al. (2002) Science 297, 1696-1700.
3. CN Chen et al. (1999) Inorg. Chem. 38, 2375-2379.
4. QT Liu et al. (1990) Inorg. Chem. 29, 4131-4137.
5. YH Deng et al. (1997) Inorg. Chem. 36, 214-219.
6. QT Liu et al. (1990) Scientia Sinica 33, 1446.
7. ZH Li et al. (2004) J. Chem. Soc. Dalton Trans., 2438-2443.

A NON-METAL SYSTEM FOR NITROGEN FIXATION UTILIZING BUCKMINSTERFULLERENE

Yoshiaki Nishibayashi[a], Sakae Uemura[a], Shin-ichi Takekuma[b] and Zen-ichi Yoshida[b]

[a]Department of Energy and Hydrocarbon Chemistry, Kyoto University, Kyoto 615-8510, Japan

[b]Department of Applied Chemistry, Kinki University, Higashi Osaka, Osaka 577-8502, Japan

In all known nitrogen-fixation processes, which range from the industrial Haber-Bosch process (1) to the biological fixation by enzymes (nitrogenases; ref. 2) and the known homogeneous synthetic systems (3), the direct transformation of the inert dinitrogen (N_2) molecule into ammonia (NH_3) makes use of the redox properties of metals. During our work with the buckminsterfullerene (C_{60}) molecule, we discovered that NH_3 is formed from N_2 in the presence of the water-soluble C_{60}/γ-cyclodextrin (1:2) complex (γ-cyclodextrin-bicapped C_{60}; **1**; ref. 4) and light under a N_2 atmosphere. Here, we report the first metal-free system achieving nitrogen fixation using light irradiation by making use of the redox properties of a fullerene derivative (5).

Treatment of the complex **1** with 100 equiv of $Na_2S_2O_4$ as a reducing reagent in water under 1 atm of N_2 with visible light at 60°C for 1 h gave NH_3 in 33% yield based on the C_{60}; a prolonged reaction time (24 h), did not increase the yield of NH_3 (Scheme 1).

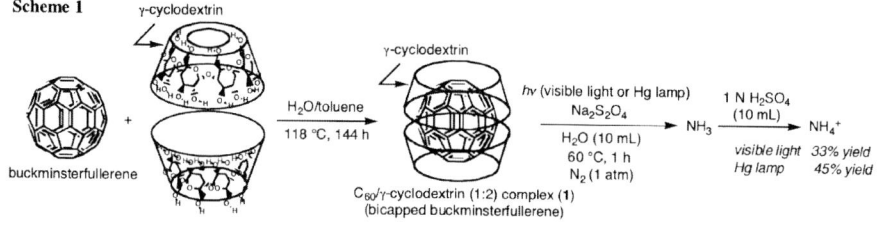

Scheme 1

The combination of $Na_2S_2O_4$, **1** and N_2 is essential for the reaction to proceed because controls missing any of these reactants yield no NH_3. Replacing **1** by γ-cyclodextrin, C_{60}, or a simple mixture of the two also failed to result in NH_3 production. Also, no NH_3 was formed at 25°C and the conditions must remain neutral because acidic or basic conditions lower the NH_3 yield. Under $^{15}N_2$, the formation of $^{15}NH_4^+$ (-363 ppm in D_2O) was confirmed by ^{15}N NMR after acidification of the crude product, providing crucial evidence for nitrogen fixation by this system.

Visible light (fluorescent lamp) is essential to produce NH_3; in the absence of light, no formation of NH_3 was observed. A slightly higher yield of NH_3 (45%) was obtained when the reaction mixture was irradiated with a high-pressure Hg lamp (450W) without a slit. Strong absorptions in the UV region and weaker but significant bands in the visible region with characteristic features of the C_{60} moiety were observed at 212(s), 259(s), 333(s), 408(m), 536(w), and 601(w) nm in the UV/VIS spectrum of **1** in H_2O. These results indicate that the charge-transfer interaction between N_2 and the reduced and electronically excited **1** (exciplex formation) are likely involved in the reduction of N_2, because it is well known that C_{60} and its derivatives are very easily excited by low-energy light, and that their excited states promote electron- and energy-transfer more readily than the ground-state species (6). Thus, light may assist electron- and energy-transfer from C_{60} in **1** to N_2 (probably coordinated to **1**). A similar light-assisted effect was previously reported by using a transition metal complex (7). All of the electrons required for the formation of NH_3 are likely supplied from the reduced C_{60} in **1** because the C_{60} in **1** was readily reduced to the corresponding C_{60}^{1-} and C_{60}^{2-} species by using $Na_2S_2O_4$ (8). The cyclic voltammogram of **1** (in H_2O under both argon and N_2 atmospheres) showed one reversible and one irreversible wave at $E_{1/2}$= -0.61 and E_{pc} = -1.09V (vs SCE), respectively, assignable to one-electron redox process, C_{60}/C_{60}^{1-} and C_{60}^{1-}/C_{60}^{2-}, respectively. Thus, the reduced complexes of **1** have sufficiently low reduction potentials (6) to reduce the coordinated N_2.

References
1. Appl M (1999) Ammonia. Wiley-VCH, Weinheim.
2. Postgate J (1998) Nitrogen Fixation. Cambridge University Press, Cambridge.
3. Chatt J et al. (1975) Nature 253, 39-40; Nishibayashi Y et al. (1998) Science 279, 540-542; Yandulov DV and Schrock RR (2003) Science 301, 76-78.
4. Yoshida Z et al. (1994) Angew. Chem. Int. Ed. Engl. 33, 1597-1599.
5. Nishibayashi Y et al. (2004) Nature 428, 279-280.
6. Guldi DM and Prato M (2000) Acc. Chem. Res. 33, 695-703.
7. Solari E et al. (2001) Angew. Chem. Int. Ed. Engl. 40, 3907-3909.
8. Takekuma S et al. (2000) Tetrahedron Lett. 41, 2929-2932.

CHEMICAL SIMULATION OF THE MOLYBDENUM COORDINATION MODE IN THE FeMo COFACTOR BY A SERIES OF MOLYBDENUM- OR TUNGSTEN-POLYCARBOXYLATE COMPLEXES WITH SULFIDE BRIDGES

Dong-Mei Li, Ji-Qing Xu*, Yong-Heng Xing, Ying-Jie Lin, and Xiao-Hua Zhou
College of Chemistry, Jilin University, Changchun 130023, P. R. China

The reduction of dinitrogen to ammonia by nitrogenases at ambient pressure and mild temperature is one of the most fascinating catalytic reactions in nature. The nitrogenase active site, the FeMo-cofactor, is composed of two cuboidal subunits Fe_4S_3 and Fe_3MoS_3 linked by three μ_2-S bridges (1). In addition, an atom (most likely nitrogen) is at the center of FeMo-cofactor as shown by a recent high-resolution crystallographic analysis (at 1.16Å) (2). With the continuing development of theoretical chemistry and Mo coordination chemistry, interest in the direct participation of both Mo and (R)-homocitrate in the process of nitrogen fixation has increased (3,4). Relevant structures with a six-coordinate Mo atom should include three sulfur atoms, a nitrogen atom from histidine, and a pair of oxygen atoms from homocitrate, in which the carboxylate oxygen could provide an open coordinate site for substrate binding (5). Therefore, much attention has been given to Mo specifically and to the functionality of homocitrate in the catalytic process. We have been focusing on the chemistry of the molybdenum's coordination mode in FeMo-cofactor for two decades, especially those aspects related to the coordination of homocitrate.

Here, some Mo(W)-polycarboxylate complexes with sulfide(oxo) bridges, which were prepared from $(NH_4)_2MS_4$ (M= Mo or W) and alternative polycarboxylate ligands under N_2, have been investigated in detail. The products obtained had infrared spectra, which suggested the presence of intense C=O vibrations (~1600 cm^{-1}), M-oxo moiety (Mo=O ~920cm^{-1}, W=O ~950cm^{-1}), and weak Mo-S-Mo bond (~470cm^{-1}). In the binuclear Mo-S complexes with citrate, homocitrate, or thioproline ligands, each molybdenum atom adopts a distorted octahedral arrangement with a terminal oxygen atom, two bridging sulfur atoms, and three atoms from the ligand (i.e., hydroxyl, α-, β-carboxylates, sufide and amine groups), as shown in Figure 1.

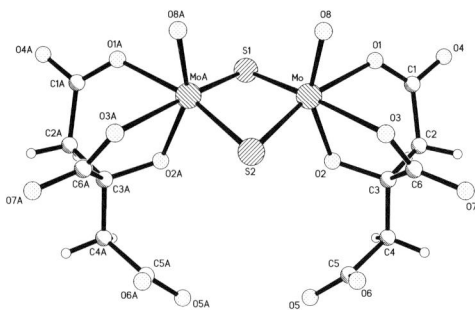

Figure 1. A perspective view of the anion of $K_4(NH_4)_2[Mo_2S_2O_2(C_6H_4O_7)_2]\cdot 10H_2O$

Other Mo-S and W-O complexes with amino acids have also been prepared and studied. With the aid of the known data, we propose that Mo atom in the homocitrate complex $[Mo_2S_2O_2(C_7H_6O_7)_2]^{6-}$ adopts a similar coordination mode as in Mo-citrate complexes. These compounds have been characterized by infrared, UV-Visible, ^{1}H and ^{13}C NMR spectroscopies, electrochemistry, XPS, X-ray crystallography. Based on numerous experiments, several factors were found to influence the formation of final products. These include pH value, the temperature, concentration of the reactants and the solvents used. Interestingly, a variety of Mo-polycarboxylate complexes exhibit catalytic activity in the reduction of C_2H_4 to C_2H_2, while the corresponding W-complexes do not. In the near future, we shall attempt to prepare a series of chiral Mo(W)-polycarboxylate complexes, which should provide a very good understanding of the coordination capacity of the central metal atom.

Acknowledgements
This work is supported by the National Science Foundation of China (No. 20271021 and No. 20333070), and the state Key Basis Research project of China (No. 2001CB108906).

References
1. Kim J and Rees DC (1992) Science 257, 1677-1682.
2. Einsle O et al. (2002) Science 297, 1696-1700.
3. Dean DR et al. (2004) Biochemistry 43, 1401-1409.
4. Demadis KD and Coucouvanis D (1995) Inorg. Chem. 117, 436-448.
5. Imperial J and Ludden PW (1989) Biochemistry 28, 7796-7799.

ACTIVATION OF HOMOCITRATE LIGAND OF FEMO-COFACTOR AS MIMICKED BY THE COORDINATION OF HOMOCITRATO AND ITS HOMOLOGUES

Zhou Z Hui*, Hou S Ya, Cao Z Xing, Deng Y Fu, Wan H Lin, and Tsai K Rui
Department of Chemistry and State Key Lab. Physics and Chemistry of Solid Surfaces, Xiamen University, Xiamen, China.

Recently, high resolution X-ray crystallographic analysis of nitrogenase has revealed a previously unrecognized light atom coordinated to six iron atoms and located at the center of FeMo-cofactor (FeMoco) (1). Density functional calculations and electrostatics calculations, as well as ENDOR and ESEEM observations, have concluded that this light atom is most likely nitrogen (2). The presence of the μ_6-N atom would make the 6 central Fe atoms coordinatively saturated. This strengthens the notion that the Mo atom directly participates in dinitrogen binding and reduction (3). It has been proposed that homocitrate may facilitate the binding of dinitrogen through the loss of the bound α-carboxyl group from [Mo] (4) with the formation of intra-molecular hydrogen bonding (5-6). However, there is no evidence for the protonation of the coordinated carboxyl group prior to its loss. As a part of our study related to coordination chemistry of hydroxycarboxylato-molybdates (7), we have systematically compared the interaction between molybdenum and hydroxycarboxylic acids, trying to obtain more information about the activation and functioning of the homocitrate ligand, which is known to be required for strong binding of N_2 to the FeMo-cofactor.

For a series of synthetic α-hydroxycarboxylato-molybdates, such as homocitric, citric, malic, tartaric, mandelic, lactic and glycolic acid, the Mo-O bond distances for both the α-alkoxyl and α-carboxyl are all similar with the average Mo-O bond for the twelve molybdate(VI) complexes listed in Table 1 being 1.944 Å and 2.210Å, respectively. There is no obvious relationship in the Mo-O (α-carboxyl) bond distances with oxidation state changes of Mo. In comparison, the Mo-O (α-carboxyl) is 2.162 Å in FeMo-co where we have likely Mo(III)~Mo(IV). In contrast, the bond distances of both Mo(VI) and Mo(V)-O α-alkoxyl, which are respectively 1.944Å and 2.028Å, are shorter for the higher Mo oxidation state and also shorter than those of citric and homocitric acid bound to FeMo-co (2.253Å and 2.212Å, respectively). This shows that the bond strength of α-alkoxyl group bound to Mo increases with the increase in Mo oxidation state and is consistent with the observation that strong reduction of glycolato, mandelato

Table 1. Comparions of Mo-O distances (Å) in hydroxycarboxylato complexes and the assignments of their absolute configurations.

Complex	Mo-O (α-alkoxyl)	Mo-O (α-carboxyl)	Abs. Config.	Ref.
Mo(VI)				
$K_2[MoO_2(glyc)_2]\cdot H_2O$	1.929(3),1.953(2)	2.202(2),2.204(2)	Δ/Λ	[7]
$\{Na_2[MoO_2(S\text{-lact})_2]_3\cdot 13H_2O$	1.947(4),1.969(5)	2.181(4),2.201(5)	$\Lambda_{S,S}$	[7]
$(NH_4)_3[GdMo_6O_{15}(S\text{-lact})_6]$	1.919(6)	2.192(6)	$\Delta_{S,S}$	[8]
$(NH_4)_2[MoO_2(R\text{-mand})_2]\cdot 3H_2O$	1.932(3),1.954(3)	2.187(4),2.228(3)	$\Delta_{R,R}$	[7]
$(NH_4)_2[MoO_2(S\text{-mand})_2]\cdot 3H_2O$	1.927(2),1.951(2)	2.183(3),2.232(3)	$\Lambda_{S,S}$	[7]
$(NH_4)_2[MoO_2(RS\text{-mand})_2]\cdot 4H_2O$	1.950(3),1.951(3)	2.205(3),2.229(3)	Δ/Λ_{RS}	[7]
$(NH_4)[MoO_2(OOCCOPh_2)_2]\cdot 2H_2O$	1.966(2),1.977(2)	2.177(2),2.166(3)	Δ/Λ	[9]
$Cs_2[MoO_2(Hmal)_2]\cdot H_2O$	1.939(8)	2.243(9)	$\Delta_{S,S}$	[10]
$Na_3[MoO_2H(mal)_2]$	1.935(2)	2.182(3)	$\Lambda_{S,S}$	[11]
$(NMe_4)[MoO_2(H_2tart)_2]\cdot EtOH\cdot 1.5H_2O$	1.94(1),1.921(9)	2.227(9),2.226(8)	$\Lambda_{R,R}$	[12]
$(NH_4)\{[Gd(H_2O)_6]Mo_2O_4(tart)_2\}\cdot 4H_2O$	1.944(4),1.946(4)	2.205(4),2.227(4)	$\Delta_{R,R}$	[13]
$Na_2[MoO_2(H_2cit)_2]\cdot 3H_2O$	1.944(4),1.946(4)	2.205(4),2.227(4)	Δ/Λ	[14]
Average	**1.944**	**2.210**		
Mo(V)				
$K_6[Mo_8O_{16}(glyc)_6(Hglyc)_2]\cdot 10H_2O$	2.062(7)av	2.162(6)av		[7]
$K_5(NH_4)[Mo_2O_2(\mu S)_2(cit)_2]\cdot CH_3OH\cdot 5H_2O$	1.993(5),1.994(5)	2.174(5),2.211(5)		[15]
Average	**2.028**	**2.177**		
Mo(III~IV)		2.120		
$[MoFe_7S_9(cit)(his)]$ in $Kp1$	2.253	2.269	Δ	[16]
$[MoFe_7S_9N(R\text{-homocit})(his)]$ in $Av1$	2.212	2.162	Δ_R	[1]
Mo(II)				
$[Mo(II)_2(glyc)_4]\cdot H_2O$		$2.121(4)_{av}$		[17]
$[Mo(II)_2(R\text{-Hmand})_4]\cdot 2THF$		$2.119(8)_{av}$		[18]
$[Mo(II)_2(OOCCOPh_2)_4]\cdot 4THF$		$2.119(2)_{av}$		[19]
Average		**2.120**		

(H_2glyc = glycolic acid, H_2lact = lactic acid, H_2mand = mandelic acid, H_3mal = malic acid; H_4tart = tartaric acid; H_4cit = citric acid, H_4homocit = homocitric acid, Kp = *Klebsiella pnemomiae*, Av = *Azotobacter vinelandii*).

and benzilato molybdates results in no coordination of the hydroxyl group. Because an α-carboxylic acid is considerably more acidic than an α-hydroxyl group, it should be considerably easier to protonate an α-alkoxyl than an α-carboxyl bound to Mo. Thus, a model for the creation of a binding site at Mo by protonation and dissociation of the α-alkoxyl group in *R*-homocitrate ligand upon reduction of the FeMo-cofactor should be taken into consideration (7).

References
1. Einsle O et al, (2002) Science 297, 1696-1700.
2. Noodleman L et al (2004) Chem. Rev. 104, 459-508.
3. Leigh GJ et al (2003) Science 301, 55-56.
4. Seefeldt LC et al (2004) Biochemistry 43, 1401.
5. Grönberg KLC et al (1998) J. Am. Chem. Soc. 120, 10613-10621.
6. Tsai KR et al (1995) J. Clust. Sci. 6, 485-501.
7. Zhou ZH et al (2004) J. Inorg. Biochem. 1037-1044, 1110–1116, 1795-1802.
8. Wu CD et al (2001) J. Chem. Soc., Dalton Trans. 3202-3204.
9. Cervilla A et al (1995) J. Chem. Soc., Dalton Trans. 3891-3895.
10. Knobler CB et al (1983) J. Chem. Soc., Dalton Trans. 1299-1303.
11. Zhou ZH et al (2002) J. Inorg. Biochem. 90, 137-143.
12 Robinson WT et al (1986) Trans. Metal Chem. 11, 86-89.
13. Wu CD et al (2003) Chem. Comm. 1284-1285.
14. Zhou ZH et al (1999) J. Chem. Soc., Dalton Trans. 4289-4290.
15. Xu JQ et al (1999) Inorg. Chim. Acta 285, 152-154.
16. Mayer SM et al (2002) J. Biol. Chem. 277, 35263-35266.
17. Cotton FA et al (2002) Inorg. Chem. Comm. 5, 527-532.
18. Cotton FA et al (2002) Inorg. Chem. 22, 382-387.
19. Liwporncharoenvong T et al (2002) Inorg. Chim. Acta 329, 51-58.

DIFFERENT ROLES OF THE SUBUNIT RESIDUE HISTIDINE 194 AND GLUTAMINE 190 OF MoFe PROTEIN AND HOMOCIREATE OF FeMoco IN CATALYSIS OF *KLEBSIELLA PNEUMONIAE* NITROGENASE

Dehua Zhao, Feng Guan, Miao Pan, Jilun Li
State Key Lab for Agrobiotechnology and Dept. Microbiology and China Agricultural University, Beijing 100094, China

The homocitrate in FeMoco and environmental α-Histidine 194 and α-Glutamine190 residues in MoFe protein play crucial roles in substrate reduction. Where the substrates interact with the FeMoco remains unknown and the definite functions of homocitrate, -Histidine 194 and -Glutamine 190 in the enzyme catalysis are undetermined. A new strategy to alter the -His194 and homocitrate or -Gln190 and homocitrate was used to block the proton and electron transfer in nitrogenase from *Klebsiella pneumoniae*, so that we can gain the mechanism details by activity analysis of the mutants.

Comparison of substrate-reduction properties of nitrogenases in *nifD* single mutants (with His-194 or Gln-190 substitutions), *nifV* single mutant , *nifD nifV* double mutants and wild type indicates that: (i) association of -Glutamine 190 and homocitrate is essential to H_2 evolution of nitrogenase; (ii)-Histidine 194 acts as a regulatory site of electron distribution between H_2 and NH_3 or H_2 and C_2H_4; (iii)-Glutamine 190 is important in the interaction of CO with the nitrogenase. Therefore we suggest that -Glutamine 190 is directly involved in the electron transfer to FeMoco and the -Histidine 194 is in close proximity to the N_2 binding site of FeMoco. The implications obtained in this study can be integrated into our two-site H_2 evolution model with the Mo atom site and the central six Fe atom region of FeMoco for the N_2-independent and N_2-dependent H_2 evolution respectively.

This work was supported by the National Natural Science Foundation of China (Grant No.30270019 and No.39970006) and the National Key Basic Research Development Program (Grant No.001CB108904).

PURIFICATION AND IDENTIFICATION OF TWO HOMOPOLYMERS IN PARTIALLY PURE NITROGENASE MoFe PROTEIN

Hui-Na Zhou, Ying Zhao, Shao-Min Bian, Jian-Feng Zhao, Fei Ren, Huang-Ping Wang and Ju-Fu Huang*
Key Laboratory of Photosynthesis and Environmental Molecular Physiology, Institute of Botany, The Chinese Academy of Sciences, Beijing 100093, China

The partially pure nitrogenase MoFe protein (Av1 preparation) was obtained from *Azotobacter vinelandii* by chromatography on DEAE-52, Q-Sepharose and Sephacryl S-200 columns (Huang et al. 2001). The Av1 preparation was shown to have two main bands at the position of α and β subunits of crystalline Av1 on SDS-gel. But on the native-gel, the preparation was shown to have other three main bands besides that of Av1. Among these three main bands, the protein with fastest migration was identified as bacterioferritin (Zhao et al. in press). The other two, termed Upper and Middle bands, were determined to be two different homopolymers, whose subunits almost migrated to the same position as those of α (62 kD) and β (59 kD) subunits of Av1 in SDS-PAGE, respectively. The molecular sizes of Upper and Middle bands were estimated to be about 880 kD and 610 kD by native-gel, respectively. It suggested that Upper is composed of 14 subunits of 62 kD and the Middle 10 subunits of 59 kD. By analysis of MALDI-TOF (matrix-assisted laser desorption/ionization time-of-flight) mass spectrometry, the Upper band was identified as GroEL which belongs to the Hsp 60 family, and the Middle band was PGI (glucose-6-phosphate isomerase).

References
Huang JF et al.(2001) Acta Bot. Sin. 43, 918-22.
Zhao JF et al.(2004) Acta Bot. Sin. in press

ACTIVATION *IN VITRO* OF FeMoco-DELETED MoFe PROTEIN FROM A NIFB⁻ MUTANT OF *AZOTOBACTER VINELANDII*

Ying Zhao, Shao-Min Bian, Jian-Feng Zhao, Hui-Na Zhou, Huang-Ping Wang, Fei Ren and Ju-Fu Huang*
Key Laboratory of Photosynthesis and Environmental Molecular Physiology, Institute of Botany, The Chinese Academy of Sciences, Beijing 100093, China

FeMoco and P-cluster are two unique metalloclusters which play pivotal roles during nitrogen fixation in nitrogenase MoFe protein. A FeMoco-deleted MoFe protein (NifB⁻ Av1) was purified from *Azotobacter vinelandii* Lipmann mutant UW45. After anaerobically treated with *o*-phenanthroline (*o*-phen), NifB⁻ Av1 became NifB⁻ Av1©, whose P-cluster was partially deficient (Huang et al 1989). The NifB⁻ Av1©, rather than NifB⁻ Av1, could be significantly activated *in vitro* by a reconstituting solution containing M (RS-M, M=Mo or Cr) in the presence of both Fe protein (Av2) and MgATP regeneration system. The highest specific proton-reduction activities of activated NifB⁻ Av1© were 1792 nmol H_2-formed·mg^{-1}·min^{-1} (reconstituted by RS-Mo) and 819 nmol H_2-formed·mg^{-1}·min^{-1} (reconstituted by RS-Cr), respectively. However, the activation could not happen in the absence of MgATP or Av2. It could be concluded that 1), activation of NifB⁻ Av1 requires pretreatment by *o*-phen, which leads to looseness and substantially conformational rearrangement of NifB⁻ Av1 (Lechenko & Sadkov 1984); 2), MgATP and Av2 are necessary during the progress of reconstitution; 3), Cr could take the place of Mo to synthesize FeCrco, which is a new evidence for the presence of nitrogenase CrFe protein.

References
Huang JF et al. (1989) Acta Bot. Sin. 31, 785-791.
Lechenko LA and Sadkov AP (1984) Dokl. Akad. Nauk. 277, 1003-1005.

CRYSTAL GROWTH AND CHARACTERIZATION OF RESIDUAL BACTERIOFERRITIN IN PARTIALLY PURIFIED NITROGENASE CrFe PROTEIN FROM A MUTANT UW3 OF *AZOTOBACTER VINELANDII*

Jian-Feng Zhao[1], He-Li Liu[2], Hui-Na Zhou[1], Zhi-Ping Wang[3], Ying Zhao[1], Shao-Min Bian[1], Shu-Xing Li[2], Ru-Chang Bi[2] and Ju-Fu Huang[1]
Key Laboratory of Photosynthesis & Environmental Molecular Physiology, Inst.Botany, The Chinese Academy of Sciences, Beijing 100093, China[1]. Inst. Biophysics, The Chinese Academy Sciences, Beijing 100101, China[2]. Inst. Nuclear Agriculture Sciences, Zhejiang Univ., Hangzhou 310029, China[3].

Attempting to obtain large crystals of nitrogenase CrFe protein, brown crystals and brick red crystals were simultaneously or independently formed from CrFe protein preparation, which was partially purified from a mutant UW3 of *Azotobacter vinelandii* Lipmann (Bishop et al. 1982; Huang et al. 2002). SDS-PAGE and anoxic native-PAGE analysis consistently showed that the protein of the brown crystal was mainly composed of subunits similar to those of OP Av1, while the protein of the brick red crystal was composed of ~20 kD subunits. Detection by 3, 5-diaminobenzoic acid of native-PAGE gels showed that the proteins forming the brick red crystal and the brown crystal were two kinds of iron-containing proteins with different electrophoretic mobility on the gel. The analysis of matrix-assisted laser desorption ionization time-of-flight (MALDI-TOF) mass spectrometry (MS) proved that the protein forming the brick red crystal was bacterioferritin of *A. vinelandii* (AvBF). The detailed structural analysis published in the near future (Liu et al. 2004) has confirmed that the brick red crystal is that of 24-meric bacterioferritin. Since AvBF was only a contamination protein in the partially purified CrFe protein, it was confirmed that minor protein component could also be crystallized in some cases.

Reference
Bishop PE et al. (1982) J. Bacteriol. 150, 1244-1251.
Huang JF et al. (2002) Acta Bot. Sin. 44, 297-300.
Liu HL et al. (2004) FEBS Lett. in press.

SECTION II.

GENETICS AND REGULATION IN HETEROTROPHS AND PHOTOSYNTHETIC BACTERIA

GENETIC REGULATION OF NITROGEN FIXATION: INTEGRATION OF MULTIPLE SIGNALS

Richard Little, Isabel Martinez-Argudo, Neil Shearer, Philip Johnson and Ray Dixon
Department of Molecular Microbiology, John Innes Centre, Norwich NR4 7UH, UK

1. Plasticity of *nif* gene regulation
The necessity to respond to the level of fixed nitrogen, external oxygen concentrations and to provide sufficient energy for nitrogen fixation imposes common regulatory principles amongst diazotrophs. However, although common regulatory components are present in nitrogen-fixing bacteria, there is considerable plasticity in the regulatory networks, which differ from organism to organism, dependent upon host physiology (Dixon and Kahn, 2004). In the proteobacteria, the universal nitrogen fixation-specific regulatory protein is NifA, a member of the enhancer binding family that activates transcription at *nif* promoters by the σ^{54}-RNA polymerase holoenzyme. NifA proteins have a conserved domain architecture consisting of a central AAA^+ ATPase domain, flanked by an amino terminal regulatory GAF domain and a C-terminal DNA binding domain. However, the activity of NifA is controlled at the post-translational level by different mechanisms in various diazotrophs. This may involve intrinsic sensing by NifA or interaction with additional regulatory proteins. For example, in most diazotrophic representatives of the α- and β-subdivisions of the proteobacteria, NifA activity is intrinsically oxygen sensitive and this property is apparently conferred by an invariant $Cys-X_4-Cys$ motif present within a linker region between the AAA^+ and C-terminal domains. However, in the γ-subdivision of the proteobacteria and in *Azoarcus* spp, this motif is absent and oxygen regulation of NifA activity is provided by the flavoprotein NifL, an anti-activator that functions to inhibit NifA under unfavourable conditions for nitrogen fixation (Martinez-Argudo et al., 2004c).

PII-like signal transduction proteins have been shown to have a common role in controlling NifA activity in response to the nitrogen source in various proteobacteria. However, despite this universal role, precise mechanisms for nitrogen control of NifA activity differ widely. For example, in *Rhodosprillim rubrum*, *Herbaspirillum seropedicae* and *Azospirillum brasilense*, the uridylylated form of the PII-like protein GlnB seems to be required to activate NifA under nitrogen-limiting conditions. In contrast, in *Rhodobacter capsulatus* and *Azorhizobium caulinodans,* the non-uridylyated forms of GlnB and GlnK are apparently required for inhibition of NifA activity under nitrogen-replete conditions (reviewed in Dixon and Kahn, 2004). In *Klebsiella pneumoniae* and *Azotobacter vinelandii*, the anti-activator NifL is required to regulate NifA activity in response to the nitrogen source. However, the mechanism by which GlnK signals the nitrogen status to the NifL-NifA systems in these organisms differs considerably. In *A. vinelandii*, the non-modified form of GlnK interacts with NifL to promote the formation of an inhibitory GlnK-NifL-NifA complex, thus preventing nitrogen fixation under conditions of nitrogen excess. In contrast, in *K. pneumoniae*, GlnK is required to prevent inhibition of NifA activity by NifL, under conditions appropriate for nitrogen fixation (reviewed in Martinez-Argudo et al., 2004c).

The GAF domain of NifA has a common role in regulating the activity of the AAA^+ domain in response to environmental cues. In organisms that lack NifL, the GAF domain seems to be the target for control by the PII-like regulatory proteins. In the examples given above, this regulation can be either "positive", in which case the interaction activates NifA, or "negative", whereby the interaction apparently leads to repression of the AAA^+ domain by the GAF domain. The GAF domain of *A. vinelandii* NifA binds 2-oxoglutarate, a metabolic signal linking carbon and nitrogen status, and this interaction modulates the ability of NifL to inhibit NifA. Therefore although GAF domains may have similar structural features, they respond to a variety of signals to regulate NifA activity. The above examples illustrate the plasticity of NifA regulation in different diazotrophs, underlying the ability of these organisms to adapt to a wide range of physiological constraints.

2. Signal integration: the case of the *A. vinelandii* NifL-NifA system

The *A. vinelandii* NifL-NifA system can integrate signals of the oxygen, nitrogen and carbon status to ensure that *nif* genes are only expressed under conditions appropriate for nitrogen fixation. The NifL protein is comprised of at least three discrete domains: an N-terminal PAS domain containing an FAD co-factor that senses the redox status, a C-terminal adenosine nucleotide-binding (GHKL) domain, similar to that found in the histidine protein kinases and a central region (or H box) that by analogy with the histidine kinases might serve as a dimerisation interface (Fig. 1).

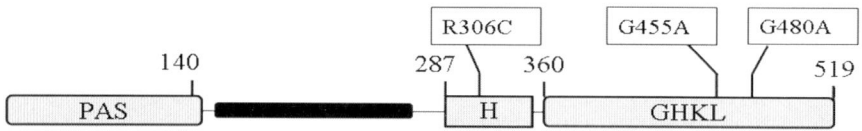

Fig. 1 domain structure of *A. vinelandii* NifL showing mutations discussed in the text.

The inhibitory activity of NifL towards NifA is promoted by three factors: (i) oxidation of the FAD moiety in the PAS domain in response to excess oxygen (Hill et al., 1996) (ii) binding of adenosine nucleotide to the GHKL domain (Söderbäck et al., 1998) and (iii) the interaction of the signal transduction protein GlnK with the GHKL domain in response to excess fixed-nitrogen (Little et al., 2002; Rudnick et al., 2002). Under conditions appropriate for nitrogen fixation, when NifL is reduced and uridylylation of GlnK prevents its interaction with NifL, adenosine nucleotide binding to the GHKL domain (signal ii) still favours inhibition of NifA activity. However, the influence of adenosine nucleotides on the NifL-NifA interaction is antagonised by the binding of 2-oxoglutarate, to the amino-terminal GAF domain of NifA, thus enabling NifA to escape from inhibition by NifL, under nitrogen-fixing conditions (Little and Dixon, 2003; Martinez-Argudo et al., 2004a).

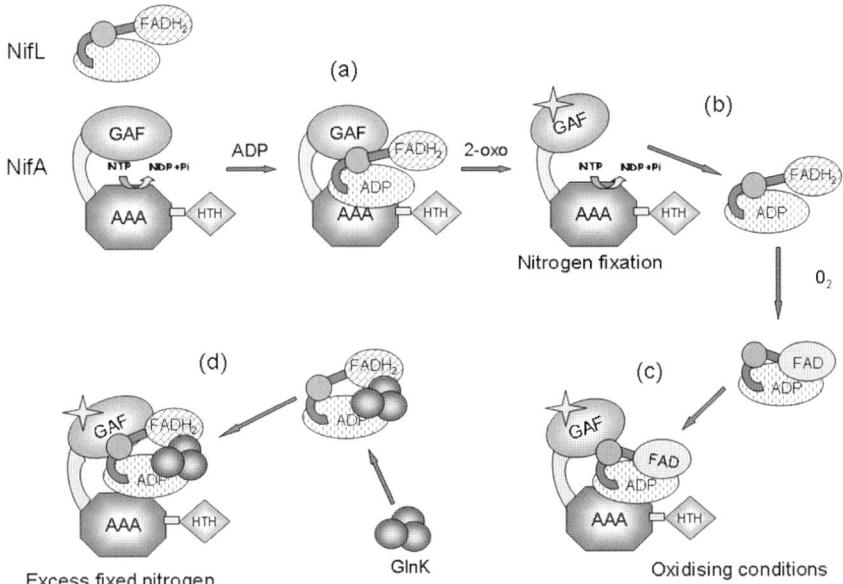

Fig. 2. Schematic model for regulation of NifA activity by NifL.

Domains of NifA are depicted in grey and domains of NifL are shown as stippled ovals connected by a central region. Binding of 2-oxoglutarate (2-oxo) to the GAF domain of NifA is shown as a star.

The regulatory dialogue between the NifL and NifA proteins involves reciprocal conformational changes in which the binding of 2-oxoglutarate to the GAF domain of

NifA plays a major role in modulating the response to the NifL (Fig 2). In the absence of 2-oxoglutarate, even the reduced form of NifL is competent to inhibit the activity of NifA, provided that adenosine nucleotide is bound to NifL (Fig. 2a). Binding of 2-oxoglutarate to the GAF domain of NifA induces a conformational change that releases NifL inhibition, allowing NifA to activate transcription under conditions appropriate for nitrogen fixation (Fig. 2b). We presume that oxidation of the flavin in NifL causes a reciprocal conformational change that re-establishes the interaction with NifA even in the presence of 2-oxoglutarate and hence NifA is inactivated by this form of NifL (Fig. 2c). Binding of GlnK to the C-terminal domain of NifL, enables the reduced form of NifL to interact with NifA, through the formation of a GlnK-NifL-NifA ternary complex (Fig. 2d).

3. Evidence for conformational changes in NifL

The above model suggests that NifL undergoes conformational changes enabling interaction with NifA in the presence of 2-oxoglutarate, either when NifL is oxidised or when GlnK is bound to the reduced form of NifL (Fig.2 c and d). These conformational changes may not necessarily be identical, since we have isolated mutant forms of NifA that can discriminate between the oxidised form of NifL and the form of NifL activated under nitrogen-excess conditions by the interaction with GlnK (Martinez-Argudo et al., 2004a; Reyes-Ramirez et al., 2002). In order to identify regions of NifL that are essential for such conformational changes we have isolated NifL mutants that constitutively inhibit NifA activity, thus preventing nitrogen fixation in *A. vinelandii*. These studies have identified an arginine residue (R360) in NifL that is required to release NifA from inhibition by NifL. Substitutions at this position result in inhibition of NifA activity under nitrogen-fixing conditions irrespective of whether the mutations introduce polar, hydrophobic or charged-side chains. Biochemical studies with one of the mutant proteins NifL-R306C, reveals that the mutation induces a conformational change in NifL that may represent the inhibitory "anti-activation" conformation. This mutant protein is apparently locked in a form that inhibits NifA in the presence of 2-oxoglutarate, irrespective of environmental signals. Arginine 306 is located in the central H box region of NifL, adjacent to a conserved histidine residue that functions as the autophosphorylation site in the histidine protein kinases (Fig. 1). However, this histidine residue is functionally redundant in *A. vinelandii* NifL, which does not exhibit kinase activity. Our results suggest that arginine 306 is critical for the transmission of the redox and fixed nitrogen signals perceived by NifL and is required for a conformational switch that prevents NifL from inhibiting NifA under conditions appropriate for nitrogen fixation (Martinez-Argudo et al., 2004b).

We have also observed that conformational changes occur in NifL upon binding of adenosine nucleotides to the C-terminal GHKL domain (Söderbäck et al., 1998). Mutations in this domain which decrease the affinity for ADP, impair the response of NifL to the redox and fixed nitrogen status *in vivo*. Biochemical studies with the GHKL substitution mutants G455A and G480A indicate that this domain plays an important role in counteracting the response of NifA to 2-oxoglutarate under conditions that are inappropriate for nitrogen fixation. Surprisingly, the NifL-R306C mutation suppresses the null phenotype of the G455A and G480A mutations to restore regulation of NifA activity by NifL (Martinez-Argudo et al., 2004b). This implies that communication between the GHKL and the H box regions of NifL is important for signal transduction.

4. References

Dixon R and Kahn D (2004) Nature. Rev. Microbiol. 2, 621-631.
Hill S et al. (1996) Proc. Natl. Acad. Sci. USA 93, 2143-2148.
Little R et al. (2002) J. Biol. Chem. 277, 15472-15481.
Little R and Dixon R (2003) J. Biol. Chem. 278, 28711-28718.
Martinez-Argudo I et al. (2004a) Mol. Microbiol. 52, 1731-1744.
Martinez-Argudo I et al. (2004b) Proc. Natl. Acad. Sci. USA in press.
Martinez-Argudo I et al. (2004c) J. Bacteriol. 186, 601-610.
Reyes-Ramirez F et al. (2002) J. Bacteriol. 184, 6777-6785.
Rudnick P et al. (2002) J. Bacteriol. 184, 812-820.
Söderbäck E et al. (1998) Mol. Microbiol. 28, 179-192.

THE STRUCTURAL BASIS FOR *nif* GENE ACTIVATION

Wigneshweraraj SR, Burrows PC, Bordes P, Schumacher J, Rappas M[1], Finn RD[2], Cannon WV, Zhang X[1] and Buck M[3].
Dept Biological Sciences, Imperial College London, Sir Alexander Fleming Building, London SW7 2AZ, UK. [1] Centre for Structural Biology, Imperial College London, The Flowers Building, London SW7 2AZ, UK. [2]The Wellcome Sanger Inst., Wellcome Trust Genome Campus, Hinxton CB10 1SA, UK. [3]Author for Correspondence

1. Introduction

The *nif* genes from *Klebsiella pneumoniae* were amongst the first bacterial genes described whose activation required an ATP hydrolysing AAA (ATPases Associated with various cellular Activities) transcriptional activator proteins (also called Enhancer Binding Protein or EBPs, for example the NifA protein). The EBPs act in concert with the RNA polymerase (RNAP) containing the sigma (σ) 54 promoter specificity factor. Here, we present some recent advances in understanding how ATP hydrolysis is used to change the functional state of the σ^{54}-RNAP closed promoter complex in order to allow DNA opening (open promoter complex formation) and transcription initiation (Fig. 1). These results help us understand how the ATP induced conformational changes in the AAA activator are used to induce alterations in the σ^{54}-RNAP closed complex. Evidently, ATP hydrolysis is used by the AAA activator to induce a series of directed conformational changes (i.e. remodelling) in σ^{54}-RNAP closed complex, rather than to allow a simple ATP conditioned binding of the interacting partners.

2. Functional Studies

The use of the ATP hydrolysis transition state analogue ADP·AlF$_x$ enabled use to show that NifA and other AAA activators of the σ^{54}-RNAP stably engage the σ^{54}-RNAP at the transition point of ATP hydrolysis [1]. For technical simplicity, we use the AAA domain of the *Escherichia coli* AAA activator Phage shock protein F (PspF), which is closely related to the NifA protein, for most experiments [2]. The AAA domain of PspF (referred to as PspF$_{1-275}$) alters the promoter DNA binding functionality of the σ^{54}-RNAP in the presence of ADP·AlF$_x$ [3]. In particular, the AAA activator appears to change the interaction between the σ^{54}-RNAP and a DNA fork junction structure (which is present adjacent to the transcription start site proximal consensus promoter element – see Fig. 1) that persists to retain the closed promoter complex and prevent its isomerisation in the absence of activation [4]. During open complex formation, the interaction between a regulatory σ^{54} domain (referred to as Region I) and the DNA fork junction structure appears to be changed, as revealed by tethered iron chelate analysis of protein-DNA proximity relationships within promoter complexes formed with the σ^{54}-RNAP [5]. The AAA activator appears to engage the σ^{54}-RNAP from the so called upstream face of the RNAP, and drive a conformational change involving the "jaw" domain of the β' subunit, as evidenced by the insensitivity of σ^{54}-dependent transcription to the bacteriophage T7 encoded RNAP inhibitor, gene protein 2 (gp2), which binds to the β' jaw domain [6].

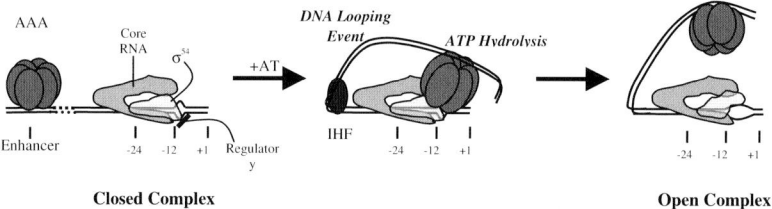

Fig. 1 Transcription Initiation by the σ^{54}-RNAP.
Cartoon showing the three major steps involved in transcription initiation by the σ^{54}-RNAP. Closed complexes formed by σ^{54}-RNAP persist, unless isomerised to open complexes by an AAA activator bound to enhancer DNA sequences located ~150 nucleotides upstream from the transcription start site at position +1. For open complex formation, the AAA activator loops out the intervening DNA (in the cartoon the DNA looping event is aided by IHF; see text) and interacts with the closed complex. One major interaction target for the AAA activator is the regulatory centre at the conserved -12 promoter recognition motif. The AAA activator converts the energy derived from ATP hydrolysis into a mechanical force used in a binding interaction with σ^{54} in order to re-model the regulatory centre and trigger open complex formation.

3. Structural Studies
We have used cryo electron-microscopy and single-particle analysis to determine the organisation of a complex formed between σ^{54} and $PspF_{1-275}$ in the presence of ADP·AlF$_x$ [7]. Analysis of this structure shows a hexameric $PspF_{1-275}$ oligomer in a ring-like configuration with σ^{54} contacted and held above the ring by extension of a loop from $PspF_{1-275}$ hexamer (Fig. 2). This so called "GAFTGA loop" is involved in the energy coupling step of transcription activation, at which the energy derived from ATP hydrolysis is coupled to the remodelling of the σ^{54}-RNAP closed complex. In the structure shown in Fig. 2, σ^{54} appears to be engaged with two adjacent $PspF_{1-275}$ monomers and the density associated with σ^{54} suggests it is quite well organised and folded in the absence of the RNAP subunits.

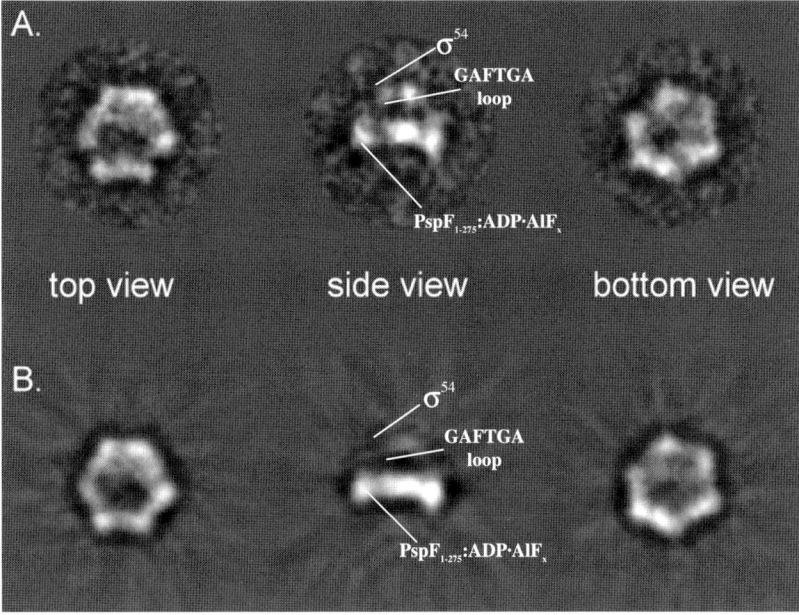

Fig. 2 Cryo-electron Microscopy Images of PspF1-275-σ54 Complex in the Presence of ADP·AlFx.
In A, (from left to right), cryo-electron microscopy class averages representing a top, side and bottom view of the PspF1-275-σ54-ADP·AlFx complex are shown. The density corresponding to σ54 is clearly visible in the side view, just above the hexameric ring of PspF1-275. The weak densities connecting the hexameric ring of PspF1-275 to σ54 represent the extended GAFTGA motifs (marked as the GAFTGA loop). In B, (from left to right), re-projections of the 3D model in the same orientations as the class averages in A are shown in order to emphasize the fidelity of the model.

We have determined the structure of $PspF_{1-275}$ to 1.75Å using X-ray crystallography [7]. In so doing, we have determined a conformational signalling pathway from the ATP binding pocket to the GAFTGA loop. Importantly, we have defined a role for a loop, called L2 (residues 130-139), in controlling the location of a GAFTGA loop, called L1 (residues 78-94), which contains the GAFTGA motif (residues 83-88) at its tip and protrudes from a helix referred to as helix3 (Fig. 3). The significance of this pathway has been tested by site directed mutagenesis and functional assays [7]. The pathway in $PspF_{1-275}$ involves several conserved residues. In the absence of ATP, interactions between residues E81 in L1 and R131 and Q136 in L2 appear to keep the L1 loop in an unfavourable conformation for interaction with σ^{54}. During ATP hydrolysis the above interaction is disrupted and novel interactions between the stem part of L1 and helix3 appear to be strengthened, hence resulting in a more "flexible" L1 tip. The latter interaction appears to be part of a large network of interactions that involves residues R95 and R98 of helix3 and S62 of a β sheet which contains the Walker B motif which is responsible for ATP hydrolysis. Thus, it is likely that during the ATP hydrolysis cycle, movement(s) of the β sheet originating from the Walker B motif is transferred to helix3, which allows L1 tip to be more flexible and adopt a conformation favorable for interaction with σ^{54}. Interestingly, the work uncovers one significant mutant class of PspF. Mutant forms of $PspF_{1-275}$ containing the R131A or V132A are able to form the ADP·AlF$_x$ dependent complex with σ^{54} but cannot hydrolyse ATP, suggesting loss of a post-transition state conformation important for functionality.

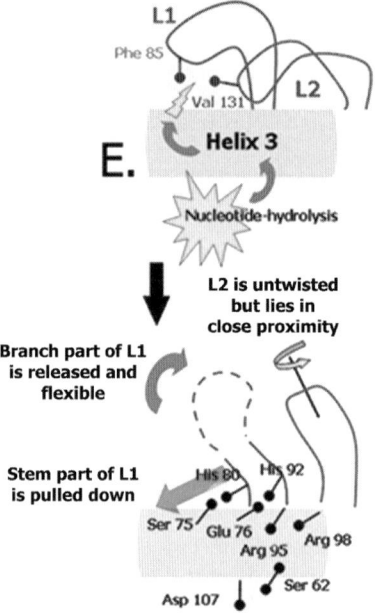

Fig. 3 Cartoon representation of the proposed mechanism for nucleotide-dependent relocation and exposure of L1 loop (GAFTGA loop) for interaction with σ 54. See text for details.

4. Future Work

Structures of AAA activator in complex with the promoter-bound σ^{54}-RNAP will provide detailed insights into the "energy coupling" mechanism during σ^{54} dependent transcription. Recent DNA-protein crosslinking studies have revealed that the AAA activator is in close proximity to promoter DNA downstream of the transcription start-site proximal consensus promoter element within the ADP·AlF$_x$ dependent complex with σ^{54}-RNAP closed complex [5, 8]. Understanding the significance and structural basis of this proximity relationship is also required.

5. Acknowledgements

The work in MB laboratory was supported by the BBSRC and Wellcome Trust. The work in XZ laboratory by the BBSRC.

6. References

1. Chaney M et al (2001). Genes Dev. 15, 2282-2294.
2. Bordes P et al (2003). Proc. Natl. Acad. Sci. USA 100, 2278-2283.
3. Cannon W et al (2003). J. Biol. Chem. 278, 19815-19825.
4. Guo Y et al. (2000). Genes Dev. 14, 2242-2255.
5. Burrows PC et al (2004). EMBO J. Advance Online Publication.
6. Wigneshwararaj SR et al (2004) EMBO J. Advance Online Publication.
7. Rappas M et al. Submitted to Science.
8. Cannon WV et al. (2004). Nucleic Acids Res. 32, 4596-4608.

HETEROCYST DIFFERENTIATION AND NITROGEN FIXATION IN *ANABAENA*

Robert Haselkorn
Department of Molecular Genetics & Cell Biology, University of Chicago,
920 East 58 Street, Chicago IL 60637 USA

The filamentous cyanobacterium *Anabaena* 7120 fixes nitrogen in specialized cells called heterocysts, which differentiate from oxygenic vegetative cells at regular intervals along the filaments under conditions of nitrogen deprivation. The differentiation requires about the same time as one vegetative cell division. The *Anabaena* genome contains at least 7,000 genes (Kaneko et al, 2001), of which more than 1,000 are differentially expressed in order to make a heterocyst. The latter genes are turned on and off according to a program that starts with the expression of a transcription factor gene, *hetR*. The product of that gene, HetR, contains a single Cys residue that can lead to dimerization under oxidizing conditions. HetR also contains several Ser residues that can be phosphorylated. The active HetR dimer binds to DNA sites upstream of certain genes and promotes their transcription (Huang et al, 2004). This transcription leads to a cascade that eventually turns on the remaining sets of genes needed to make a functional heterocyst. We will first describe the mature heterocyst and then return to the transcription program.

The heterocyst is essentially a factory for reduction of atmospheric nitrogen gas (Haselkorn, 2005). To do this, the cell synthesizes large amounts of the nitrogenase complex in addition to electron carriers that feed the enzyme with low-potential electrons and protons. The normal pathways of carbon fixation are inactivated in these specialized cells. Instead, carbon stores such as glycogen are first consumed, and then sucrose is brought in from neighboring vegetative cells, where photosynthesis continues. The sucrose is converted to hexoses by invertase (Vargas et al, 2003) and these are metabolized to provide reduced pyridine nucleotide for several purposes: the synthesis of glycolipids and polysaccharides to make the protective outer envelope and the reduction of the electron carriers to nitrogenase. Carbon is also needed for the carriers that shuttle newly fixed nitrogen to the vegetative cells. These carriers are 2-oxo-glutarate, glutamate, glutamine, aspartate and arginine, in that order. The product of nitrogen fixation, ammonia, is added to glutamate to make glutamine—that much is certain. How the heterocyst acquires enough glutamate is still a mystery (Haselkorn, 2005). Some of the glutamine is probably transported directly to the neighboring vegetative cells. Most of it, particularly at high light intensity when nitrogen fixation is rapid, is converted to arginine which is then polymerized into the storage polymer cyanophycin. The latter insoluble material has a backbone of poly-aspartate; the arginine is bound to the polymer by isopeptide linkage to the carboxyl group of each aspartate side-chain. The cyanophycin accumulates at the poles of the heterocyst where, in electron micrographs, it appears to form polar plugs. However, these plugs go through the membranes separating the two cell types. Enzymes exist that specifically break down cyanophycin to yield dipeptides and another enzyme that degrades the dipeptides to aspartate and arginine, so it is possible that cyanophycin provides the major mechanism for transport of fixed nitrogen from the heterocyst to the vegetative cell (Ziegler et al, 1998; Richter et al, 1999).

Although it was possible to obtain a great deal of information about the genes involved in heterocyst differentiation by cloning and sequencing candidates such as nitrogenase and genes encoding elements of the photosynthetic apparatus, carbon fixation enzymes, glutamine synthetase, electron transporters and transcription components, it was not until a systematic genetic approach was taken that the key regulatory genes were uncovered. The genetic approach was introduced by Wolk and Elhai (1984), who showed how to construct vectors that could transfer *Anabaena* DNA from *E. coli* to *Anabaena*. Their system was modified by Buikema and Haselkorn (1991a) to permit the isolation of genes by complementation of mutants. Using chemical mutagenesis, he isolated numerous mutants with well-defined phenotypes, such as inability to differentiate or inability to maintain stable filaments. In this way, the first master regulator, HetR was isolated (Buikema and Haselkorn 1991b). Some of the properties of HetR, along with the properties of other regulators, are summarized in Table 1. The phenotype of a *hetR* mutant is simple: it cannot differentiate and it cannot fix nitrogen. The phenotype of a strain in which *hetR* is over-expressed (knock-in, by virtue of a Cu^{++}-regulated promoter) is called MCH, for Multiple Contiguous Heterocysts. The knock-in strain can differentiate nearly 30% of its cells, even in the presence of levels of ammonia normally sufficient to completely repress differentiation (Buikema and Haselkorn, 2001).

The HetR protein can be countered by either of two negative regulators of differentiation. One is HetN, which looks as though, based on its sequence, ought to be an enzyme of fatty acid elongation, but nevertheless whose phenotype when inactivated is the precocious differentiation of heterocysts (Table 1). On the other hand, over-expression of *hetN* suppresses differentiation (Black TA and Wolk CP, 1994; Callahan SM and Buikema W, 2001). PatS is similar in phenotype, with over-expression suppressing differentiation and a knockout differentiating wildly (Yoon and Golden, 2001). Here, however, there is more information on mechanism: PatS is a peptide whose C-terminal five amino acids (RSGSR) bind to HetR and prevent HetR from binding to DNA and activating transcription. Thus, over-expression of HetN or PatS gives the same phenotype as a HetR knock-out. A fantastic result is obtained when both *hetN* and *patS* are inactivated: nearly all of the cells differentiate (Callahan SM, personal communication). This is the case even when the cells are growing on nitrate. Of course the cells die, but before they pass on, they should reveal some of the details of the early transcription cascade.

Table I A few of the many genes involved in heterocyst differentiation in *Anabaena*

Gene product	Function	Expression	Phenotype of Knock-out	Phenotype of Knock-in
NtcA	Transcription regulator	V and H	Het-, Fix-	unknown
GlnB (PII)	Transcription regulator	V and H	Het-, Fix-	unknown
HetR	autoprotease, transcription regulator	H	Het-, Fix-	MCH
HetF	not known, but phenotype similar to HetR	H	Het-, Fix-	MCH
HetN	regulator, anti-HetR	H	MCH	Het-
PatA	response regulator	V and H	terminal hets only	unknown
PatS	peptide, anti-HetR	H	MCH	Het-
PatB	redox-sensitive transcription	H	Fox-	bizarre

See text for details

There are other proteins that have been shown to play important roles in regulating transcription in *Anabaena*. NtcA can be either a negative or positive factor. It is required for transcription of *hetR*, but not directly. NtcA is likely to be the real sensor of the nitrogen status of the cell, because it binds 2-oxo-glutarate and must do so to promote transcription. Bacteria in general monitor the ratio of 2Ooxo-glutarate to glutamine to determine whether they have to activate nitrogen fixation or other nitrogen-acquiring systems. PatA is a kinase required, directly or indirectly, for HetR to be active in all cells except terminal ones (Liang J et al 1992). PatB is a redox-sensitive transcription factor probably needed for oxygen protection of nitrogenase in the heterocyst (Jones et al 2003).

With the complete genome sequence for Anabaena 7120 available at Kazusa, it became possible to construct full genome arrays and to use them to study transcription during heterocyst differentiation. This is a fairly difficult task because only ten percent of the cells differentiate. Nevertheless, Ehira et al (2003) have done this with very interesting results. There are at least 19 sets of genes that increase in transcript abundance, the sets defined by the time of increase following nitrogen step-down. Most of these genes have unknown functions. One striking result is that the genes are clustered on the chromosome according to their transcription program, in what the authors call "expressed islands". Anabaena has an abundant DNA-binding protein, HU, which disappears during heterocyst differentiation and is replaced by a different "histone-like" protein (Nagaraja & Haselkorn 1994), which might have something to do with the program. Time will tell.

Acknowledgement
I am grateful to Sean Callahan for communicating unpublished results.

References
Black TA and Wolk CP (1994) J. Bacteriol. 176, 2282-2292.
Buikema W and Haselkorn R (1991a) J. Bacteriol. 173, 1879-1885.
Buikema W and Haselkorn R (1991b) Genes & Develop. 5, 321-330.
Buikema W and Haselkorn R (2001) Proc Natl Acad Sci USA 98, 2729-2734.
Callahan SM and Buikema W (2001) Mol. Microbiol. 40, 941-950.
Ehira S et al (2003) DNA Res. 30, 97-113.
Haselkorn R (2005) In: Associative and Endophytic Nitrogen-fixing Bacteria, Elmerich C and Newton WE (eds) The Netherlands, Kluwer Publishers, in press
Huang X et al (2004) Proc Natl Acad Sci USA 101, 4848-53.
Jones K et al (2003) J. Bacteriol. 185, 2306-2314.
Nagarajah R and Haselkorn R (1994) Biochimie 76, 1082-1089.
Kaneko T et al (2001) DNA Res. 8, 205-253.
Liang J et al (1992) Proc. Natl. Acad. Sci. USA 89, 5655-5659.
Richter R et al (1999) Eur. J. Biochem. 263, 163-169.
Vargas W et al (2003) Planta 216, 951-960.
Wolk CP (1984) Proc. Natl. Acad. Sci. USA 81, 1561-1565.
Yoon HS and Golden JW (2001) J. Bacteriol. 183, 2605-2613.
Ziegler K et al (1998) Eur. J. Biochem. 254, 154-159.

SIGNAL TRANSDUCTION IN HETEROCYST DIFFERENTIATION OF THE CYANOBACTERIUM *ANABAENA* SP. PCC 7120

Yinhong Zhao, Yunming Shi, Guohua Yang, Ying Zhang, Xu Huang and Jindong Zhao
State Key Lab of Protein Engineering, College of Life Sciences, Peking University, Beijing, 100871, China

The cyanobacteria are a group of gram-negative bacteria that perform oxygenic photosynthesis. Some filamentous cyanobacteria are able to form specialized cells, called heterocysts, specialized in nitrogen fixation under nitrogen limiting growth conditions (Adam, 2000). Heterocyst properties allow them to maintain a low internal oxygen concentration for optimal nitrogenase activity (Wolk et al., 1994). Heterocysts have a thick envelope consisting of a polysaccharide layer and a glycolipid layer that limits oxygen penetration. Heterocysts lack photosystem II and do not evolve oxygen. A high respiratory activity of heterocyst helps consuming oxygen inside the cells. Heterocysts also lack Rubisco and cannot fix CO_2. The reducing power for nitrogenase comes from sugar supply from adjacent vegetative cells. In return, heterocysts provide fixed nitrogen to vegetative cells probably in the form of amino acids. Because the metabolite exchange between vegetative cells and heterocysts is required for a normal function of heterocysts, chains of heterocysts are rarely seen on cyanobacterial filaments. In fact, heterocyst distribution along the filaments is often present in a semi-regular pattern. In *Anabaena* sp. PCC 7120, there is one heterocyst per approximately 10-15 vegetative cells. According to fossil record, heterocyst might the earliest example of differentiation in biological world (Gobulic et al, 1995).

There are many genes involved in heterocyst differentiation and in heterocyst formation at regular intervals. Genes responsible for heterocyst formation could be divided into three groups: (1) the sensory genes that perceive environmental stimuli such as nitrogen starvation and plant signal in symbiosis; (2) the genes involved in initiation of the differentiation such as *hetR* and *ntcA*. This group of genes controls the initiation process and is also required for up-regulation of the expression of the genes involved in heterocyst structure and function (3) the genes that are required for heterocyst morphogenesis and function. This group of genes includes *hepK*, *devR* and the *nif* genes, responsible for heterocyst envelope and nitrogenase synthesis, respectively. While progresses are made in the understanding of heterocyst differentiation initiation and pattern formation, the first step of the signal transduction pathway remains largely unknown. In this communication, we provide evidence that calcium, a key messenger in eukaryotic cell signal transduction, is involved in heterocyst formation.

1. *hetR* gene and its function
Identification of the *hetR* gene was first reported by Buikema and Haselkorn (1991). They showed that *hetR* was required for heterocyst formation. By using *lux* gene as a reporter gene, Black and Wolk (1994) showed that *hetR* was a positively autoregulates its expression. This suggests that *hetR* could be a master switch gene for heterocyst formation. The primary sequence of *hetR* gene product, however, did not reveal any biochemical function of the HetR protein. Analysis of recombinant *Escherichia coli* strains that overproduced HetR (Zhou et al., 1998) showed that HetR was a serine-type protease. Dong et al. (2000) showed that the active site of HetR was the Ser152 residue. Replacing Ser152 with an Ala residue resulted in a non-functional HetR, which showed no auto-degradation and was unable to initiate heterocyst formation. Recently, Huang et al. showed (2004) that HetR forms homodimer through a disulphide bond and the dimerization is required for heterocyst differentiation. They also showed that HetR is a DNA-binding protein and dimerization is required for the DNA-binding activity. It has been demonstrated that an intact *hetR* gene is required for up-regulation of *hepA*, *hetR* itself and *patS*, and Huang et al. (2004) showed that HetR dimer binds promoters of these genes. The DNA-binding activity of HetR is inhibited *in vitro* by PatS pentapeptide RGSGR, which inhibits heterocyst differentiation when present in growth medium in micromolar concentration (Yoon and Golden, 1998). Thus, HetR and PatS interaction plays an important role in controlling heterocyst differentiation and spacing pattern formation.

2. Calcium and heterocyst differentiation
In studying recombinant HetR, Zhou et al. (1998) showed that Ca^{2+} induced HetR conformation change as revealed by Ca^{2+}-dependent gel mobility shift. Circular dichroism spectra also showed that Ca^{2+} could influence HetR structure. Other reports indicate that Ca^{2+} could play a role in heterocyst differentiation. Smith and Wilkins (1988) showed that the concentration of calcium in growth medium influenced heterocyst frequency in *Nostoc* 6720. Torrecilla et al. (2000) showed that *Anabaena* sp. PCC 7120 had a mechanism for Ca^{2+} homeostasis and Onek et al. (1994) reported isolation of a calmodulin-like protein from *Nostoc* 6720. However, a role of Ca^{2+} in heterocyst differentiation was not firmly established.

As shown in Figure 1, Ca^{2+} is required for heterocyst formation in *Anabaena* sp. PCC 7120. Removal of Ca^{2+} from the growth medium resulted in complete loss of heterocyst formation. We then partially purified a Ca^{2+} binding protein from *Anabaena* sp. PCC 7120. $^{45}Ca^{2+}$ overlay experiment showed that the protein was indeed binding Ca^{2+}. The protein was named cyanobacterial calcium binding protein, or CcbP. The gene encoding CcbP was cloned and it was found present in other heteroystous cyanobacteria. The predicted molecular mass of CcbP is 15 kDa and a pI of 4.1. CcbP is present in soluble form when overproduced in *E. coli* and it is resistant to heat treatment at 80°C for 10 min. Gel mobility shift assay demonstrated that CcbP, like calmodulin (CaM), showed characteristic calcium-dependent mobility shift. One molecule of CcbP could bind 18 Ca^{2+}.

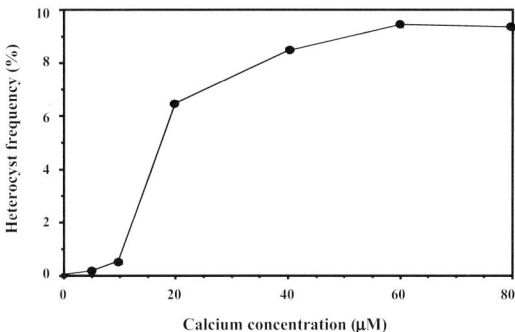

Figure 1. The relationship between calcium concentration in growth medium and heterocyst frequency of *Anabaena* sp. PCC 7120

We constructed a vector for conditional overproduction of CcbP in *Anabena* sp. PCC 7120. The *ccbP* gene was cloned in a way that it was controlled by the copper-inducible promoter of *petE* on the shuttle plasmid pRL25C and transformed into *Anabaena* sp. PCC 7120. It was found that in the absence of copper, *Anabaena* sp. PCC 7120 could produce heterocyst with a normal differentiation pattern. In the presence of copper, heterocyst differentiation under nitrogen limiting condition was completely inhibited. We also constructed a *ccbP* mutant by deleting the gene. The strain *ccbP-* showed normal morphology under nitrogen replete condition. In the absence of combined nitrogen, *ccbP-* produced multiple contiguous heterocysts (MCH) with a much higher frequency of heterocysts. The *patS-* strain (which shows MCH phenotype under nitrogen limiting condition) did not produce heterocyst when *ccbP* was overproduced, suggesting that the effect of CcbP was on early steps of heterocyst differentiation. Western blotting showed that the initial HetR up-regulation, which could be dependent upon NtcA (Fries et al., 1993) could still be observed when *ccbP* was overproduced. However, the up-regulation of *hetR* could not be sustained in the strain that over-expressed *ccbP*, suggesting that a HetR-dependent *hetR* up-regulation was suppressed by CcbP.

There are two possibilities for CcbP's function in suppression of heterocyst differentiation. One is that CcbP interacts with some proteins involved in heterocyst differentiation and the other is that CcbP regulates cellular free calcium concentration which in turn regulates heterocyst differentiation. To determine which hypothesis was correct, we constructed a CaM expression system under control of *petE* promoter. We found that when CaM was overexpressed in *Anabaena* sp. PCC 7120, heterocyst differentiation under nitrogen limiting condition was inhibited. We also found that overexpression of CaM in another species *Anabaena variabilis* inhibited heterocyst formation. These results strongly suggest that CcbP functions as calcium sequester in *Anabaena* cells. Expression of the green fluorescent protein gene under control of the *ccbP* promoter (P_{ccbP}-GFP) showed that mature heterocyst contained much less GFP, suggesting that the expression of *ccbP* in differentiating cells was turned off. This suggestion was confirmed by Western blotting, which showed that CcbP was not present in isolated heterocysts. Further analysis of intracellular free calcium content with showed that calcium concentration in heterocysts was several times higher than that in vegetative cells. A calcium concentration increase in differentiating cells could be observed 12 h after nitrogen step-down. These results strongly suggest that calcium play an important role in regulation of heterocyst differentiation.

3. Conclusion

While calcium is one of the most important molecules in signal transduction in eukaryotic cells, its role in prokaryotic cellular signal transduction has not been well established. Our experimental results showed that sequestration of intracellular calcium prevented heterocyst formation and that calcium plays an important role in initiation of heterocyst differentiation. Calcium could also be important to heterocyst function since mature heterocyst accumulates calcium. The mechanism of calcium in regulation of heterocyst formation is not known. Calcium could regulate activities of some key proteins through calcium-dependent kinase/phosphatase. The observation that HetR binds calcium implies that calcium could also influence HetR's activity directly. Further study is needed to understand calcium's role in cyanobacterial cell differentiation.

4. References

Adam DG, 2000, Curr Opin Microbiol 3, 618-624.
Black TA and Wolk CP 1994, J Bacteriol 176, 2282-2292.
Buikema WJ and Haselkorn R 1991, Genes Dev 5, 321-330.
Dong Y et al. 2000, J Bacteriol 182, 1575-1579.
Fries JE et al. 1994, Mol Microbiol 14, 823-832.
Golubic S et al. 1995, Lethaia, 179, 285-298.
Huang X et al.2004, Proc. Natl Aca Sci USA 101, 4848-4853.
Markova SV et al. 2002, Biochemistry 41, 2227-2236.
Onek S et al.1994. Arch Micriol 161, 352-358.
Smith RJ and Wilkins A 1988, New Phytol 109, 157-161.
Torrecilla I et al. 2000, Plant Physiol 123, 161-175.
Wolk CP et al. 1994, in Bryant DA, ed. The molecular biology of cyanobacteria, Kluwer Academic, pp769-823.
Yoon HS and Golden JW 1998, Science 282, 935-938.
Zhou R et al.1998, Proc. Natl Aca Sci USA 95, 4959-4963.

P_{II} SIGNAL TRANSDUCTION IN CYANOBACTERIA: NOVEL MECHANISMS OF GLOBAL NITROGEN CONTROL

Karl Forchhammer, Annette Heinrich, Nicole Kloft, Mani Maheswaran and Ulrike Ruppert
Institut für Mikrobiologie und Molekularbiologie der Justus-Liebig
Universität Giessen, Heinrich-Buff-Ring 26-32; D-35392 Giessen, Germany

P_{II} signal transduction plays pervasive roles in microbial nitrogen control. Among all of the various P_{II} signalling systems, that in cyanobacteria is so far unique: as demonstrated in the unicellular species *Synechococcus elongatus* and *Synechocystis* sp. strain PCC 6803 and recently in the heterocystous, nitrogen-fixing strain *Nostoc* (*Anabaena*) sp. PCC 7120, the P_{II} proteins are modified by serine phosphorylation in response to the cellular nitrogen status (Forchhammer 2004a; Laurent et al. 2004). Moreover, P_{II} modification is affected by the CO_2 supply and modulated by conditions, which alter the redox-status of the cells (Hisbergues et al. 1999; Forchhammer et al. 2004b). The environmental conditions are sensed by the intracellular levels of metabolites that act as effector molecules through their binding to P_{II}. The most important effectors are 2-oxoglutarate and ATP (Forchhammer 2004, Ruppert et al. 2002) since their binding alters the P_{II}-reactivity towards kinase and phosphatase and thus, controls the phosphorylation state of P_{II}. Recent advances have been the identification of the P_{II} phosphatase (termed PphA) in *Synechocystis* (Irmler and Forchhammer 2001), the resolution of the crystal structure of P_{II} proteins from *Synechococcus* and *Synechocystis* (Xu et al. 2003) and the identification of novel functions of P_{II} regulation (see below).

Global nitrogen control at the transcriptional level in cyanobacteria operates through the CRP-like transcription factor NtcA (Herrero et al. (2001). *In vivo* analyses showed that in the absence of P_{II}, activation of NtcA-dependent gene expression is strongly impaired under conditions of nitrogen deprivation (Aldehni et al. 2003), suggesting that P_{II} signals nitrogen limitation to NtcA. NtcA activation is also impaired (although less severe) in a mutant in which Ser 49, the site of phosphorylation, had been replaced by alanine. This implies that activation of NtcA is mediated by the phosphorylated form of P_{II} (P_{II}-P) (Paz-Yepez et al. 2003; Forchhammer et al. unpublished results). In contrast to these nitrogen-downshift conditions, nitrogen-upshift (addition of ammonium to nitrogen-limited cultures) seems not to depend on P_{II}-phosphorlyation/ dephosphorylation. In a PphA-deficient mutant of *Synechocystis* PCC 6803 (unable to rapidly dephosphorylate P_{II}), up-regulation of the *gif*-genes (encoding glutamine synthetase inhibitor factors) is not impaired following ammonium treatment. The *gif* genes are repressed by active NtcA (absence of ammonium) and are derepressed following ammonium addition through inactivation of NtcA. Since this is not impaired in the absence of P_{II} dephosphorylation, it follows that P_{II}-P does not permanently activate NtcA. The mode of signal transduction from P_{II}-P to NtcA remains to be established.

To identify molecular receptors of the P_{II} signal, a yeast-two hybrid screening was conducted with P_{II} as bait. By this means, the first P_{II}-receptor in cyanobacteria could be identified: N-Acetyl glutamate kinase (NAGK), the key enzyme of the arginine biosynthesis pathway, forms a tight complex with P_{II} (Heinrich et al. 2004). NAGK is the first biosynthetic enzyme with P_{II} receptor function. Titration and ultracentrifugation experiments revealed that one P_{II} trimer binds to one hexameric NAGK protein (Maheswaran et al. 2005). In contrast to the binding of proteobacterial P_{II} proteins to their receptors, complex formation of *Synechococcus* P_{II} with NAGK does not require the presence of effector molecules. Binding of P_{II} to NAGK depends on seryl-residue 49 of P_{II} being non-phosphorylated.

Binding of P_{II} to NAGK strongly affects the catalytic activity of NAGK (Maheswaran et al. 2005): The K_M for N-acetylglutamate decreases approximately four-fold and V_{max} increases 10-fold. Feedback inhibition of NAGK activity by arginine is strongly reduced upon P_{II} binding. Together, this results in a tight control of NAGK activity by P_{II}. In accord with the data that complex formation was impaired when P_{II} Ser49 was mutated, NAGK activity in *S. elongatus* extracts correlated with the phosphorylation status of P_{II}. High NAGK activities corresponded to non-phosphorylated P_{II} (nitrogen-excess condition) and low activities corresponded to increased levels of P_{II} phosphorylation (nitrogen–poor conditions), thus subjecting the key enzyme of arginine biosynthesis to global nitrogen control.

The stability of the NAGK-P_{II} complex was analyzed by BIACORE surface plasmon resonance analysis. In the absence of divalent cations, ATP and ADP had a strong negative effect on complex formation. Increasing concentrations of Mg^{2+} restored complex formation in presence of ATP, but not ADP. In presence of Mg^{2+}/ ATP, increasing concentrations of 2-oxoglutarate abolished complex formation. Ca^{2+} was able to restore the NAGK-P_{II} complex in the presence of ATP and 2-oxoglutarate. Together, this suggests that NAGK activity is controlled by P_{II} in multiple manners: in response to the nitrogen status via phosphorylation and immediately by the 2-oxoglutarate effect; furthermore, in response to the energy status via the adverse effect of ADP on complex formation. The control of arginine synthesis by P_{II} is corroborated by deregulated accumulation of the arginine storage compound cyanophycin in mutants of the P_{II}

signalling pathway in the heterocystous cyanobacterium *Nostoc* (*Anabaena*) PCC 7120 (Laurent et al. 2004).

Acknowledgments:
The work in the author's lab was supported by a grant from the "Deutsche Forschungsgemeinschaft" (Fo 195/4) and by the "Fonds der Chemischen Industrie"

References
Aldehni M-F et al. (2003) J. Bacteriol. 185, 2582-2591.
Forchhammer K (2004a) FEMS Microbiol. Rev. 28, 319-333.
Forchhammer K et al. (2004b) Physiol. Plantarum 120, 51-56.
Heinrich A et al. (2004) Mol. Microbiol. 52, 1303-1314.
Herrero A et al. (2001) J. Bacteriol. 183, 411-425.
Hisbergues M et al. (1999) FEBS Lett. 463, 216-220.
Irmler A and Forchhammer K (2001). Proc. Natl. Acad. Sci. USA. 98, 12978-12983.
Laurent S et al. (2004) FEBS Lett. 576, 261-265.
Maheswaran M et al. (2005) J. Biol. Chem. *In press*
Paz-Yepez J et al. (2003) FEBS Lett. 543, 42-46.
Ruppert U et al. (2002) Mol. Microbiol. 44, 855-864.
Xu Y et al. (2003) Acta Cryst.D 59, 2183-2190.

MOLECULAR SIGNAL OF NITROGEN STARVATION, INSIGHT FROM THE CYANOBACTERIUM *ANABAENA* PCC 7120

Cheng-Cai Zhang[1,2], Sophie Laurent[1], and Sylvie Bédu[1]
[1]Laboratoire de Chimie Bactérienne, CNRS, 31 chemin Joseph Aiguier, 13402 Marseille cedex 20, France
[2]National Key Laboratory of Agricultural Microbiology, Huazhong Agriculture University, Wuhan 430070, China

The cyanobacterium *Anabaena* PCC 7120 responds to starvation of combined nitrogen source in the growth medium by the differentiation of cells, called heterocysts, devoted to molecular nitrogen fixation. The two cells types are interdependent to ensure the growth of filament in the absence of combined nitrogen in the growth medium. Heterocysts provide nitrogen source, likely in the form of aminoacids, to surrounding vegetative cells and receive from the later carbon synthates and reductants derived from photosynthetic activity (Herrero et al 2004). A close interdependence between these two cell types exists for the assimilation of nitrogen. Glutamine synthetase (GS) is active in heterocysts, but glutamate synthase (GOGAT) is active in vegetative cells. As a consequence, glutamate has to come to heterocysts from vegetative cells, while glutamine and 2-oxoglutarate (2- OG) from heterocysts should be sent to vegetative cells (Martin-Figueroa et al, 2000)

How filaments of *Anabaena* interpret nitrogen availability? The intracellular pool of 2-OG is thought to reflect nitrogen status within cells of cyanobacteria (Muro-Pastor et al, 2001). Cyanobacteria have no 2-oxoglutarate dehydrogenases, thus an incomplete Krebs cycle. In this case, this metabolite is mainly used as a carbon skeleton to incorporate ammonium and synthesize aminoacids. Our results show that the intracellular pool of 2-OG rapidly increases by 2-2.5 folds following nitrogen starvation (unpublished data). To modulate the intracellular pool of 2-OG *in vivo*, we have expressed the *kgtP* gene encoding 2-oxoglutarate permease from *E. coli* under the control of a regulated promoter. Our results show that uptake of extracellular 2-OG into cells promotes heterocyst development, thus mimicking nitrogen starvation status (Li et al, 2003). These results are consistent with those obtained from unicellular cyanobacterial strains (Herrero et al 2004). However, it should be stressed that no formal evidence in vivo is available to ascertain the signalling function of 2-OG since it is rapidly assimilated into a variety of metabolites. It remains therefore to determine that the effect of 2-OG is a direct one.

What are the receptors of 2-OG in cyanobacteria? Two proteins whose activity is influenced by the level of 2-OG are known in cyanobacteria, NtcA and PII. NtcA is required for heterocyst development, and its transcription is regulated by the presence of 2-OG *in vitro* (Herrero et al, 2004). The role of PII in heterocyst development remains unknown. PII responds to the level of 2-OG by reversible phosphorylation/ dephosphorylation on a serine residue in *Synechococcus* PCC 7942. Nitrogen limitation favours the phosphorylation of PII, while nitrogen sufficient condition leads to its dephosphorylation (Arcondéguy et al, 2001; Forchhammer, 2004). We found that PII phosphorylation in *Anabaena* is cell type specific. In the presence of ammonium PII was dephosphorylated. In filaments grown under N-fixation conditions, both forms of PII are found in vegetative cells, but only the dephosphorylated one is present in heterocysts (Laurent et al, 2004). In heterocystous cyanobacteria, the *glnB* gene encoding PII is likely to be essential. That was shown in *Nostoc punctiforme* (Hansen et al 1998), and our own screening of about a hundred of exconjugants has not allowed us to get one single mutant with complete segregation in *Anabaena*. We thus constructed several mutants in which the modification state of PII is affected. In one mutant (PII S49A), the serine 49 is changed to alanine, while in the other, a putative PII phosphatase (*prpS*) was deleted and in which the phosphorylated form of PII is present even under ammonium regime. No mutant in which serine 49 is changed to glumamate, mimicking constitutively phosphorylated form of PII, was obtained (Laurent et al 2004).

We conclude that: 1) the unphosphorylated form of PII is likely essential; 2) the phosphorylated form of PII inhibits the accumulation of the nitrogen reserve cyanophycin in heterocysts, and favours instead its accumulation in vegetative cells; 3) PII is unlikely to be involved in the early step of heterocyst development, since in the PII S49A mutant where PII is never phosphorylated, therefore unable to transmit the signal of nitrogen starvation, heterocysts are still normally formed. It remains to determine however, the essential function of PII in heterocystous cyanobacteria. PII has a high affinity towards 2-oxoglutarate in unicellular cyanobacteria (Forchhammer, 2004), thus very sensitive to the small variation of 2-OG pool. PII modification/demodification is likely to be involved to tune metabolism subtly. For NtcA, we propose that it is less sensitive to the variation of 2-OG level, and contributes to the initiation of heterocyst development after a big jump in the level of 2-OG.

Arcondéguy T et al. (2001)   Microbiol. Mol. Rev. 65, 80-105.
Forchhammer K. (2004) FEMS Microbiol. Rev. 28, 319-333.
Hansen TE et al.   (1998) Microbiol. 144, 1537-1547.
Herrero et al. (2004) FEMS Microbiol. Rev. 28, 469-487.
Laurent S et al. (2004) FEBS Lett. 576, 261-265.
Li JH et al. (2003) Microbiol. 149: 3257-3263.
Martin-Figueroa E et al. (2000) FEBS Lett 476, 282-286.
Muro-Pastor MI et al. (2001) J. Biol. Chem. 276, 38320–38328.

ELECTRON TRANSPORT TO NITROGENASE IN *RHODOSPIRILLUM RUBRUM*

Tomas Edgren and Stefan Nordlund
Department of Biochemistry and Biophysics, Arrhenius Laboratories,
Stockholm University, SE-106 91 Stockholm, Sweden

Rhodospirillum rubrum is a purple non-sulfur photosynthetic bacterium and was shown to fix nitrogen in 1949 by Gest and Kamen. This discovery was followed by a number of studies seminal to our understanding of the physiology of nitrogen fixation in this group of diazotrophs (Ormerod, et al., 1961). These phototrophs meet the high ATP-demand of nitrogen fixation through photophosphorylation at the expense of the proton motive force generated by light–driven cyclic electron transport. In addition to the demand for ATP, the high energy requirement of nitrogen fixation is also manifested in the need for low-potential reductant in a form that can donate electrons to the Fe-protein of nitrogenase.

The identification of pathways for electron transfer from metabolic intermediates to nitrogenase has been given surprisingly limited interest and today only the pathway utilized in *Klebsiella pneumoniae* has been characterized in great biochemical and genetic detail (Shah et al., 1983). In *K. pneumoniae* electrons are transferred from pyruvate to the Fe-protein, with the *nif*–specific flavodoxin NifF as the direct electron donor. NifF is reduced and pyruvate oxidized to acetylCoA in a reaction catalyzed by NifJ, a *nif*–specific pyruvate:flavodoxin oxidoreductase. In most other diazotrophs this pathway does not seem to be the major means by which reductant is transferred to nitrogenase. The common denominator for the organisms studied is that the direct electron donor to nitrogenase is either a flavodoxin or a ferredoxin.

In purple, non-sulfur photosynthetic bacteria electron transport to nitrogenase has mainly been studied in *R. rubrum* and *Rhodobacter capsulatus*. A number of ferredoxins have been characterized in *Rb. capsulatus*, four of these as well as a flavodoxin are *nif*–specific (Hallenbeck and Gennaro, 1998; Jouanneau et al. 1995; Saeki et al., 1996). The main direct electron donor to nitrogenase in this phototroph is believed to be either ferredoxin I (FdI), encoded by *fdxN* or the flavodoxin (encoded by *nifF*). A major question is the identity of the reaction(s) by which these proteins are reduced. In *Rb. capsulatus* two possible systems have been identified, one is a pyruvate:ferredoxin oxidoreductase which has been shown to have the capacity to catalyze the reduction of the ferredoxins and the flavodoxin (Yakunin and Hallenbeck, 1998). The other system is comprised of the proteins encoded by the *rnf* operon (Jeong and Jouanneau, 2000; Kumagai et al., 1997). In this case FdI is suggested to be reduced in a reaction driven by the membrane potential across the cell membrane.

In *R. rubrum* two soluble ferredoxins, FdI and FdII, were shown to support nitrogenase activity in vitro using an illuminated thylakoid system (Yoch and Arnon, 1975). The gene encoding FdI was later sequenced and named *fdxN* (von Sternberg and Yoch, 1993) and the protein levels were shown to be up-regulated under *nif*-conditions. No *nif*–specific flavodoxin has been identified in this diazotroph. We have identified and characterized a pyruvate:oxidoreductase, which together with a mixture of *R. rubrum* ferredoxins could support nitrogenase activity in vitro (Brostedt and Nordlund, 1991). This enzyme which is quite similar to NifJ from *K. pneumoniae* is however not *nif*–specific. In addition, a mutant lacking the enzyme has no phenotype with respect to growth under diazptrophic conditions (Lindblad et al.,1996). It is thus obvious that this NifJ-like enzyme is not part of the major electron transport pathway to nitrogenase in *R. rubrum*. We have not been able to identify *rnf* genes in the *R. rubrum* genome draft.

Studies in our laboratory have shown that nitrogenase activity in *R. rubrum* is dependent on an operational TCA-cycle, as addition of fluroacetate leads to inhibition (Brostedt et al., 1997). This would suggest that reductant is transferred to nitrogenase in reactions involving NADH. Furthermore we have shown that when the proton motive force is affected, either by uncouplers or by darkness, there is a decrease in nitrogenase activity which is not due to a decrease in the ATP concentration (Lindblad and Nordlund, 1997).

Recently we identified the *fixABCX* genes in *R. rubrum* and showed that a mutation in *fixC* caused lower growth rate and a 75% decrease in nitrogenase activity in vivo (Edgren and Nordlund, 2004). Furthermore, we demonstrated that the expression of the *fix* genes was repressed under nitrogen sufficient conditions indicating that the genes are *nif*-regulated. FixA and FixB were shown to be soluble proteins whereas most likely FixC forms a complex with FixX, localized to the membrane.

In our continued studies we have identified a new ferredoxin in *R. rubrum*, which shows high similarity to FdI. A mutant devoid of this ferredoxin showed a 25% decrease in nitrogenase activity, whereas there was no effect when the *fdxN* gene encoding FdI was mutated. Furthermore using Northern blotting we could not detect an *fdxN* transcript under any of the growth conditions tested. We have chosen to name the new ferredoxin FdN and the gene *fdxN2* to distinguish it from the previously reported ferredoxin. *fdxN2* is located within a cluster of putative *nif* genes including *nifB* and *nifZ*. We are convinced that FdN is the major direct electron donor to nitrogenase in *R. rubrum* although it is also expressed under nitrogen sufficient conditions. As we have not yet been able to reveal the conditions under which *fdxN* is expressed at high levels, we believe that in most of the experiments previously reported, FdN was in fact present and probably the dominating ferredoxin. We have also constructed a double mutant, lacking both FdI and FdN, which has only about 15% in vivo activity. We suggest that this remaining activity is due to ferredoxin II. The double mutant could not grow diazotrophically.

In order to further clarify the role(s) of the ferredoxins, the Fix proteins and the NifJ-like protein we have constructed a number of mutants in which we have combined mutations in each of these proteins. The mutant in which both FixX and the NifJ-like protein are lacking does not grow diazotrophically and nitrogenase activity *in vivo* is less than 10% of wild type.

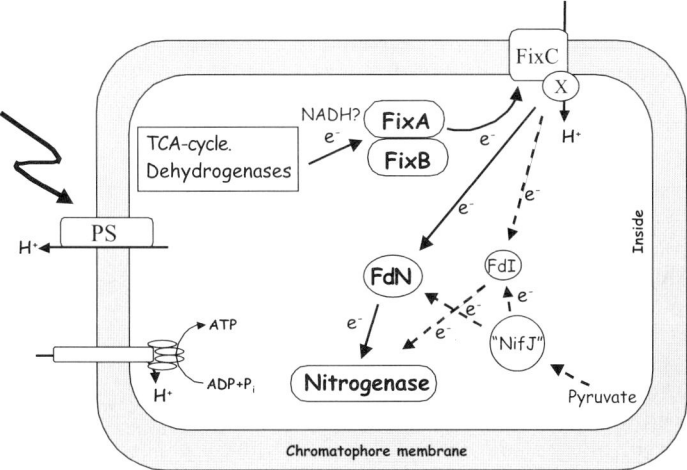

Figure 1. Model for the electron transfer pathway to nitrogenase in *R. rubrum*. Dashed arrows indicate reactions not part of the major pathway

One of the interesting characteristics of nitrogen fixation in *R. rubrum* is the metabolic regulation, manifested as the "switch-off" effect (Nordlund and Ludden, 2004). The molecular basis for this regulatory mechanism is reversible ADP-ribosylation of the Fe-protein in reactions catalyzed by DRAT and DRAG. A major question has been how the activities of these proteins are controlled. The redox-state of the Fe-protein has been suggested to be one important factor in this control (Halbleib et al., 2000). We have also investigated the "switch-off" effect in the various mutants, but have not been able to determine any significant differences from wild type, neither in response to ammonium ions nor when switching to darkness. We therefore conclude that the redox-state of the Fe-protein probably is important in the metabolic regulation of nitrogenase but it is not the factor triggering the increase in DRAT activity.

In summary we propose that the major electron transfer pathway to nitrogenase in *R. rubrum* is from either NADH or a dehydrogenase to the FixAB dimer, which in turn can reduce FdN in a reaction catalyzed by FixCX being dependent on the proton motive force (Figure 1). We further suggest that FdI can substitute for FdN although not as efficiently. An alternative pathway is comprised of the NifJ-like protein, which catalyzes the reduction of FdI or FdN with pyruvate as electron donor. This route is however less efficient at least under photoheterotrophic conditions. It still remains to demonstrate the role of FixAB, as well as the relation between the proton motive force and the function of FixCX at the molecular level.

References
Brostedt, E and Nordlund, S (1991) Biochem. J. 279, 155-158.
Brostedt, E et al. (1997) FEMS Microbiol. Lett. 150, 263-267.
Edgren, T and Nordlund, S (2004) J. Bacteriol. 186, 2052-2060.
Gest, H and Kamen, MD (1949) Science 109, 558-559.
Halbleib, CM et al. (2000) J. Biol. Chem. 275, 3493-3500.
Hallenbeck, PC and Gennaro, G (1998) Biochim. Biophys. Acta 1365, 435-442.
Jeong, HS and Jouanneau, Y (2000) J. Bacteriol. 182, 1208-1214.
Jouanneau, Y et al. (1995) Biochim. Biophys. Acta 1232, 33-42.
Kumagai, H et al. (1997) Biochemistry 36, 5509-5521.
Lindblad, A et al. (1996) Mol. Microbiol. 20, 559-568.
Lindblad, A and Nordlund, S (1997) Photosynth. Res. 53, 23-28.
Nordlund, S and Ludden, PW (2004) in: Genetics and regulation of nitrogen fixation in free-living bacteria, (Klipp, W et al., Eds.) Kluwer Academic Publisher, the Netherlands, pp. 175-196.
Ormerod, JG et al. (1961) Arch. Biochem. Biophys. 94, 449-463.
Saeki, K et al. (1996) J. Biol. Chem. 271, 31399-31406.
Shah, VK et al. (1983) J. Biol. Chem. 258, 12064-12068.
von Sternberg, R and Yoch, DC (1993) Biochim. Biophys. Acta 1144, 435-438.
Yakunin, AF and Hallenbeck, PC (1998) Biochim. Biophys. Acta 1409, 39-49.
Yoch, DC and Arnon, DI (1975) J. Bacteriol. 121, 743-746.

REGULATION OF NITROGEN FIXATION IN THE PHOTOSYNTHETIC BACTERIUM *RHODOSPIRILLUM RUBRUM*: ROLES OF GLND AND P_{II} PROTEINS

Yaoping Zhang, Edward L. Pohlmann and Gary P. Roberts
Department of Bacteriology and the Center for the Study of Nitrogen Fixation, University of Wisconsin-Madison, Madison, WI 53706, USA

1. The various roles of nitrogen regulatory protein P_{II} in *R. rubrum*
In *Rhodospirillum rubrum*, nitrogen fixation is tightly regulated at both the transcriptional and posttranslational levels, with GlnD and P_{II} homologs playing very important roles. Three P_{II} homologs, GlnB, GlnK, and GlnJ, have been identified in *R. rubrum* (Johansson and Nordlund 1996; Zhang et al. 2000; 2001b), and they play both distinct and overlapping functions in the cell (Fig. 1).

NifA activation.
Unlike that seen in *Klebsiella pneumoniae*, transcription of *nifA* is not regulated in *R. rubrum*, but NifA activity is tightly controlled in response to NH_4^+ (Zhang et al. 2000). This regulation is probably through the direct interaction between NifA and the uridylylated form of GlnB, which is required for the activation of NifA activity under nitrogen-limiting conditions. Neither GlnK nor GlnJ can replace GlnB to activate NifA (Zhang et al. 2000; 2001b). Recently, we found that the requirement for GlnB in NifA activation reflects properties of P_{II} proteins, rather than the level/time of protein synthesis. GlnJ and GlnK are still unable to activate NifA even when they are expressed from the *glnB* promoter. With site-directed mutagenesis, we identified a few residues in GlnB critical for its interaction with NifA, by converting GlnJ to GlnB for NifA activation. Two (residues 45 and 54) are in the T-loop, but other critical residues lie in the C-terminus (residues 95-97 and 109-112) and near the N-terminus (residues 3-5 and 17). These results lead to the hypothesis that the T-loop of GlnB is flexible enough to come into the proximity of the C- or N-terminal regions of the protein in order to bind NifA (Zhang et al. 2004).

Posttranslational regulation of nitrogenase activity.
The nitrogenase activity of *R. rubrum* is also regulated in response to exogenous NH_4^+ or to energy limitation. This regulation involves the reversible ADP-ribosylation of

dinitrogenase reductase, catalyzed by the DRAT/DRAG regulatory system (Ludden 1994). Dinitrogenase reductase ADP-ribosyl transferase (referred to as DRAT) carries out the transfer of the ADP-ribose from NAD to the Arg-101 residue of dinitrogenase reductase, resulting in inactivation of that enzyme. The ADP-ribose group attached to dinitrogenase reductase can be removed by another enzyme, dinitrogenase reductase activating glycohydrolase (referred to as DRAG), thus restoring nitrogenase activity. The activities of DRAT and DRAG are regulated by GlnB and GlnJ (Zhang et al. 2000; 2001a; 20001b).

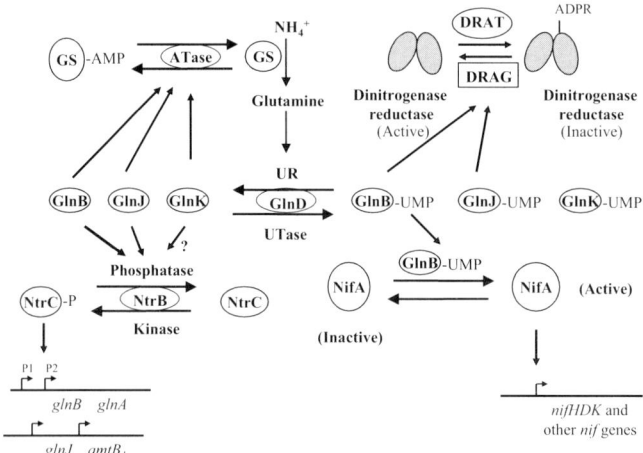

Fig. 1. The various roles of P_{II} and GlnD in *R. rubrum*

Adenylylation of glutamine synthetase.
Similar to the case in *Escherichia coli* and other bacteria, glutamine synthetase of *R. rubrum* is modified by adenylylation in response to NH_4^+. This is catalyzed by ATase and P_{II}, and any of the three P_{II} homologs can support this function in vivo (Zhang et al. 2001b).

Cell growth.
Strains lacking both *glnB* and *glnJ*, irrespective of the presence of *glnK*, grow poorly on both minimal and rich media (Zhang et al. 2001b). It is unknown if these P_{II} homologs have another target affecting cell growth or if this is an indirect effect.

Regulation of NtrB/NtrC regulatory system.
NtrB/NtrC is a two-component regulatory system. NtrB is a histidine kinase that phosphorylates NtrC or acts as a phosphatase to dephosphorylate NtrC. Evidence

suggests that in *R. rubrum,* as in *E. coli,* P interacts with NtrB to stimulate its phosphatase activity under nitrogen-excess conditions, resulting in the dephosphorylation of NtrC. However, under nitrogen-limiting conditions, the P_{II} proteins could be uridylylated by GlnD, preventing their interaction with NtrB, so that NtrB is dominated by its kinase activity to phosphorylate NtrC. The phosphorylated form of NtrC acts as a transcriptional activator of *glnBglnA, glnJamtB₁* and other operons. In *R. rubrum, glnB* is expressed from both σ^{54}- and σ^{70}-dependent promoters, so that GlnB is present under all conditions. However, the expression of *glnJ* is tightly regulated by NtrB/NtrC system in response to nitrogen status. All three P_{II} homologs accumulate to different levels in *R. rubrum.* GlnJ is the most abundant and GlnK is the least abundant in the cell under nitrogen-limiting conditions.

2. The various roles of nitrogen sensing protein GlnD in *R. rubrum.*
GlnD is a bifunctional uridylyltransferase/uridylyl-removing enzyme (UTase/UR) and is believed to be the primary sensor of nitrogen status in the cell (Fig. 1). It plays an important role in nitrogen assimilation and metabolism by reversibly regulating the modification of P_{II} proteins. By searching the Pfam protein family database, *R. rubrum* GlnD appears to have four conserved domains. A nucleotidyltransferase domain in the vicinity of residues 105 to 184 has some residues in common with many proteins in an ancient nucleotidyltransferase family, such as kanamycin nucleotidyltransferase and DNA polymerase β. This domain probably represents the UTase activity (Holm and Sander 1995). An HD domain located at residues 488 to 610 has some residues in common with proteins of a new superfamily of metal-dependent phosphohydrolases, and possibly represents the UR activity (Aravind and Koonin 1998). Two putative ACT domains have been identified in the extreme C-terminus residues (residues 727-797 and 839-917). ACT domains have been found in functionally diverse proteins and may serve as amino acid-binding sites in some metabolic enzymes (Chipman and Shaanan 2001). These ACT domains in GlnD might be involved in binding glutamine and 2-ketoglutarate.

UTase is located in the N-terminal region of GlnD.
A *glnD* mutant with a Gm^r insertion at codon 393 (where *aacC1* is transcribed in the opposite orientation as *glnD*) showed negligible nitrogenase activity. Similarly, *glnD* deletion mutants (deleted residues 23-864) also showed no nitrogenase activity. Complementation experiments with N-terminal domain of GlnD confirmed the UTase is located in this region. These results also indicate that UTase activity of GlnD is essential for *R. rubrum* NifA activity, certainly through GlnB-UMP.

Effect of *glnD* mutation on the DRAT/DRAG regulatory system.
All *glnD* mutants responded normally to darkness shifts in their regulation of nitrogenase activity. However, no recovery of nitrogenase activity was seen in *glnD* deletion mutants after shifting back to light, though a complete recovery was seen in *glnD* insertion mutants. This indicated that UTase activity of GlnD is essential for

DRAG activation. The fact that a *glnD* insertion mutant had sufficient UTase activity for DRAG activation, but not for NifA activation, implies a hierarchy of effects of the levels of GlnD on the different regulatory systems.

Effects of *glnD* mutation on GS modification and cell growth.
R. rubrum glnD deletion mutations are not lethal, and such mutants grow reasonably well in both rich and minimal malate-glutamate medium. However, there were growth defects on minimal medium with no added nitrogen source and with NH_4^+ as the sole nitrogen source. The slow growth in minimal NH_4^+ medium is apparently due to low glutamine synthetase activity, because a Δ*glnD* mutant with an altered glutamine synthetase (GS-Y398F) that cannot be adenylylated can grow well in minimal NH_4^+ medium. Most GS was modified in Δ*glnD* mutants under all conditions, even in nitrogen-limiting conditions. However, there was always a small amount of unmodified GS in these *glnD* mutants, even in cells grown in nitrogen excess conditions, which might be the reason for their non-lethality.

Effects of *glnD* mutation on NtrB/NtrC system and *glnJ* expression.
In *R. rubrum*, the expression of *glnJ* is tightly regulated by NtrB/NtrC system, through the interaction with P_{II} proteins. No GlnJ was found in Δ*glnD* mutants, indicating that modified P_{II} is essential for activating NtrC activity. Interestingly, a dramatic decrease of GlnJ accumulation was found in Δ*glnD* mutants with N-terminus of GlnD from a multi-copy plasmid or a single-copy allele. These results indicate that the full function of the NtrB/NtrC system requires a higher level of UTase activity than that needed for NifA activation and DRAT/DRAG regulation.

In summary, *R. rubrum* strains deleted for *glnD* are viable, but the presence of reduced levels of UTase activity in some *glnD* insertion mutants has a differential effect on NifA activation, DRAG regulation, and NtrC-activation. Different levels of UTase activity are required for proper regulation of these three different regulatory systems, suggesting that the analysis of the roles of GlnD in other organisms might be rather more difficult to analyze than has been generally appreciated.

References:
Aravind L and Koonin EV (1998) TIBS 23, 469-472.
Chipman DM and Shaanan B (2001) Curr. Opin. Struct. Biol. 11, 694-700.
Holm L and Sander C (1995) TIBS 20, 345-347.
Johansson M and Nordlund S (1996) Microbiology 142, 1265-1272.
Ludden PW (1994) Molecular & Cellular Biochemistry 138, 123-129.
Zhang Y et al. (2000) J. Bacteriol. 182, 983-992.
Zhang Y et al. (2001a) J. Bacteriol. 183, 1610-1620.
Zhang Y et al. (2001b) J. Bacteriol. 183, 6159-6168.
Zhang Y et al. (2004) Proc. Natl. Acad. Sci. USA 101, 2782-2787.

FUNCTIONAL ANALYSIS OF THE NIFA GAF DOMAIN OF *AZOSPIRILLUM BRASILENSE:* EFFECT OF TYR-PHE MUTATIONS IN NIFA AND INTERACTION OF THE MUTATED PROTEINS WITH GLNB USING THE YEAST TWO-HYBRID SYSTEM

Sanfeng Chen[1], Li Liu[1], Xiaoyu Zhou[1], Claudine Elmerich[2], and JiLun Li[1]
[1]Department of Microbiology and National Key Laboratory for Agrobiotechnology, China Agricultural University, Beijing 100094, P. R. China; [2]Institut des Sciences du Végétal, UPR 2355-CNRS, 91198 Gif-sur-Yvette and Institut Pasteur, 75724 Paris Cedex 15, France.

NifA is synthesized both in conditions compatible and incompatible with nitrogen fixation in *Azospirillum brasilense* (Arsène et al., 1996). The mechanism by which NifA is converted from an active form to an inactive form is still unknown. However, it was shown that GlnB is required to maintain NifA in its active form and the target of GlnB activity is the N-terminal part of NifA. The NifA N-terminal domain contains a GAF domain found in sensory proteins (Studholme and Dixon, 2003). Using the yeast two-hybrid system, it was shown that GlnB binds to NifA. In particular, GlnB interacts with the N-terminal GAF domain of NifA, but not with the central or C-terminal domains (Du et al., 2002; Chen et al., 2003).

1. Mutagenesis of Tyr residues in the GAF domain of NifA: Properties of a double mutant carrying Tyr-Phe substitutions at positions 18 and 53

Deletion of most of the NifA GAF domain resulted in a NifA protein that was active independently of the presence or absence of GlnB and in the presence of ammonia (Arsène et al., 1996). Mutagenesis of the Tyr residues at positions 18, 43 and 53 in the NifA GAF domain led also to active NifA proteins. In particular, the mutated NifA-Y18F was active in a *glnB* mutant, but nitrogen fixation was still repressed by ammonia (Arsène et al., 1999). A double mutant that carried Tyr-Phe substitutions at positions 18 and 53 was then constructed. The doubly mutated NifA was active in *nifA* and *glnB* mutated backgrounds. The increase of the nitrogenase activity was greater than that observed with the wild type or single NifA-mutants (Table 1). This increase was correlated to an increase of *nifH* expression, but *nifH* expression was still repressed by ammonia (not shown).

2. Interaction of mutated NifA with GlnB

Yeast two-hybrid experiments showed that wild type NifA GAF domain binds GlnB (Chen et al., 2003). Similar experiments were performed with the NifA mutants. We observed binding between GlnB and NifA-Y18F as well as with NifA-Y53F. The interaction between GlnB and NifA-Y18/53F was not detected (data not shown).

Table 1. Nitrogenase activity of the wild type Sp7, nifA (6067) and glnB (6006) strains complemented wild plasmids carrying wild type or mutated nifA genes.

	No plasmid	pNifA	pNifA-Y18F	PNifA-Y18/53 F
		% of nitrogenase activity		
Wild type	100	140	180	220
nifA	<1	110	180	220
glnB	<1	<1	90	120

3. Discussion

Data of Chen et al. (2003) and those reported here suggest a mechanism of protein-protein interaction between NifA GAF domain and GlnB. As the binding was detected in yeast, it can be concluded that in *Azospirillum* non-uridylylated GlnB binds the NifA terminal domain, both under ammonia assimilation and nitrogen-fixation conditions. We know that GlnB is required in its uridylylated form to antagonize the inhibition of NifA terminal domain on NifA activity. This is based on the finding that a *glnD* mutant and a *glnB* mutant (with a mutation of the Tyr 51 codon susceptible of uridylylation) were Nif⁻ (Arsène et al., 1999; van Dommelen et al., 2002). It thus appears that the binding between NifA and GlnB is not sufficient to convert NifA into its active form, unless GlnB is in the GlnB-UMP form. Indeed, some mutants independent of GlnB (Y18) still bind NifA whereas others (Y18/53) did not. Moreover, it could not be demonstrated that the Tyr residues in NifA were the target of a covalent modification controlled by GlnB. The NifA mutants deleted from a large part of their GAF domain differ from those with mutations in Tyr residues. The first ones displayed low nitrogenase activity both independently from GlnB and in the presence ammonia (Arsène et al., 1996). In contrast, mutants in Tyr residues, in particular Tyr 18 and Tyr 18/53, display high nitrogenase activity, whose synthesis is still regulated by ammonia, but not by GlnB. This supports control by ammonia and by GlnB using different mechanisms

This work was supported by the Development Program of China ("973" Program, Grant No. 001CB1089004).

4. References

Arsène F et al. (1996) J.. Bacteriol. 178, 4830-4838 .
Arsène F et al. (1999) FEMS Microbiol. Lett. 179, 339-343.
Chen SF et al. (2003) Chinese Science Bulletin 48, 170-174.
Du J et al. (2002) in T Finan et al (eds). Nitrogen Fixation: Global Perspectives. CAB International, Wallingford, UK, p. 374.
Studholme D and Dixon R (2003) J. Bacteriol. 185, 1757-1767.
van Dommelen A et al. (2002) Mol. Genet. Genomics 266, 813-820.

EFFECTS OF 2-OXOGLUTARATE AND THE PII HOMOLOGUES NIFI$_1$ AND NIFI$_2$ ON NITROGENASE ACTIVITY IN CELL EXTRACTS OF *METHANOCOCCUS MARIPALUDIS*

J. A. Dodsworth and J. A. Leigh
Dept. of Microbiology, University of Washington, Box 357242, Seattle, WA 98195, USA

The post-translational regulation of nitrogen fixation in response to a preferred nitrogen source such as NH_4^+ is termed ammonia switch-off, and has been observed in many diazotrophic prokaryotes. The best understood mechanism of this process occurs in *Rhodospirillum rubrum*, and involves reversible covalent modification of the *nifH* gene product, dinitrogenase reductase (Ludden and Roberts, 1989). Alternative mechanisms for switch-off apparently exist in some other diazotrophs (e.g. Yoch et al., 1988), however these mechanisms are not as well understood. This is the case in the methanogenic archaeon *Methanococcus maripaludis*, where switch-off does not seem to involve covalent modification of NifH (Kessler et al., 2001). Utilizing the genetic tools available for this organism, it was determined that two open reading frames directly downstream of *nifH* were required for switch-off. These genes, *nifI*$_1$ and *nifI*$_2$, are divergent members of the PII family of nitrogen regulatory proteins, which also include *glnB* and *glnK*. PII proteins are present in most prokaryotes, and play a key role in regulation of nitrogen metabolism by sensing intracellular nitrogen status and affecting regulation by interaction with other proteins (Arcondéguy et al., 2001).

To better understand switch-off in *M. maripaludis*, we assayed nitrogenase activity in crude cell extracts. Similar levels of activity, approximately 0.4 nmol C_2H_2 reduced min^{-1} mg protein^{-1}, were obtained in extracts from nitrogen fixing and switched-off cells. To determine if the NifI proteins had an effect on *in vitro* nitrogenase activity, the same experiments were performed with extracts from a Δ*nifI*$_1$*nifI*$_2$ strain. Activity in this switch-off deficient strain was approximately 5-fold higher than in extracts from wild type cells, suggesting that the NifI proteins have a negative effect on nitrogenase activity. Similar results were obtained with extracts from single *nifI* deletion mutant strains, and extracts from a strain where these *nifI* deletion mutations were complemented had activity similar to extracts from wild type strains. This indicates that switch-off deficiency is correlated with increased *in vitro* nitrogenase activity. To determine if potential signals of cellular nitrogen status might have an effect on wild type or mutant extracts, activity was determined in the presence of 7.5 mM ammonium, glutamine, glutamate, or 2-oxoglutarate. None of these additions had an effect on extracts from the $\Delta nifI_1 nifI_2$ strain. In extracts from wild type cells, only 2-oxoglutarate had an effect, resulting in an increase in activity to levels similar to those in the $\Delta nifI_1 nifI_2$ strain extracts. These results suggest that 2-oxoglutarate counteracts the negative effect of the NifI proteins on nitrogenase activity *in vitro*.

The effect of 2-oxoglutarate on *in vitro* nitrogenase activity suggested that it was functioning as a signal of nitrogen limitation. To test this hypothesis, 2-oxoglutarate was quantitated in cell extracts from nitrogen fixing and switched-off cultures. Before switch-off, levels of 2-oxoglutarate were 3.5 nmol/mg protein, approximately ten-fold higher than in cells ten minutes after switch-off. Furthermore, when cultures were transiently switched-off by addition of 50 µM NH_4^+, 2-oxoglutarate increased to levels similar to those before switch-off after *in vivo* nitrogenase activity had resumed. These results are consistent with 2-oxoglutarate acting as a signal of nitrogen limitation.

Based on these results, we propose a model of switch-off where 2-oxoglutarate, operating as a signal of nitrogen limitation, prevents NifI-dependent inhibition of nitrogenase activity under conditions of nitrogen limitation (Fig. 1). Addition of ammonium to cultures results in a decrease in intracellular 2-oxoglutarate to levels, which can no longer relieve the negative effect that the NifI proteins have on nitrogenase activity.

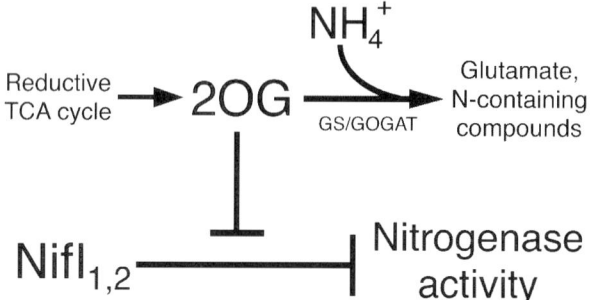

Figure 1. Working model for regulation of nitrogenase activity in *M. maripaludis*. 2OG, 2-oxoglutarate.

References
Arcondéguy T et al. (2001) Microbiol. Mol. Rev. 65, 80-105.
Kessler P et al. (2001) J. Bacteriol. 183, 882-889.
Ludden PW and Roberts GP (1989) Curr. Top. Cell. Regul. 30, 23-56.
Yoch D et al. (1988) Arch. Microbiol. 150, 1-5

MECHANISM OF REGULATION OF NITROGEN FIXATION BY NIFLA IN *PSEUDOMONAS STUTZERI* A1501

Zhihong Xie[1,2], Jie Pan[1], Shuzheng Ping[1], Ming Chen[1], Guoying Wang[2], Yi Yang[3], Min Lin[1], and Claudine Elmerich[4]
[1]Biotechnology Research Institute, CAAS, Beijing; [2]Biology College, Chinese Agricultural University, Beijing; [3]Sichuan University, Chengdu, Sichuan, P. R. China; [4]ISV, CNRS UPR-2355, Gif-sur-Yvette, France

Pseudomonas stutzeri A1501 (formerly *Alcaligenes faecalis* A1501), which was isolated from a rice field in South China in 1980, can colonize the rhizoplane of host plants and invade root tissues (You et al., 1991). The strain can fix N_2 under microaerobic conditions in the free-living state. A 30 kb DNA region with most of the *nif* and *rnf* genes has been characterized (Desnoues et al., 2003). The nitrogen-fixation genes (*nifHDK*, *nifTY*, *nifNE* and associated ORFs) are organized like *Azotobacter vinelandii*. A set of genes upstream of *nifH* shares identity with the *rnfCDGEFH* operon, involved in electron transport to nitrogenase in *Rhodobacter capsulatus*. The *nifLA* region was mapped within the main *nif* cluster. Using appropriate *lacZ* fusions, we observed that *nif*-gene expression is repressed by O_2 and ammonia and that the *nif* and *rnf* genes are under the control of *ntrBC*, *nifLA* and *rpoN* (Desnoues et al., 2003). Mutations in the *nif* and *rnf* genes led to Nif⁻ strains showing that they are essential for nitrogen fixation in *Pseudomonas*. The objective of this work was to further document the role of NifLA.

1. Properties of a *nifL* mutant non-polar on *nifA* expression.

The *nifL* gene is found essentially in nitrogen fixers belonging to the gamma subgroup of proteobacteria and in a strain of *Azoarcus* (Egener et al., 2002). Deduced translation products of different *nifL* genes showed that the A1501 NifL shared a high level of identity with that of *A. vinelandii* (72%). To construct a non-polar *nifL* mutant, we inserted a KIXX kanamycin cassette within the *nifL* coding sequence so that *nifA* expression would be under the control of the Km promoter. The resulting strain displayed some residual nitrogenase activity, whereas a *nifA-Km* mutant was Nif⁻. This result supports NifL being an inhibitor of NifA activity as described in other nitrogen fixers containing a *nifL* gene.

2. Effect of the overexpression of *nifA*

A 1.62 kb DNA fragment containing only the *nifA* coding sequence was cloned in both orientations under the control of the Km promoter of the vector. The two constructions were then introduced into the wild type A1501 and into the *nifA* and *nifL* mutants. The resulting plasmid, which expressed *nifA* from the Km promoter, restored only 30% nitrogenase activity to the *nifA* mutant strain, but the complementation reached 60% after induction by kanamycin. In addition, this plasmid increased nitrogenase activity in both the wild type and the *nifL* mutant. In contrast, the plasmid in which *nifA* was cloned in the opposite orientation abolished nitrogen fixation. It, thus, appears that modulation of *nif* genes expression through an antisense *nifA* RNA may play an important role in *P. stutzeri*.

3. Physical interaction between NifL and NifA.

It has been established that NifL modulates the activity of NifA by protein-protein interaction in *Klebsiella pneumoniae* and *Azotobacter vinelandii* (Money et al., 1999; Lei et al., 1999). The yeast two-hybrid system is a useful tool to detect protein-protein interactions (Fields, 1993). This technique was used to assay for an interaction between the NifA and NifL proteins from *P. stutzeri*. NifA and NifL are organized into domains. The *nifA* and *nifL* genes, as well as fragments encoding domains, were amplified by PCR and cloned in-frame in the yeast-vector system based on GAL4 transcriptional activator. Only the C-terminal domain of NifL displayed binding activity to the entire NifA protein and to the NifA central AAA+ domain, as reported for *K. pneumoniae* (Lei et al., 1999). This suggested that the NifL inhibitory effect on NifA activity both in air and in the presence of ammonia is mediated through the C-terminal domain of NifL. The purification of NifA and NifL to study their interaction *in vitro* is in progress.

We are grateful for support from Programs 973 and 863, Ministry of Sciences and Technology of China and Program Franco-Chinois de Recherches Avancées (PRA-2B).

References
Desnoues, N et al. (2003) Microbiology 149, 2251–2262.
Egener, T et al. (2002). Microbiology 148, 3203-3212.
Fields, S. (1993) Meth. Enzymol. 5, 116-124.
Lei, S. et al. (1999) J. Bacteriol. 181, 6535–6539.
Money, T (1999) J. Bacteriol. 181, 4461–4468.
You, C. B., et al. (1991) Plant Soil 137, 81-85.

RHIZOBIAL FIXL/FIXJ SYSTEM IS A PARADIGMN OF THE TWO-COMPONENT SIGNAL TRANSDUCING SYSTEMS: A TWO-CYLINDER RECIPROCATING ENGINE MODEL OF PHOSPHORYLATION REACTIONS

Hiro Nakamura
RIKEN and Yokohama City University, Tsurumi, Yokohama, 230-0045 Kanagawa, Japan

Rhizobial FixL and FixJ proteins are the O_2-sensing two-component signal-transducing apparatus, which directs the expression of nitrogenase genes in leguminous root nodules. FixL is a heme-based sensory histidine kinase, and the O_2-free form undergoes autophoshorylation at low O_2 tensions. But O_2 binding to the heme moiety causes inactivation of kinase activity at high O_2 tensions. Because the oxy and deoxy FixLs are spectroscopically analyzable, the FixL/FixJ system serves as a paradigm of the two-component systems (Saito et al., 2003; Nakamura et al., 2004).

Here, we show that ADP reduces the O_2 binding affinity of the sensor domain in FixL by 4-5-fold when it is produced from ATP in the kinase reaction. In contrast, phosphorylation-deficient mutants, where the highly-conserved ATP-binding catalytic site of the kinase domain is impaired, showed no such allosteric effect. Although it is known that sensory histidine kinases undergo *trans*-autophosphorylation between each subunit, this discovery casts a novel light on the significance of homodimerization of the two-component histidine kinases; ADP, produced in the phosphorylation reaction in one subunit of the homodimer, enhances the histidine kinase activity of the other by reducing the ligand binding affinity in a reciprocating manner.

FixL, like other histidine kinases, may be *trans*-autophosphorylated in the homodimeric form. Although it is unknown whether the O_2-bound sensor domain in one subunit downregulates the histidine-phosphorylation of the same polypeptide (*cis*-acting repression) or that of the partner (*trans*-acting repression), the latter *trans*-acting mode is assumed in our model (Figure 1), as shown by wedges in steps I and I'. In the *trans*-acting mode, the O_2-free sensor domain would either unmask the histidine in the other subunit or promote an ATP-phosphoryl transfer reaction in the same subunit. (R, R') represent the O_2-bound resting state, where the kinase activity is repressed, whereas (I, I') represent the initial kinase-active species, where the bound O_2 is released from one subunit, resulting in the *trans*-phosphoryl transfer from ATP to the histidine in the other subunit. At (A), ADP produced at the binding site in subunit A reduces the O_2-binding affinity of subunit B in a *trans*-acting manner (larger sensor domain of subunit B). The phosphoryl group of subunit B is transferred to FixJ. At (B), phospho-FixJ and ADP are then released from subunit A. In turn, the ATP-phosphoryl transfer reaction in subunit B is enhanced due to the allosteric effect in step A. Meanwhile, FixJ and ATP are reloaded onto subunit A. At (C), the O_2-binding affinity of subunit A is decreased by the ADP that is produced in subunit B, leading to the ATP-phosphoryl transfer reaction of subunit A in step (D).

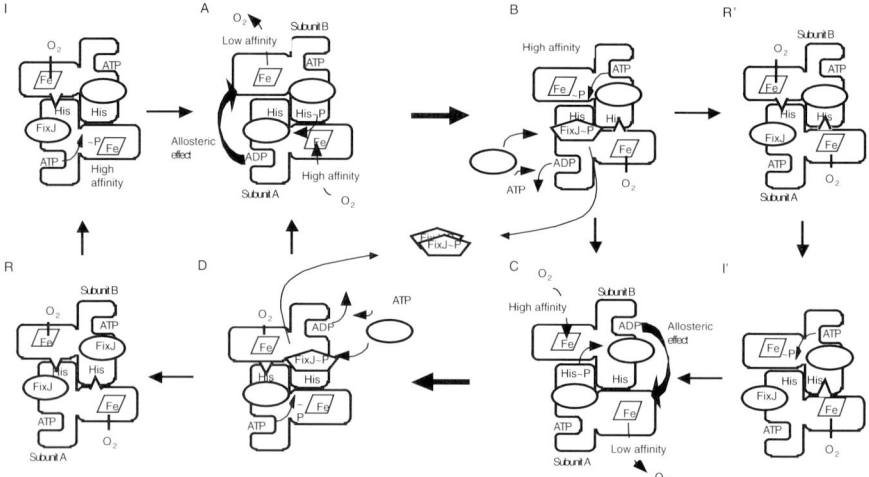

Fig 1. Two-cylinder reciprocating-engine model of FixL/FixJ phosphorylation reactions.

After the overall phosphoryl-transferring reactions, two phospho-FixJ are librated from FixL to form a regulatory active dimer (in center of figure). O_2 binding at the high affinity hemes (subunit A in step D and subunit B in step B) determines the content of kinase-active FixL and the inactive resting form in the turnover cycle. The allosteric effect facilitates the velocity of the transitions from step A to B, and from step C to D, resulting in the amplification of the phospho-FixJ production. If the O_2-bound sensor domain inhibits the histidine-phosphorylation of the same subunit (*cis*-acting repression), the model implies that the ADP allosteric effect would be exerted in the sensor domain of the same subunit (*cis*-acting manner).

References
Saito K et al., (2003) Mol. Microbiol. 48, 373-383.
Nakamura H et al., (2004) Proc. Natl. Acad. Sci. USA 101, 2742-2746.

REGULATION OF NITROGEN FIXATION IN *RHODOBACTER CAPSULATUS*

Bernd Masepohl[1], Meriyem Aktas[1], Margit Brusch[2], Thomas Drepper[3], Britta Schubert[1], Christa Sicking[1], Sven Vermöhlen[1], Jessica Wiethaus[1], and Klaus Schneider[2]

[1] Ruhr-Universität Bochum, LS Biologie der Mikroorganismen, Germany
[2] Universität Bielefeld, LS Anorganische Chemie I, Germany
[3] FZ Jülich, Institut für Molekulare Enzymtechnologie, Germany

The photosynthetic purple bacterium *Rhodobacter capsulatus* has the potential to synthesize two nitrogenases, a Mo-containing and a Fe-only nitrogenase. The synthesis and activities of both nitrogenases are regulated by ammonium on at least three levels. In the absence of ammonium, NtrC activates transcription of *nifA* and *anfA* (level 1). In turn, NifA or AnfA activate expression of all the other *nif* or *anf* genes, respectively. Addition of ammonium to nitrogen-fixing cultures leads to inhibition of NifA and AnfA activity (level 2) and to DraT-mediated modification of both nitrogenase reductases (level 3). As suggested by genetic analyses and yeast two-hybrid (Y2H) studies, ammonium inhibition of NifA activity involves direct interaction with GlnB or GlnK. Remarkably, a *glnB,glnK* double mutant synthesizes active Mo nitrogenase not only under *nif*-derepressing conditions but also under high ammonium concentrations. Similar to a *glnB,glnK* double mutant, a *glnK* (but not a *glnB*) mutant carrying a constitutively expressed *amtB* gene synthesizes active Mo nitrogenase in the presence of ammonium. Therefore, at least in a *glnK* mutant background, AmtB seems to bind and titrate out GlnB. In contrast to the situation for NifA, activity of AnfA is still inhibited by ammonium in a *glnB,glnK* mutant strain, and AnfA does not interact with either GlnB or GlnK in Y2H studies. These findings suggest that at least one other protein is involved in ammonium control of AnfA. In order to identify such protein(s) we performed Y2H studies based on an *R. capsulatus* genomic library with AnfA as a bait. Screening of this library identified Orf1818 as interactor of AnfA. Orf1818 exhibits clear similarity to metallopeptidases of the M20 family. Studies on the putative role of Orf1818 in control of AnfA activity are currently under investigation. In a parallel Y2H approach we identified two new genes, *ranR* and *ranT* (for genes **r**elated to **a**lternative **n**itrogenase), coding for products that interact with Anf1. Although both genes play important roles for diazotrophic growth via Fe nitrogenase, expression of neither *ranR* nor *ranT* is regulated by ammonium.

HYDROGEN PRODUCTION BY NITROGENASE IN *RHODOSPIRILLUM RUBRUM*

Ruiyan Zhu[1], Yaoping Zhang[2,1] and Ji-Lun Li[1]
[1]State Key laboratory for Agrobiotechnology and College of Biological Science, China Agricultural University, Beijing 100094, P. R. China
[2]Departments of Biochemistry and Bacteriology and the center for the study of Nitrogen Fixation, University of Wisconsin-Madison, Madison WI 53706, USA

Rhodospirillum rubrum, a non-sulfur photosynthetic bacterium, can produce pure hydrogen about 2.0±0.1 l/l by nitrogenase in the derepressed medium (MG medium) under light and anaerobic condition in the absence of dinitrogen gas. In order to increase the yield of H_2 evolution, we constructed a mutant deleted the large subunit of uptake hydrogenase (Hup). The H_2 yield and maximal rate in *hupL*⁻ mutant were 5.02 times and 3.15 times that of the wild type strain respectively (Fig. 1 and Fig. 2).

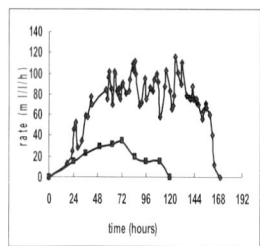

Fig.1 H_2 production rate in wild type and *hupL*⁻ strain of *R. rubrum*.Symbols: ■, wild type; □, *hupL*⁻ strain.

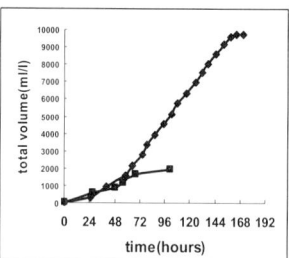

Fig. 2 H_2 yield in cultures of the wild type and *hupL*⁻ strain of *R. rubrum*. Symbols: ■, wild type; □, *hupL*⁻ strain. ●

The *draT*⁻ *hupL*⁻, a double mutant, could produce H_2 in low level under continuous darkness, while the wild type strain could not. This mutant is benefit for hydrogen production by nitrogenase during day and night. Hydrogen production rate was increased in presence of nickel and iron in malate-glutamate medium. These results show that other NiFe hydrogenase probably exists in *R. rubrum*, and it should be verified by further experiments.

Acknowledgment:
This work was supported by a grant of National Basic Research Program of China.(No.2001CB108904)

ACTIVITY OF DINITROGENASE REDUCTASE IS REGULATED BY ADP-RIBOSYLATION IN *MAGNETOSPIRILLUM GRYPHISWALDENSE*

Wei Jiang, Ying Li, Ji-Lun Li
State Key Laboratory for Agrobiotechnology and Department Microbiology, China Agricultural University, Beijing 100094, P. R. China

It has been reported that *Magnetospirillum magnetotacticum*, *M. gryphiswaldense* and *M.* sp. AMB-1 can fix nitrogen in N-limiting semi-solid media and the existence of structure gene of nitrogenase was confirmed by Southern blot with the *nifHDK of Rhodospirillum rubrum* [1, 2]. Our previous work indicated that *M. gryphiswaldense* can only fix nitrogen in the absence of ammonia and under microaerobic (0.4~0.8% of O_2) condition [3]. It appears to be that the activity of dinitrogenase reductase (Fe protein) of *M. gryphiswaldense* is regulated by covalent modification.

The dinitrogenase (MoFe protein) and dinitrogenase reductase were isolated and purified separately from *M. gryphiswaldense* under anaerobic condition. However there is no nitrogenase activity when the two components combined each other by C_2H_2 reduction assay. But the complementary activity of MoFe protein of *M. gryphiswaldense* with Fe protein from *K. pneumoniae* was 341.3nmol C_2H_4 formed min^{-1} mg of protein $^{-1}$, indicating that the loss of nitrogenase activity was due to Fe protein. An 1113-bp *draT* fragment, include a 327 bp promoter and 786 bp of coding sequence, was cloned by PCR and sequenced from this microorganism. The deduced amino acids sequence had 58 % identity with DraT of *R. rubrum* and 67 % with the deducted DraT sequence from *M. magnetotacticum*. These experimental results indicated that the activity of dinitrogenase reductase of *M. gryphiswaldense* could also be regulated by DRAT/DRAG regulatory system.

References
1. Bazylinski DA et al. (2000) Environ Microbiol 2(3), 266~273.
2. Bazylinski DA and Blakemore R P (1983) Current Microbiol 9, 305~308.
3. Jiang Wei et al (2002) *Chinese Science Bulletin* 47 (24), 2095~2099.

EXPRESSION AND PURIFICATION OF THE N-TERMINAL DOMAIN OF AZOSPIRILLUM *BRASILENSE* NIFA IN *ESCHERICHIA COLI*

Xiaoyu Zhou[1], Yaoping Zhang[2], Jilun Li[1]
[1]State Key laboratory for Agrobiotechnology and College of Biological Science, China Agricultural University, Beijing 100094, P. R. China
[2]Departments of Biochemistry and Bacteriology and the center for the study of Nitrogen Fixation, University of Wisconsin-Madison, Madison WI 53706, USA

Azospirillum brasilense is a diazotroph associated with many important agricultural crops such as corn and wheat, and shows potential as a biofertilizer by contributing fixed nitrogen and/or phytohormones to plant growth.

The NifA protein is a prokaryotic transcriptional activator that specifically activates expression of the nitrogen fixation (*nif*) genes in response to the cellular nitrogen and oxygen status. In many organisms, NifA is insoluble when over-expressed in *Escherichia coli*. Biochemical analysis of NifA has been impaired by its insolubility. It was previously reported that the N-terminal domain of *A.brasilense* NifA could inhibit its own activity by some unknown mechanisms in the presence of ammonia.

We report here that the N-terminal domain of *A.brasilense* NifA was over-expressed in the form of a His-Tag fusion protein in *Escherichia coli* and purified by metal-affinity chromatography to yield a highly purified and soluble protein. The purified NifA N-terminal protein was also used as immunogen for the generation of rabbit antiserum, and these works provide very useful tools to study the regulation of NifA activity both *in vivo* and *in vitro*.

Acknowledgements
This work was supported by the National Key Fundamental Research Program of China (973 Program, Grant No.2001CB108904).

CRP-CAMP MEDIATED *TRANS* AND *CIS* REGULATORY EFFECTS ON *GLNA*P2 DERIVED PROMOTERS IN *E. COLI*

Zhe-Xian Tian and Yi-Ping Wang
National Laboratory of Protein Engineering and Plant Genetic Engineering, College of Life Sciences, Peking University, Beijing 100871, P. R. China

*E. coli glnA*p2 is σ^{54}-dependent promoter, controlling expression of glutamine synthetase. Our previous works reported that *glnA*p2 is repressed by CRP-cAMP complex, which provide coordination between nitrogen and carbon metabolism (Tian et al., 2001). In order to investigate the role of CRP-binding site in the regulation, a series of *glnA*p2 derived promoters (named as cc-promoters) were constructed by introducing a CRP binding site at different locations between the activator bonding site and the core promoter without changing the distance in-between. In the *E. coli cya crp* double mutant TP2339-1, containing plasmids harboring the wild type *crp* gene, *glnA*p2 activities from each cc-construct were measured. At high exogenous cAMP concentrations, all promoters were permanently downregulated, regardless the locations of CRP binding site. This result confirms our previous observation that CRP downregulation at *glnA*p2 is dominant *in trans* (Tian et al., 2001). This *trans*-dominant effect by wild type CRP can be reversed by either lowering exogenous cAMP concentrations, or replacing wild type CRP with either CRP AR1 or AR3, but not AR2 mutant. In these cases, a *cis* dominant, and binding site dependent up- and down-regulatory pattern at these promoters was observed. CRP-cAMP downregulates cc-41.5, cc-50.5, cc-61.5 and cc-72.5 constructs, and upregulates cc-54.5, cc-55.5 and cc-66.5 constructs. The length of the helix-pitch on these *glnA*p2 derivatives fits in with the general believed 10.4 bp. Furthermore, at cc-61.5 construct, 5 bp or 10 bp insertions were used to control the relative phase of DNA helix of each protein. Only when both the NtrC binding sites and RNA polymerase binding site were placed to the opposite phase of the DNA helix, it results in complete abolishment of *cis* dominant downregulation of CRP. Therefore, CRP mediated *cis* regulatory effects on *glnA*p2 are based on DNA bending and relative phase between Eσ^{54} and activators.

Reference: Tian ZX et al. (2001) Mol. Microbiol. 41, 911-924.

Acknowledgements: Work supported by 973 (No. 2001CB108903) and 863 (No. 2001AA214021) from MOST, and NNSF (No. 39925017, No. 30200047) in China.

PROTEIN-INDUCED-DNA-BENDING REVEALS THAT BACTERIAL ENHANCER-BINDING PROTEIN APPROACHES σ^{54}-RNA POLYMERASE FROM THE OPEN-FACE OF THE PROMOTER DNA HELIX DURING TRANSCRIPTION INITIATION

Yi-Xin Huo[1,2], Zhe-Xian Tian[1], Jin Wen[1], Martin Buck[3], Yi-Ping Wang[1,2] * and Annie Kolb[2]
[1]College of life Sciences, Peking University, Beijing 100871, P. R. China.
[2]Unité des Régulations transcriptionnelles, Institut Pasteur, Paris75724, France. [3]Imperial College of Science, London SW72AZ, UK.
* Corresponding author.   E-mail: wangyp@pku.edu.cn

In bacteria, σ^{54} and σ^{70} bind to the same core RNA polymerase and presumably are located at similar places on the core structure even though they bear no more than a passing resemblance at the level of amino acid sequence. Transcription by σ^{54}-RNA polymerase can be viewed as a hybrid of prokaryotic and eukaryotic mechanisms. σ^{54}-RNA polymerase (Eσ^{54}) is restricted predominantly to one face of the DNA helix in the closed promoter complex, and requires contact with upstream Enhancer-like sequence (ES) bound proteins (EBP) mediated by DNA looping to make open complexes. Effective DNA looping depends on the correct orientation of RNA polymerase and EBP binding sites. Up to now the precise face of the σ^{54}-holoenzyme that contacts the activator to convert the closed complex to an open one remains unclear. In this study, by introducing a protein induced DNA bend at precise locations between the ES and the core promoter of the σ^{54}-dependent *glnA*p2 promoter without changing the distance in-between, we observed a strong up and down regulation of promoter activity. The relative orientations of the Eσ^{54} and EBP were determined by *in vitro* footprinting. Intriguingly the locations from which the DNA bending protein exerted its optimal stimulatory effects were all found on the opposite face of the DNA helix as the DNA bound σ^{54}-holoenzyme in the closed complex. We suggest that the target site of activator in the σ^{54}-holoenzyme is accessible from the open face of the σ^{54}-holoenzyme closed complex to enable triggering of the isomerization step. Protein induced DNA bending appears a novel approach to explore the effect of the "on and off" contacts between two prokaryotic or eukaryotic regulatory factors.

Acknowledgements:
Work supported by 973 (No. 2001CB108903) and 863 (No. 2001AA214021) programs from the MOST, and the NNSF (No. 39925017, No. 30200047) in China.

COPG INDUCED DNA BENDING AND ITS REGULATION ON σ^{54}-DEPENDENT *GLNA*P2 PROMOTER IN *E. COLI*

Yan-Cheng Chen, Zhen-Feng Liu, En-Ce Yang, Zhong-Yu Liu and Yi-Ping Wang
National laboratory of Protein Engineering and Plant Genetic Engineering, College of Life Sciences, Peking University, Beijing 100871, P. R. China

CopG is a classical DNA bending protein, and when it is a monomer, it consists a 45 amino acids polypeptide. In our study, a series promoters were constructed by replacing different fractions of the intervenient region between upstream Enhancer-like Sequence (ES) and the core promoter at σ^{54}-dependent *glnA*p2 with a consensus CopG binding sequence, and the length of promoter DNA remains unchanged. When this series of promoter's activity were measured in the presence/absence of CopG in *E. coli*, upon CopG-mediated DNA binding and bending, a binding site dependent up-and-down regulatory pattern was observed. In these cases, CopG downregulates the promoter when its bending center is located at -50, -61 and -72 respectively, and upregulates the promoter when its bending center is located at -56, -66 and -77 respectively. And the length of the helix-pitch on *glnA*p2 fits in with the general believed 10.4 bp. Therefore, we demonstrate that the activity of σ^{54}–dependent *glnA*p2 promoter can be regulated by a classical protein induced DNA bending, in a DNA helical dependent fashion. We propose that CopG mediated regulatory effects on *glnA*p2 are based on DNA bending and relative phase between Eσ^{54} and activators.

Acknowledgements:
Work supported by 973 (No. 2001CB108903) and 863 (No. 2001AA214021) programs from the MOST, and the NNSF (No. 39925017, No. 30200047) in China.

CRP-CAMP INHIBITS STABLE CLOSED-COMPLEX FORMATION AND DOWN-REGULATES TRANSCRIPTION OF THE σ^{54}-DEPENDENT *PU* PROMOTER IN *ESCHERICHIA COLI*: BY AN UNUSUAL MECHANISM

Zhi-Ting Li[1], Yuan-Tao Zhang[1], Zhi-Hong Zhang[1], Zhe-Xian Tian[1], Yi-Cheng Sun[1], Annie Kolb[2], Yi-Ping Wang[1]
[1]National Laboratory of Protein Engineering and Plant Genetic Engineering, College of Life Sciences, Peking University, Beijing 100871, P. R. China.
[2]Unite des Regulations Transcriptionnelles, Institut Pasteur, 75724 Paris cedex 15, France

The σ^{54}-dependent *Pu* promoter of the TOL plasmid pWW0 of *Pseudomonas putida* is activated by the enhancer-binding protein XylR in response to a variety of aromatic inducers. In this study, the XylR-mediated activation of the *P. putida Pu* promoter was recreated in the heterologous host *Escherichia coli*. Here we show that the cAMP receptor protein (CRP) had an inhibitory effect on the *Pu* promoter in a cAMP-dependent manner. When *Pu* was activated by the alternative regulator NtrC-phosphate, a similar inhibitory effect was observed, indicating that the result observed was not activator specific. $KMnO_4$ and DMS *in vivo* footprinting indicated that the inhibitory effect was at the transcriptional level and that CRP-cAMP complex prevented stable closed-complex formation at *Pu*. Moreover, surfaces of CRP involved in the inhibition of *Pu* could be pinpointed by analyzing various mutants that fail to interact with different subunits of $E\sigma^{70}$. In this case, only the CRP mutants defective in AR1, but not AR2 or AR3, lost the capability of inhibiting *Pu*. Thus, in contrast to previously studied CRP-cAMP-mediated inhibition of open-complex formation at the *dctA* promoter, CRP-cAMP may hinder the interaction between the αCTD of $E\sigma^{54}$ RNA polymerase and the UP-like element, which is essential for stable closed-complex formation at the *Pu* promoter. All these data point to the conclusion that formation of a stable closed-complex is the main regulatory check point of the *Pu* promoter *in vivo*.

Acknowledgements:
Work supported by 973 (No. 2001CB108903) and 863 (No. 2001AA214021) programs from the MOST, and the NNSF (No. 39925017) in China.

CRP-CAMP MEDIATED DOWN-REGULATION AND THE ROLE OF UP-LIKE ELEMENT AT THE σ^{54}-DEPENDENT *PU* PROMOTER

Yuan-Tao Zhang, Zhi-Ting Li, Zhe-Xian Tian and Yi-Ping Wang
National Laboratory of Protein Engineering and Plant Genetic Engineering, College of Life Sciences, Peking University, Beijing 100871, P. R. China.

Recruitment of RNA polymerase is a rate-limiting step for the XylR-mediated activation of the σ^{54} promoter *Pu* of *P. putida* (Carmona et al. 1999). This recruitment requires the interaction of the αCTD of Eσ^{54} RNA polymerase with the UP-like elements at *Pu* (Bertoni et al. 1998). Previous studies indicate that CRP-cAMP complexes had an inhibitory effect *Pu* promoter in *E. coli*. In this study, the mechanism by which CRP-cAMP inhibits transcription of *Pu* was investigated. 1) The UP-like $_{Pu}{}^1$ element was replaced by progressive sequences scrambling, and the maximum activity of *Pu* was not affected. 2) The effect of CRP-cAMP on the UP-like mutated *Pu* was examined in *E. coli*. In this case, under the lower concentration of cAMP, CRP had a strong inhibitory effect on the mutated *Pu*, in contrast, the wild type *Pu* was hardly affected. 3) When CRP AR1 mutant defective in interaction with the αCTD of RNA polymerase (CRP-H159L) was used in the assay, CRP-H159L had no effect on both *Pu* promoters. 4) *in vivo* DMS footprinting indicated that CRP-cAMP prevented stable closed-complex formation at the mutated *Pu* more markedly than that at the wild type *Pu*. Taken together, the results suggest that disruption of UP-like $_{Pu}{}^1$ element made αCTD of RNA polymerase apt to interact with CRP-cAMP. A model can be proposed, therefore, that CRP-cAMP inhibits expression of *Pu* through directly interaction with αCTD of RNA polymerase, which in turn, may cause instability of closed-complex formation at *Pu*.

References: Carmona M et al. (1999) J. Biol. Chem. 274, 33790-33794.
Bertoni G et al. (1998) EMBO J. 17, 5120-5128.

Acknowledgements: Work supported by 973 (No. 2001CB108903) and 863 (No. 2001AA214021) programs from the MOST, and the NNSF (No. 39925017) in China.

CRP-CAMP REPRESSES *KLEBSIELLA PNEUMONIAE NIFU* PROMOTER BY COMPETING WITH NIFA FOR ITS UPSTREAM BINDING SITES

Zhi-Ting Li, Bei-Yan Nan, Yi-Xin Huo, Xian-Jun Mao and Yi-Ping Wang*
National laboratory of Protein Engineering and Plant Genetic Engineering, College of Life Sciences, Peking University, Beijing 100871, P. R. China.

There are three NifA Upstream Activator Sequences (UASs) at the *Klebsiella pneumoniae nifU* promoter. Sequence analysis indicates that UAS2 completely overlaps with a strong CRP-binding site. In this study, the role of this CRP-binding site on *nifU* promoter is investigated. First, UAS2 is mutated for CRP binding, and maintained its function for NifA. Secondly, the expression of this CRP binding site mutated *nifU* promoter is tested under various conditions, together with its wild type counterpart, in *Escherichia coli* background. When the activator NifA is constitutively over-expressed and exogenous cAMP concentration is high, CRP-cAMP mediated inhibition effect is observed on both *nifU* promoters (Li et al. 2002); In contrast, when wild type CRP is replaced by its AR1 mutant CRP-H159L, the effect is diminished on both *nifU* promoters, indicating that the inhibitory effect is *trans*-dominant. When NifA is expressed under lower- and inducible-conditions, and cAMP is titrated to simulate the physiology condition, we find that at some specifically CRP-cAMP/NifA ratio, CRP-cAMP mediated inhibition on *nifU* promoter is dependent on the CRP binding site. In this case, when wild type CRP is replaced by its AR1 mutant CRP-H159L, the effect remains, indicating that CRP can out-complete with NifA for binding on the *nifU* promoter. This competition can be also demonstrated by DNaseI footprints. Taken together, under the relative high concentrations of CRP-cAMP, the inhibitory effect is *trans*-dominant (1); while under relative low concentrations of CRP-cAMP, this effect mainly depends on the competition between CRP and NifA with their DNA binding sites upstream of the *nifU* promoter.

Reference: Li ZT et al. (2002) Chinese Sci. Bul. 47, 1821-1825.

Acknowledgements: Work supported by 973 (No. 2001CB108903) and 863 (No. 2001AA214021) programs from the MOST, and the NNSF (No. 39925017) in China.

STRUCTURAL AND FUNCTIONAL ANALYSIS OF C4-DICARBOXYLATE TRANSPORT (*DCT*) GENES IN THE N2-FIXING *PSEUDOMONAS STUTZERI*

Hongquan Li[1,2], Shuzhen Ping[1], Liying Wang[1], Ming Chen[1], Claudine Elmerich[3], and Min Lin[1]
[1]Biotechnology Research Institute, CAAS, Beijing, 2 College of Life Science, Hebei University, Baoding, P. R. China; [3]Institut des Sciences du Végétal, CNRS UPR-2355, Gif-sur-Yvette, France.

The nitrogen-fixing strain of *Pseudomonas stutzeri* A1501, isolated from a rice field in China, colonizes rice endophytically. The strain can use a variety of carbon sources including dicarboxylic acids, usually present in large amounts in the root exudates. A DNA region containing the C4-dicarboxylates transport (*dct*) genes was isolated from a gene bank of *P. stutzeri* A1501. Nucleotide sequencing of a 9408-bp DNA region (AJ313422) led to the identification of 5 *loci* with translation products sharing high similarity to DctB-DctD-DctP-DctQ and DctM. The organization of the *dct* gene cluster resembles that of *Rhodobacter capsulatus*. DctB and DctD shared high similarity to the two-component sensor-regulator pair that controls expression of *dct* genes in other bacteria. DctP, which did not include membrane-spanning regions, is likely a periplasmic C4-dicarboxylate-binding receptor. Structural analysis was consistent with DctQ and DctM being integral membrane proteins. They contained 5 and 12 transmembrane segments, respectively, and they likely form a secondary transporter to transport C4-dicarboxylate using eletrochemical ion gradient.

A *dctB* insertion mutant could not utilize fumarate and succinate as the sole carbon sources, but the strain grew with malate suggesting that another transporter different from that encoded by DctPQM is present in *P. stutzeri*. The strain had a Nif+ phenotype when lactate or malate was the carbon source. Nitrogenase activity was not detected with succinate or fumarate as the carbon source. Introduction of a broad host range plasmid carrying the *dctBD* region into the *dctB* mutant restored growth and nitrogenase activity in succinate and fumarate containing media. In addition the nitrogenase activity of the A501 strain containing the *dctBD* plasmid was increased. These data are in favor of *dctD* being the transcription regulator that controls *dctPQM* expression. It also suggests that DctBD could play a role in maximal nitrogenase activity.

Supported by the Chinese 863 program and French-Chinese PRA-2B.

NADH: FERREDOXIN OXIDOREDUCTASES IN GLUTAMATE FERMENTING BACTERIA, SIMILAR TO RNF PROTEINS IN *RHODOBACTER CAPSULATUS*

Clara Dana Boiangiu,[1] Elamparithi Jayamani,[1] Daniela Brügel,[1] Marcus Hans,[1] Jihoe Kim,[1] Irini Vgenopoulou,[2] Julia Fritz-Steuber[2] and Wolfgang Buckel.[1]
[1]Laboratorium für Mikrobiologie, Fachbereich Biologie, Philipps-Universität, 35032 Marburg, Germany.
[2]Institut für Biochemie, Universität Zürich, Zürich, Switzerland.

In *R. capsulatus* the six *rnf* genes probably encode a NADH:ferredoxin oxidoreductase, which catalyses the thermodynamic uphill reduction of ferredoxin ($E_0' = -420$ mV) by NADH ($E_0' = -320$ mV) driven by $\Delta\mu H^+$ or $\Delta\mu Na^+$. The reduced ferredoxin is used as electron donor for nitrogen fixation. In membranes from *Clostridium tetanomorphum* and *Fusobacterium nucleatum* NADH dehydrogenases have been detected, which have been characterised as Rnf-related proteins, each containing one covalently and two non-covalently bound flavins as well as six iron-sulfur clusters. Both enzymes are composed of six subunits each, which were identified via their N-termini as homologues of RnfABCDEG. The dehydrogenases catalyse the reduction of ferricyanide by NADH as well as the reduction of NAD by reduced ferredoxin. In *C. tetanomorphum* the enzyme is used for energy conservation, whereas in *F. nucleatum* the sodium ion pump glutaconyl-CoA decarboxylase generates $\Delta\mu Na^+$, which drives the reduction of ferredoxin by NADH necessary to activate the 2-hydroxyglutaryl-CoA dehydratase. The activation of this enzyme is similar to the electron transfer in nitrogenase. The activator contains a [4Fe-4S] cluster located at the interface between its two identical subunits. The cluster is reduced by one electron derived from ferredoxin. Hydrolysis of two ATP bound to the activator, one to each subunit, transfers the electron to the dehydratase, where a low potential $[4Fe-4S]^{2+}$ cluster is probably reduced. Further transfer to the substrate (*R*)-2-hydroxyglutaryl-CoA generates a substrate-derived ketyl radical anion, which expels the adjacent hydroxyl group. The resulting enoxy radical is deprotonated to a product-related ketyl radical anion. Finally the electron is removed by the next incoming substrate leading to the product glutaconyl-CoA and starting a new turnover.

References:
Schmehl M et al. (1993) Mol. Gen. Genet. 241, 602-615.
Kim J et al. (2004) FEMS Microbiol. Rev. 28, 455-468.
Buckel W et al. (2004) Curr. Opin. Chem. Biol. 8, 462-467.

FLAGELLAR GENES ARE FOUND IN *AZOSPIRILLUM BRASILENSE*

Yanqi Chang, Xiaoyu Zhou, Jilun Li
State Key laboratory for Agrobiotechnology and College of Biological Science, China Agricultural University, Beijing 100094, P. R. China

The nitrogen-fixing bacterium *Azospirillum brasilense* has the potential to increase the yield of economically important cereals and grasses. Their capability of adaptation to the root is remarkably improved because of the production of the polar flagellum and lateral flagella in media of certain density. In *A. brasilense* we have known only one flagellar gene, the structural *laf1* gene for flagellin of the lateral flagella. Recently, we cloned a 5.9-kb DNA fragment from *A. brasilense* Yu62 which contain 5 homologs of known flagellar genes. We conclude that it is a part of flagellar gene cluster in *A. brasilense*. A putative sigma-54-type promoter region was identified by similarity to known promoter consensus in upstream of the *fliF* gene.

In *Salmonella* and *E. coli*, FliF proteins assemble into the MS ring, a central motor component of the bacterial flagellum that anchors the structure in the inner membrane. FliH is a negative regulatory component for FliI, the ATPase that is responsible for driving flagellar protein export. FliN is a major structural protein of the C-ring. As a part of stator, MotA has four membrane-crossing segments and is responsible for flagellar rotation. The *flbD* codes a 53-kDa protein homologous to the σ^{54}-dependent transcriptional activators NtrC (NR1), NifA, DctD and HydG. In *Caulobacter crescentus*, FlbD is a transcriptional activator of class III and IV flagellar genes. Previously, our experiments indicated that *flbD* plays an important role in regulation of both polar and lateral flagella biosynthesis in *A. brasilense*.

Acknowledgements
This work was supported by the National Key Fundamental Research Program of China (973 Program, Grant No.2001CB108904).

SECTION III.

FROM GENOME TO FUNCTION

A REPORT ON THE GENOME OF *HERBASPIRILLUM SEROPEDICAE* STRAIN Z78.

Fábio O. Pedrosa and Genopar Consortium
Department of Biochemistry and Molecular Biology ,Universidade Federal do Paraná UFPR, C. Postal 19046, CEP 81531-990 Curitiba-PR, Brazil. fpedrosa@ufpr.br

The sequence and annotation of the genome of *Herbaspirillum seropedicae* strain Z78 is being completed by The Paraná State Genome Programme, GENOPAR (www.genopar.org) a consortium of 10 Institutions (UFPR, UEL, UEM, UEPG, UNIOESTE, UNIPAR, IAPAR, EMBRAPA-SOJA, PUC-PR, UFSC). *H. seropedicae* is an endophytic nitrogen-fixing bacterium found inside tissues of important crops such as corn, sugar cane, rice, wheat and sorghum (Baldani et al. 1986). It fixes nitrogen under microaerobic conditions and can express *nif* promoters *in planta*. It has been found to promote plant growth probably through the production of phytohormones. The regulation of nitrogen fixation by *H. seropedicae* has been reviewed recently (Souza et al., 2000b, Pedrosa et al., 2001, Pedrosa and Elmerich, 2005 in press).

We have sequenced the genome of *H. seropedicae* using a shotgun strategy and the automatic DNA sequencer MegaBace 1000. The number of non-redundant bases has reached a plateau close to 5,511,539 bp. Closure of sequence gaps and the refining of sequences is being completed. Potential protein coding regions were identified by an automatic annotation platform (GAnM: Genome Annotation Module) using Glimmer2, RBS finder and Blast softwares.

Of the suggested 51,895 orfs, manual annotation has, so far, revealed, 2,892 with assigned function, 1,433 with no known function but conserved in other organisms and 754 hypothetical orfs, adding to a probable total of 5,079 genes in the genome of *H. seropedicae*. These orfs have been categorized into 4 main functions according to the Cluster of Orthologous Groups of proteins (COG). These functions are: Information storage and processing 631 (12.4%), Cellular processes 1,169 (23.0%), Metabolism 1,523 (30.0%) and Poorly characterized proteins of unknown function 1756 (34.6%).

Information storage and processing.
In this category the number of orfs is as follows: J - Translation, ribosomal structure and biogenesis (188), K – Transcription (423) and L - DNA Replication, recombination, and repair (20).

Cellular processes.
The cellular processes category comprises the following orfs: B - Chromatin structure and dynamics (2), D - Cell cycle control, cell division, chromosome partitioning (35), M - Cell wall/membrane/envelope biogenesis (258), N - Cell motility (71), O - Post translational modification, protein turnover, chaperones (128), P - Inorganic ion transport and metabolism (242), T - Signal transduction mechanisms (307), U - Intracellular trafficking, secretion, and vesicular transport (68), V - Defense mechanisms (58).

Metabolism
In Metabolism the numbers are distributed as follows: C - Energy production and conversion (284), E - Amino acid transport and metabolism (419), F - Nucleotide transport and metabolism (73), G - Carbohydrate transport and metabolism (320), H - Coenzyme transport and metabolism (136), I - Lipid transport and metabolism (199), Q - Secondary metabolites biosynthesis, transport and catabolism (92).

Poorly characterized proteins
The orfs in the 4th group are, R - General function prediction only (411) and S –

Function unknown (1,345).
The nitrogen fixation genes (*nif*) of *H. seropedicae* including *nifA, nifB, nifZ, nifH, nifD, nifK, nifE, nifN, nifX, nifQ, nifW, nifV, nifU, nifS*, were found in a region spanning 40 kbp interspersed with mod, *hes, fdx, hsc* and other genes. The *modABC* and *modE* genes are present in this region, as well as *fixA, fixB, fixC,* and *fixX* and four ferredoxin genes.

Nitrogen regulatory genes.
The genes involved in regulation of nitrogen metabolism of *H. seropedicae*, *glnAntrBntrC* (single operon) are present, as are also *orfglnKamt*, a separate single *amtB* gene and two probable *ntrYX* –like operons, one in the region of the *glnKamtB* operon.

Nitrate metabolism.
Single copies of nitrate and nitrite reductases are present, but no genes involved in nitrification or dissimilatory nitrate reduction were found.
Ammonia assimilation. Genes for low affinity (glutamate dehydrogenase, *gdhA*) and high affinity (glutamine synthetase, *glnA* and D-glutamate synthase, *gltB* and *gltD*) ammonia assimilation are present.

Auxin biosynthesis.
The plant growth promoting bacteria probably owe their effect to their ability to synthesize and secrete phytohormones. In *H. seropedicae* we have found a collection of genes possibly involved in four pathways for the synthesis of indole acetic acid from tryptophan. The most probable route is from tryptophan to indolepyruvate and then indole acetic acid, catalyzed by tryptophan transaminase and indolepyruvate ferredoxin oxidoreductase, respectively. Enzymes for the other possible pathways are present: i) tryptophan to indoleacetate via indoleacetamide, ii) from indoleacetamide to indoleacetate via indoleacetonitrile and iii) tryptophan to indoleacetate via tryptamine and indoleacetaldehyde.

Oxidative phosphorylation and the respiratory chain.
Genes for the respiratory chain and oxidative phosphorylation of *H. seropedicae* have been identified. Genes for the synthesis of NADH dehydrogenase, succinate dehydrogenase, cytochrome c reductase, coenzyme Q biosynthesis, cytochrome c, cytochrome c oxidase aa3, three alternative terminal oxidases (bd, cbb3 and o) are present as well the complete set of genes for ATP synthase.

Iron transport and metabolism.
H. seropedicae has at least 27 genes involved in iron transport and metabolism. It can apparently transport either iron 2^+ and 3^+. Genes for siderophores and siderophore receptors have been identified. Presumably iron uptake is via active transport involving ABC-type systems and TonB/ExbB/ExbD. We have also detected the presence of the global iron regulator gene *fur*. In a preliminary search we found two *fur* boxes: one in the fhuA and the other in *tpb* genes.

Genes of protein secretion in *H. seropedicae*.
Genes for both Sec-dependent and Sec-independent protein secretion pathways (Hueck, 1998; Thanassi & Hultgren, 2000) have been found in the *Herbaspirillum seropedicae* genome. The evidence suggests that the Sec-dependent pathways: i) Autotransporter, ii) Single accessory, iii) Chaperone/usher and iv) Type II secretion / Type 4 pili homolog are present. Similarly the Sec-independent pathways: i) Type I secretion/ABC transporter, ii) Type III secretion/flagella homolog; and iii) Twin arginine translocase, were found. No evidence for Type IV secretion pathway was detected.

Genes of plant-microbe interaction in *H. seropedicae*.
The main family of genes for plant-microbe interaction found in *H. seropedicae* was similar to the hypersensitivity response and pathogenicity (*hrp*) genes of plant pathogens and to the *Yersinia pestis ysc* genes. The whole complement of genes for the assembly of a protein secretion apparatus similar to the Type III secretion system of *Y. pestis*, with the exception of *yopB* and *yscF*, are present in the *H. seropedicae* genome. The genome also contains three *nod*-like genes, *nodB*, *nodD* and *nodN*; however the

common *nod* genes, *nodABC*, responsible for the synthesis of the oligochitin backbone, have not being found.

Finally, the genome the *H. seropedicae* has revealed a high degree of similarity with those of other β-Proteobacteria, in particular the plant pathogens *Ralstonia solanacearum* and *Pseudomonas syringae*, two plant pathogens.

Acknowledgments
We would like to thank the Secretary of Science Technology and Higher Education of Paraná State and the Ministry of Science and Technology of Brazil/The Brazilian Research Council CNPq/MCT for financial support, and all researchers, students and staff involved in this project. Their names are listed as Genopar Consortium, see www.genopar.org.

References:
Baldani J I et al. (1986) Int. J. Syst. Bacteriol. 36, 86-93.
Hueck CJ (1998) Microbiol. Mol. Biol. Rev. 62, 379-433.
Pedrosa FO and Elmerich C (2005) In Associative Nitrogen-fixing Bacteria and Cyanobacterial Associations (Elmerich C and W.E. Newton, eds.), The Netherlands, Kluwer Academic Publishers. In press
Souza E M et al. (2000). In Nitrogen fixation: from molecules to crop productivity (Pedrosa FO et al. eds.) (pp. 83-86). Dordrecht: Kluwer Academic Publishers.
Thanassi DG and Hultgren SJ (2000) Curr Opin Cell Biol. 12, 420-30.

FUNCTIONAL ANALYSIS OF GENES OF UNKNOWN FUNCTIONS IN *SINORHIZOBIUM MELILOTI* 1021

J. Cheng, J. Fowler, A. Cowie, R. Zaheer, C. Patten, C. Sibley, P. Chain, Z. Yuan, C. Baron, T. Charles[b], B. McCarry[a], P. Summers, J. Xu, E. Weretilnyk, G. B. Golding and T. M. Finan
Department of Biology and [a]Department of Chemistry, McMaster University, Hamilton, Ontario L8S 4K1, Canada. [b]Department of Biology, University of Waterloo, Waterloo, Ontario, Canada

1. Introduction

Sinorhizobium meliloti, a free-living gram-negative bacterium, induces N_2-fixing root nodules on legume alfalfa (*Medicago sativa*). *S. meliloti* possesses three replicons: a 3.65-Mb chromosome and two megaplasmids of 1.35-Mb pSymA and 1.68-Mb pSymB. The *S. meliloti* genome contains some 6,204 predicted protein-coding regions, of which about 2,535 are classified as hypothetical, conserved hypothetical, or putative genes. We classify these genes as genes of unknown function. In many organisms such genes constitute about 40% of the total genome content. Here we outline a Genome Canada funded project the major objective of which is to identify biological roles for these proteins of unknown function or proteins encoded by *puf* genes. As the functions of these genes are diverse, we are employing a multidisciplinary approach including: (i) the construction of reporter gene fusions and gene disruption; (ii) examination of metabolite profiles of *S. meliloti* using GC-MS methodologies under conditions where the target gene is expressed; (iii) bioinformatics including the use of phylogenetic profiling; and (iv) population studies addressing the distribution of the *puf* genes.

Since the gene content of the bacteria will reflect its life history, we envisage that the identification of the functions and phenotypes associated with the *puf* genes will give new insight into the life of *Sinorhizobium* in the rhizosphere and in the soil. Hence, although some of the *puf* genes will be related to the legume symbiosis, other will have roles in survival in soil, in the rhizosphere, in interactions with other organisms, in osmotic stress, heat stress etc. While we envisage that many of these genes will not be expressed under routine laboratory conditions and our initial experiments are directed to addressing the validity of this assumption. We also note that 12% of the genes in *S. meliloti* are involved in transport (Galibert et al., 2001) and hence we think it likely that a significant number of the *puf* genes will be involved in catabolism of diverse compounds. We envisage using microarrays to screen for expression of genes in nodules and under routine laboratory conditions, however, to screen for conditions under which *puf* genes or *puf*-gene clusters are expressed, we wished to employ reporter gene fusions to monitor gene expression in *S. meliloti*. Here, we report on progress made in the generation of large deletion mutants and on the construction of a reporter gene fusion library of the *S. meliloti* genome.

2. Deletion formation

As an additional approach to identifying gene function, we wished to construct *S. meliloti* strains in which large gene clusters were removed by deletion. We have previously constructed *S. meliloti* strains in which large pSymB deletions were generated through IS50-directed recombination between Tn5-derivative inserts flanking 100-300 kb gene regions. Since these deletions were difficult to identify, we employed Flp recombinase and Flp recombination target sequences (FRT) (Broach and Hicks, 1980) to delete large regions of the *S. meliloti* genome. Several integration vectors, to allow integration of FRT sites into the *S. meliloti* genome and vectors to allow expression of Flp, have been made. These include the vectors pTH455 and pTH456. An outline of the method employed to construct strains carrying two FRT is in direct orientation is shown in Figure 1. Strains carrying deletions were then screened using Biolog plates to identify carbon and nitrogen source non-utilization phenotypes.

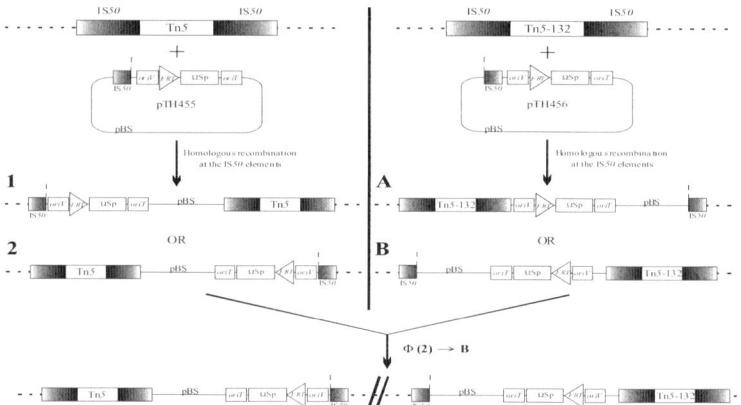

Figure 1. Construction of strains carrying FRT in direct orientation. FRT was directed to target Tn5 and Tn5-132 insertions using the vectors pTH455 and pTH456 respectively. The two orientations of the targeting plasmids are revealed by PCR as previously described (see Chain et al., 2000). Double co-integrate strains were made by transduction with ΦM12.

3. Reporter Gene Fusion Library

We wished to construct a library of the *S. meliloti* genome in which genes are fused to reporter genes that are readily assayed in microtitre plates. Our objective is to use this library to identify conditions under which the *puf*-reporter gene fusions are expressed. The pTH1522 reporter plasmid carries a ColE1 origin of replication and hence replicates at moderate copy number in *E. coli* but not in *S. meliloti*. It carries the *accC4* gene encoding gentamycin resistance and is mobilizable by the plasmids pRK2013 and pRK600. Flanking the XhoI cloning site are dual reporter genes. Thus *gusA* (β-glucuronidase) and *tdimer2(12)* (an RFP variant) are positioned in tandem in one orientation, and *lacZ* (β-galactosidase) and *gfp+* (a GFP variant) also in tandem are in the opposite orientation. The *gfp+* variant has been reported to reliably monitor gene expression *in vivo* and carries mutations that increase folding efficiency and fluorescence yield (Scholz et al., 2000). The *tdimer2(12)* gene is a nonoligomerizing derivative of an engineered red fluorescence protein from *Discosoma* coral, such that the modified DsRed protein is duplicated and the resulting dimeric protein is joined by a linker containing twelve arginine residues (Campbell et al., 2002). Translation stop codons in the three reading-frames are positioned on both sides of the *Xho*I site upstream of the reporter genes – hence this vector should only generate transcriptional gene fusions. We employed a vector in which reporter genes were positioned on both sides of the cloning site as this allowed the generation of reporter fusions regardless of the orientation of the insert DNA (Figure 2). Moreover, the use of the dual reporters also allows us to compare/confirm values from one reporter (e.g., GFP) with the second reporter activity (e.g., β-galactosidase). In preliminary studies, we evaluated the expression of *lacZ*, *gfp+*, *gusA* and *tdimer2(12)* by using well characterized *pckA*, *dme* and *nifH* promoters, and uncharacterized *chvI* and SmC00171 promoters in *S. meliloti*.

Figure 2. Expression of *gfp+* and *lacZ* in *S. meliloti* constructs containing the *pckA* and *chvI* promoters

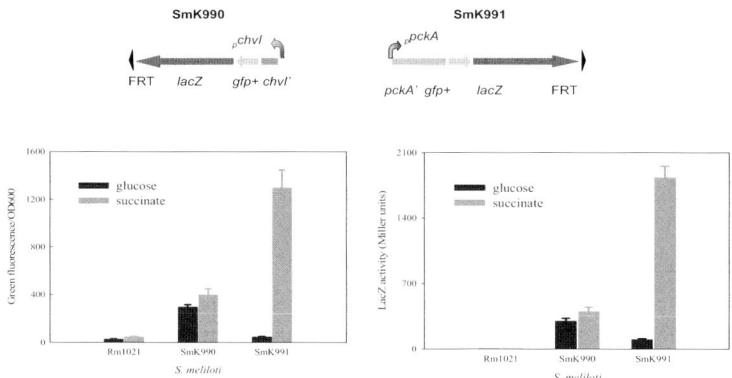

An initial genomic library with over 5000 clones has been constructed from 1.5-3 kb size fractionated DNA obtained from a partial Sau3A digest of *S. meliloti* 1021 DNA. The insert DNA filled in with G A and ligated to XhoI restricted, and T C filled in, pTH1522. The insert DNA in each of these clones was sequenced from both ends and hence the precise gene fusion junction is known. Each clone is transferred into *S. meliloti* and single-crossover co-integrates are selected as gentamycin resistant transconjugants. High throughput screening procedures for monitoring gene expression in 96 well format have been developed. Thus, standard LacZ and GusA reporter gene assays have been modified and optimized for semi-automated measurement. The modifications include eliminating the use of chloroform or toluene for cell permeabilization prior to adding substrate and using 0.01% (w/v) SDS instead, along with enzyme substrate and assay buffer. Fluorescence intensities are measured in a plate reader and to date we have screened 2100 gene fusion strains in LBmc, M9-glucose or M9-succinate medium (Figure 3). This genome-wide functional analysis of gene fusions provides an independent test of results from other gene expression analyses e.g., transcriptome, proteome, or metabolome.

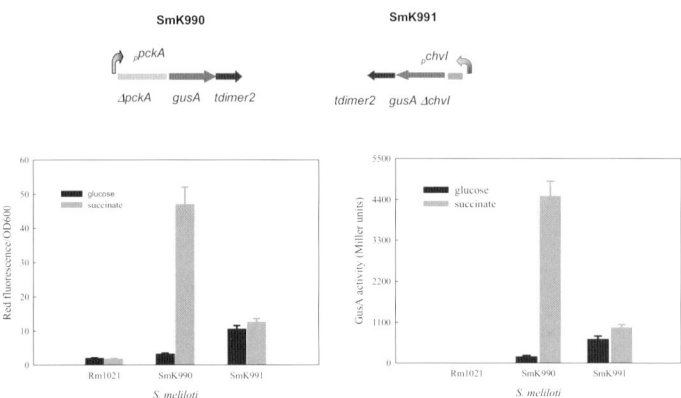

Figure 3. Expression of *gusA* and *tdimer2* in *S. meliloti* constructs containing the *pckA* and *chvI* promoters

We thank Genome Canada, The Ontario Genomics Institute and the Ontario Research and Development Challenge Fund for financial support.

Broach J and Hicks J (1980) Cell 29, 227-234.
Campbell et al. (2002) Proc. Natl. Acad. Sci. USA 99, 7877-7882.
Charles, T and Finan T (1991) Genetics 127, 5-20.
Chain et al. (2000) J. Bacteriol. 182, 5486-5494.
Galibert F et al. (2001) Science 293, 668-672.
Scholz O et al. (2000) Eur. J. Biochem. 267, 1565-1570.

WHAT CAN BACTERIAL GENOME RESEARCH TEACH US ABOUT IRON UPTAKE IN *SINORHIZOBIUM MELILOTI*?

A. Pühler, A. Becker, J. Buhrmester, T.-C. Chao, S. Rüberg, and S. Weidner
Department of Genetics, Faculty of Biology, Bielefeld University, POB 100131, D-33501 Bielefeld, Germany

This article deals with the genome based analysis of iron transport in *S. meliloti*. In particular, the use of microarrays to study the transcriptional regulation of genes involved in iron uptake is demonstrated.

1. The *S. meliloti* genome and its use for the development of PCR-fragment and oligo nucleotide based microarrays

The sequence of the *S. meliloti* genome published by Galilbert et al. in the year 2001 belonged to the first rhizobial genomes ever sequenced. The *S. meliloti* genome is composed of the chromosome and the two megaplasmids pSymA and pSymB. The chromosome carries – as expected – all the house keeping genes (Capela et al., 2001) whereas the pSymA plasmid, also designated as the symbiotic plasmid, was found to carry the genetic information for nodulation, infection and nitrogen fixation (Barnett et al., 2001). The megaplasmid pSymB is characterized by a huge amount of polysaccharide biosynthesis genes and in addition by many DNA-regions coding for ABC-transporters (Finan et al., 2001). Altogether, around 6207 genes could be identified in the *S. meliloti* genome.

The detailed information on the genetic level represents an ideal base for the functional analysis of *S. meliloti*. In particular, the construction and use of microarrays contributed a lot to our knowledge on the life style of *S. meliloti* in the free-living as well as in the symbiotic state. In the Bielefeld group two types of *S. meliloti* microarrays were developed, a PCR-fragment based (Rüberg et al., 2003) and an oligo nucleotide based one (Krol and Becker, 2004). Both types of microarrays were used in the meantime in a whole series of articles concerning the analysis of interesting biological properties of *S. meliloti*, like osmoadaptive gene expression (Rüberg et al., 2003), gene expression under microoxic and symbiotic condition (Becker et al., 2004), regulation of the *sit*-operon (Chao et al., 2004), phosphate stress response (Krol and Becker, 2004), LuxR gene expression (Hoang et al., 2004), ExoR and ExoS gene expression (Yao et al., 2004) and finally the transcriptional modulating effect of *ntrR* (Puskás et al., 2004).

2. The identification of *S. meliloti* genes involved in iron transport
It is well-known that iron supply of the microsymbiont plays an important role in the nodule symbiosis between rhizobia and legumes. Therefore, the microsymbiont should be equipped with special transport systems to ensure a sufficient iron supply. Three types of iron ABC-transporters can be identified in *S. meliloti*. In the case of the siderophore-type of iron ABC-transporters, ferri-siderophore complexes are taken up by outer membrane receptors. The ferric-type of iron ABC-transporters is characterized by its property to take up ferric iron, which is acquired from iron storage proteins like lactoferrin or transferin. Last but not least, the third type of iron ABC-transporter is the metal-type. For this type, it is not yet clear whether outer membrane receptors or the Ton-complex are parts of the system.

Using bioinformatic tools the genomic sequence of *S. meliloti* was screened for the presence of ABC-transporter gene regions. Eight DNA regions possibly coding for iron transporters were identified (Table 1). These DNA regions were designated *smFe-A* to *smFe-H*. The four DNA regions *smFe-C*, *smFe-D*, *smFe-E* and *smFe-F* most probably encode siderophore/heme-type ABC-transporters whereas the three DNA regions *smFe-B*, *smFe-G* and *smFe-H* were found to encode putative ferric-type ABC-transporters. Finally, the DNA region *smFe-A* is the only one to encode a putative metal-type ABC-transporter. In particular, it should be mentioned that *smFe-A* shows homology to *sitABCD*, *smFe-B* to *afuABC* and *smFe-C* to *hmuPSTUV*.

Table 1: Putative iron ABC-transporter genes and their regulation by iron and RirA

putative iron ABC-transporter gene region	regulation by iron	RirA
smFe-A: smc02509-06(sitABCD)[1]	-	-
smFe-B: smc04317-00(afuABC)[2]	-	-
smFe-C: smc01513-10(hmuPSTUV)[3]	+	+
smFe-D: smc1746-41	+	+
smFe-E: smc20056-58	-	-
smFe-F: smc20365-63	-	-
smFe-G: smc21432-29	+	+
smFe-H: smc21540-42	-	-

[1] *sit* from *Samonella* iron transport
[2] *afu* from *Actinobacillus* ferric uptake
[3] *hmu* from *Hemin* uptake

Table 2: Putative IROMP genes and their regulation by iron and RirA

putative IROMP gene	regulation by iron	RirA
smc01611(fhuA1)[1]	-	+
smc01657(fhuA2)[1]	-	-
sma2414(rhtA)[2]	+	+
smc2721	-	-
smc02726(shmR)[3]	+	+
smc02890	-	-
smc04205	-	-
sma1747	-	-

[1] *fhu* from ferric hydroxamate uptake
[2] *rht* from rhizobactin 1021 transport
[3] *shmR* from *Sinorhizobium* heme receptor

The DNA regions *smFe-B,C,D,E,F,G* and *H* encoding putative iron ABC-transporters were selected for a mutational analysis. Marker-free deletion mutants were constructed with deletions in the genes coding for substrate binding proteins. For all genes with the exception of *smFe-F* the appropriate mutants could be obtained. By means of iron dependent growth tests of the corresponding mutants the involvement of three siderophore/heme-type and one ferric-type transporter in iron acquisition of *S. meliloti* RM1021 was verified.

As already mentioned, outer membrane receptors play an important role in iron uptake. These outer membrane receptors are also designated IROMPs (IROMP from iron regulated outer membrane protein). Therefore, the *S. meliloti* genome was also screened for IROMP genes. The results are presented in Table 2. Eight putative IROMP genes could be identified. The genes *smc01611* and *smc01657* showed homology to *fhu*. Further on, it was recognized that *sma2414* represents the *rhtA* gene and *smc02726* the *shmR* gene. In order to learn more about the function of the putative IROMP genes, marker-free deletion mutants were generated and tested for their growth behaviour in iron depleted TY-medium containing 200 µM of the iron chelator 2,2'-dipyridyl. Only the *S. meliloti rhtA* mutant showed a severe growth defect. Since the *rhtA* gene is known to code for the outer membrane receptor involved in the uptake of the endogenous siderophore rhizobactin 1021, the observed growth defect confirms former findings.

3. The regulatory gene *fur* and its role in the regulation of the *sitABCD* operon

Amongst the DNA region coding for iron ABC-transporters a region was identified which encoded an ABC-transporter of the metal type (Table 1). The four genes located in this DNA region were designated *sitA,B,C,* and *D* since their gene products showed high identity to the gene product of the *sit* operon analyzed in the human pathogen *Salmonella typhimurium* and *Yersinia pestis*. Upstream of the *sitABCD* operon on the opposite strand the gene *fur* encoding a ferric uptake regulator (Fur) could be identified. In addition, in front of the *sitA* gene a putative Fur-box motif was found. The *S. meliloti sitABCD* operon as well as the regulatory gene *fur* were intensively analyzed (Chao et al., 2004). Applying the PCR-fragment based *S. meliloti* microarray it could be demonstrated that in the *fur*-deletion mutant 23 genes were up-regulated and 10 genes were down-regulated when compared to the wild type strain. Among the up-regulated genes only the *sitABCD* operon could be associated with metal uptake. Thus in *S. meliloti* the Fur repressor does not represent a global regulator. As a surprise it turned out that the *sitA*-promoter was strongly repressed by Mn(II) with dependence on Fur. Therefore, it has to be concluded that in *S. meliloti* the *sitABCD*-operon together with the *fur*-gene is in the first place responsible for manganese transport.

4. The global regulatory gene *rirA* and its influence on the iron metabolism of *S. meliloti*

Following the finding that *fur* is not involved in the regulation of genes coding for iron uptake we concenetrated on the *S. meliloti rirA* gene. The *rirA* gene (*rir* from rhizobia iron regulator) was first identified in *Rhizobium leguminosarum*. In *S. meliloti*, the RirA regulator is characterized by 77 % identity to the *R. leguminosarum* RirA. It could be demonstrated that a marker-free *S. meliloti rirA* deletion mutant showed the expected phenotype. The *S. meliloti rirA*-mutant is reduced in its growth behaviour if cultivated in a medium containing an enhanced level of iron. This can be explained easily since a higher cellular iron concentration causes toxic effects. In order to learn whether RirA represents the global iron regulator in *S. meliloti* a microarray experiment was carried out which compared the wild type strain with the *rirA*-deletion mutant. The results obtained were as follows. Altogether, 131 genes were up-regulated and 63 genes turned out to be down-regulated in the *rirA*-mutant. Evidently, *rirA* is involved in a huge regulatory network. In order to show that *rirA* mainly regulates genes involved in iron metabolism further microarray experiments were carried out with the *S. meliloti* wild type strain grown in minimal as well as in complex medium under low and high iron concentrations. To make the story short, genes up- or down-regulated in all three microarray experiments were of special interest. In such a way, a core set of 45 up-regulated genes and of four down-regulated genes was generated. Among the 45 up-regulated genes, three ABC-transporter gene regions (Table 1) and two IROMP genes (Table 2) could be identified. This clearly demonstrates that RirA can be considered to represent the global iron regulator in *S. meliloti*. It is now of interest to identify the RirA-binding motif and to solve the question under which conditions RirA binds to DNA.

5. Conclusions

In this article, we show that *S. meliloti* candidate genes for iron ABC-transporters and for outer membrane receptors could be successfully identified by the help of bioinformatics. Further on, the proposed functions of the candidate genes were tested by mutational analysis. Whole genome microarrays were then used for the analysis of *S. meliloti* genes involved in the regulation of iron metabolism. It could be shown that the *S. meliloti* RirA repressor controls a network of genes up-regulated during iron starvation. Evidently, genome research helps a lot to analyze the life style of *S. meliloti*.

6. Literature

Barnett MJ et al. (2001) PNAS 98, 9883-9888
Becker A et al. (2004) MPMI 17(3), 292-303
Capela D et al. (2001) PNAS 98, 9877-9882
Chao TC et al. (2004) J. Bacteriol. 186(11), 3609-3620
Finan TM et al. (2001) PNAS 98, 9889-9894
Galibert F et al. (2001) Science 293, 668-672
Hoang HH et al. (2004) J. Bacteriol. 186(16), 5460-5472
Krol E and Becker A (2004) Mol. Gen. Genomics 272, 1-17
Puskás LG et al. (2004) Mol. Gen. Genomics, in press
Rüberg S et al. (2003) J. Biotechnol. 106(2-3), 255-268
Yao SY et al. (2004) J. Bacteriol. 186(18), 6042-6049

STRUCTURAL AND FUNCTIONAL GENOME ANALYSIS OF *LOTUS JAPONICUS* AND *MESORHIZOBIUM LOTI*

S. Sato, T. Kaneko, Y. Nakamura, E. Asamizu, T. Kato and S. Tabata
Kazusa DNA Research Institute, 2-6-7 Kazusa-Kamatari, Kisarazu, Chiba, 292-0818, Japan

1. Introduction

Progress in DNA-sequencing technology has allowed us to perform systematic and comprehensive analysis of genetic information in a variety of organisms. This accumulation of large quantities of information on genomes and gene structure is changing drastically the experimental strategies in many biological fields. Furthermore, information and material resources obtained during the course of genome sequencing can be utilized to study functional aspect of the genes in the genome.

Lotus japonicus is a promising model legume for molecular genetic and physiological studies on legume-specific phenomena, such as plant-microbe interaction and symbiotic nitrogen fixation, because of its small genome size, short generation time, self-compatibility, and feasibility of genetic transformation. In addition, molecular genetic approaches, such as transposon (Schauser et al., 1999) and T-DNA (Webb et al., 2000) tagging, have been successfully performed and a number of mutants related to symbiosis and nitrogen fixation have accumulated. Recent progress in large-scale analysis of cDNA has also increased the significance and usefulness of *L. japonicus* as a model legume (Asamizu et al., 2004).

Mesorhizobium loti performs nitrogen fixation on several *Lotus* species within determinant-type nodules. Nodule formation and nitrogen fixation result from the sequential expression of a series of genes from both bacteria and host plants. To understand the genetic system required for the entire process of symbiotic nitrogen fixation from both the bacterial and host plant sides, we initiated a large-scale analysis of the *L. japonicus* and *M. loti* genomes.

2. Status of the rhizobial genome project

The complete genome sequence of *M. loti* strain MAFF303099 was determined by the whole-genome shotgun strategy, using four shotgun libraries with different insert sizes

(Kaneko et al., 2000). The length of the genome was 7,596,297 bp and consists of three circular molecules, a chromosome of 7,036,071 bp and two plasmids, designated as pMLa and pMLb, of 351,911 bp and 208,315 bp, respectively. A total of 7,281 potential protein-coding genes, two sets of rRNA genes, 50 tRNA genes, and an Rnase P RNA gene were identified on the genome. The assigned genes on the *M. loti* genome are presented in the Web database, RhizoBase, at http://www.kazusa.or.jp/rhizobase/.

For a comparative analysis of rhizobia and to provide information and material resources of agricultural importance, we continue determining the nucleotide sequence of the genome of *Bradyrhizobium japonicum*, which is a nitrogen-fixing endosymbiont of soybean (*Glycne max*). The genome of *B. japonicum* is a single circular chromosome of 9,105,828 bp (Kaneko et al., 2002). The chromosome comprises 8,317 potential protein coding genes, one set of rRNA genes, and 50 tRNA genes. On comparison with the entire genomes of *M. loti* and *Sinorhizobium meliloti* (Galibert et al., 2001), 34% of *B. japonicum* genes showed significant sequence similarity to those of both *M. loti* and *S. meliloti*, whereas 23% were unique to this species.

Comparative genomics holds the promise of identifying either common genes or genes unique to individual organisms, and of studying the process of gene and genome evolution. With the completion of the *M. loti*, *S. meliloti*, and *B. japonicum* genomes, the plasmids of *Rhizobium* sp. strain NGR234 (Freiberg et al., 1997; Streit et al., 2004), the symbiosis island of *M. loti* R7A (Sullivan et. al., 2002), and the symbiotic plasmid p42d of *Rhizobium etli* strain CFN42 (Gonzalez et al., 2003), comparative genomics among rhizobia has become feasible. In order to establish a web site suitable for the comparative analysis of the rhizobial genomes, we improved RhizoBase by storing all of the available rhizobial genome data in an in-house database and presenting them through the same interface. One of the important advantages of the improved RhizoBase is that it is possible to search simultaneously either all or selected genomes. In addition, with the aim of increasing the information on individual genes, we established (at http://www.kazusa.or.jp/rhizobase/comment.html/) a data deposit type database, RhizoGene.

3. Status of the genome project of *Lotus japonicus*

For *L. japonicus* genome sequencing, we selected the genomic clones from multiple seed points, using the expressed sequence information. The nucleotide sequence of each selected clone was determined by the bridging shotgun method. As of October, 2004, a total of 1,659 clones, including 451 walking clones from seed sequences, have been selected. Twenty-five of them are ready for sequencing, 5 are being sequenced, 1,173 are in the phase1 draft sequence stage, and 438 clones covering 44.9 Mbp region of the genome have been annotated (Kato et al., 2003). Including the sequences in the phase1 draft stage, a total of 162 Mbp of the genome has been sequenced so far.

Assignment of the protein-coding regions and gene modeling are being performed by combining similarity search results and computer predictions. At present, complete structures of 4,089 potential protein-encoding genes were assigned in the 44.9 Mbp region of the genome.

In order to establish the bases of the mapping and map-based cloning, the sequenced clones were genetically localized onto the linkage map by generating PCR-based DNA markers, simple sequence repeat length polymorphism (SSLP), and derived cleaved amplified polymorphic sequence (dCAPS). Currently, 1,310 clones have been located on the linkage map. Among them, 691 clones were mapped by SSR markers, 80 were mapped using dCAPS markers, and the remaining 539 clones were anchored by confirming the overlap with mapped seed clones. These PCR-based markers with neighboring sequence information have facilitated gene mapping and map-based cloning in *L. japonicus*. The sequence data, gene information, and mapping information are available in the web database, http://www.kazusa.or.jp/lotus/.

4. References

Asamizu E et al. (2004) Plant Mol Biol. 54, 405-14.
Freiberg C et al. (1997) Nature. 387, 394-401.
Galibert F et al. (2001) Science. 293, 668-672.
Gonzalez V et al. (2003) Genome Biol. 4, R36.
Kaneko T et al. (2000) DNA Res. 7, 331-338.
Kaneko T et al. (2002) DNA Res. 9, 189-197.
Kato T et al. (2003) DNA Res. 10, 277-285.
Schauser L et al. (1999) Nature. 402, 191-195.
Streit WR et al. (2004) J Bacteriol. 186, 535-542.
Sullivan JT et al. (2002) J Bacteriol. 184, 3086-3095.
Webb KJ et al. (2000) Mol. Plant-Microbe Interact. 13, 606-616.

ANALYZING A *SINORHIZOBIUM MELILOTI* 1021 ORFEOME IN A FUNCTIONAL GENOMIC PLATFORM

Brenda K. Schroeder, Brent L. House, Michael W. Mortimer, Scott C. Maloney, Casey A. Taylor, Kristel L. Ward, Hope T. Ziemkiewicz, Scott Clark, John J. Bovitz, Hao Jin, Svetlana Yurgel, and Michael L. Kahn.
Institute of Biological Chemistry and School of Molecular Biosciences, Washington State University, Pullman, WA 99164-6340

The Gram negative soil bacterium *Sinorhizobium meliloti* is a complex microorganism that can grow as part of the normal soil microflora or form a nitrogen-fixing symbiotic relationship with legumes such as alfalfa (*Medicago sativa*) and *Medicago truncatula* (1). When the bacterial cells perceive legume root exudates in the soil, they migrate toward and ultimately invade the root hairs. Further development leads to the formation of root nodules, which are specialized plant structures where *S. meliloti* cells can differentiate into bacteroids that can fix atmospheric dinitrogen into ammonia. N fixation is an energetically expensive reaction, requiring 16 ATP molecules and 8 low potential electrons per molecule of N_2 reduced. The plant provides carbon and energy sources, such as organic acids, to the bacteroids and benefits from the fixed nitrogen that is made available by the bacteria.

The genome sequence *S. meliloti* 1021 was determined by an international collaboration (2). Sequence annotation of the chromosome (3.7 Mb), and the two megaplasmids, pSymA (1.4 Mb), and pSymB (1.7 Mb), predicted that greater than 6200 ORFs were present. Analysis of these ORFs indicated that *S. meliloti* 1021 was genetically complex and metabolically diverse. The availability of the entire genome sequence posed an interesting dilemma, to continue genetic analysis of *S. meliloti* one gene at a time or to embark on a genetic analysis of *S. meliloti* on a genomic scale. Both avenues are important and their impact should be complementary. We report on progress with a novel approach to experimentation on a genomic scale.

Bioinformatics enables prediction of the putative products of many of the ORFs, but it is important to confirm these assignments and generate biological data based on them. There are several types of tools based on the sequence that could provide biologically relevant information about the ORFs including: nucleic acid probes for transcription analysis; operon fusions for monitoring bacterial gene expression in free-living culture and in symbiosis; mutants lacking one or more genes for phenotype analysis; constructs for overproducing proteins of interest in both *E. coli* and *S. meliloti*; and "tagged" proteins for various kinds of functional analysis. No single manipulation immediately enables all of these tools to be constructed. But strategies have been developed, such as the GATEWAY™ system (Invitrogen, Inc) where DNA is cloned with the aid of phage recombination enzymes, so that after the initial and efficient cloning of an ORF, the ORFs can be moved efficiently into other plasmids specialized for the manipulations described above. In our implementation of this strategy, all of the *S. meliloti* ORFs have been cloned into pMK2010 (3) as a first step toward generating plasmids for expressing proteins, creating reporter gene fusions or constructing mutations. Adding *oriT* sequences to all of the plasmids, which were originally brought into the design to enable DNA to be conjugated into *S. meliloti*, has also been useful in enabling recombination manipulations to be carried out *in vivo*, saving time and reagents. This technique was successfully utilized in the analysis of the denitrification gene cluster of *S. meliloti* (3).

Currently, 99.9% of the 6317 predicted ORFs have been cloned into pMK2010 via GATEWAY integrase-mediated recombination and we refer to the set of plasmids as the S. meliloti ORFeome. We have transferred greater than 68% of the ORFeome from the entry vector into pMK2030, a destination vector that allows the construction of fusions to a tandem GFP and GUS reporter operon (4). This transfer is accomplished via a penta-parental mating, which uses a five parent mating mix to mobilize the entry construct and pMK2030 into a strain that expresses integrase and excisionase thus promoting the exchange of the cloned ORF from the entry vector into the destination vector without DNA preparation or the expense of additional enzymes. We are subsequently transferring the reporter fusions carrying the ORFs into *S. meliloti* 1021 to generate fusions of the reporter genes to the native promoters of each ORF. We have in hand vectors for generating deletions (3) and for expressing histidine tagged versions of the proteins. The unique advantage to using this system is that new methods can be adopted relatively easily by creating new destination vectors. This technique will be especially useful for analyzing properties of the many protein families present in the *S. meliloti* genome.

Projects are now being completed in the laboratory that are utilizing the genetic resources described above. For example, the entry vector carrying the *agaL1* gene, coding for α-galactosidase, was recombined via penta-parental mating into pMK2030, and then recombined into the *S. meliloti* 1021 genome, generating fusions to the GFP and GUS reporter genes. The resulting strain was cultured in minimal NH_4^+ media amended with 0.2% raffinose, or raffinose+succinate and then the GUS and α-galactosidase activity were determined. The values obtained indicated that promoter activity of *agaL1* was induced when cultured in the presence of raffinose and repressed by succinate as previously reported (5). The ability to measure not only the activity of GUS but also α-galactosidase, indicates that expression of the wild type gene is not hindered by the presence of the reporter construct in the genome.

The genetic resources developed by this effort will enable the investigation of scientific questions on a genomic scale. The genetic materials will also be immediately available for gene-specific analysis. Thus, the genetic analysis of *S. meliloti* on a genomic scale will provide the genetic tools to complete any number of investigations into the complex biology of this important microorganism.

We thank the NSF Microbial Genetics Program for support of this project.

1. van Rhijn P and Vanderleyden J (1995) Microbiol. Rev. 59,124-142.
2. Galibert et al. (2001) Science. 293,668-672.
3. House B et al. (2004) Appl. and Environ. Microbiol. 70, 2806-2815.
4. Ramos et al. (2002) J. Biotechnol. 97, 243-252.
5. Bringhurst RM and Gage DJ (2002) J. Bacteriol. 184, 5385-5392.

FUNCTIONAL GENOMIC ANALYSIS OF THE SDR FAMILY IN *SINORHIZOBIUM MELILOTI*

Sirin Adham, Asha Jacob, David Capstick, Trevor C. Charles
Dept. Biology, University of Waterloo, 200 University Ave. West, Waterloo, ON, N2L 3G1 Canada

The ability of bacterial cells to grow and compete in the soil and rhizosphere is influenced by the ability to degrade a wide variety of organic compounds, thus obtaining energy and carbon. This involves oxidation, carried out by specific enzymes, which have substrate specificity. The genomes of *Sinorhizobium meliloti* and other plant associated alpha-proteobacteria are rich in dehydrogenases and oxidoreductases, reflecting the broad catabolic capacity of the organisms and the nutritional richness of the soil and rhizosphere environments. A major class of such enzymes is the SDR (**S**hort-chain **D**ehydrogenases/**R**eductases) superfamily (Kallberg et al. 2002), catalyzing NAD(H) / NAD(P)(H) dependent oxidation / reduction reactions. According to the Pfam database (Bateman et al. 2004), the *S. meliloti* genome encodes 78 members of the SDR superfamily, 35 on replicon C, 21 on replicon B and 22 on replicon A. These enzymes likely make major contributions to the nutritional diversity of the organism.

Members of the SDR family are typically proteins of 250-350 residues that can share as little as 15% residue identity but have strikingly high similarities in tertiary structure. The characterized members of the SDR family are homodimeric or homotetrameric. Surprisingly, the substrate specificities have been experimentally determined for only very few of those encoded in the *S. meliloti* genome, and in bacteria in general. Two examples of SDRs of known function in *S. meliloti* are the PHB cycle enzymes acetoacetyl-CoA reductase (encoded by *phbB*) (Aneja et al. 2004) and D-3-hydroxybutyrate dehydrogenase (encoded by *bdhA*) (Aneja and Charles 1999). Both of these genes have been characterized genetically, and the enzyme activities have been demonstrated biochemically. Another SDR is the *nodG* encoded oxoacyl-ACP reductase, involved in fatty acid biosynthesis related to the lipochitin oligosaccharide Nod factor (López-Lara and Geiger 2001). The goal of the present study is to attempt to determine the substrate specificities and metabolic functions of the remaining SDRs.

As part of the Genome Canada funded project "Genomic Analyses of Soil Microorganisms", we are developing and applying methods for the determination of the substrate specificities of the members of the SDR family in *S. meliloti*. We have designed and constructed a new expression vector for analysis of these and other genes. This vector, pSA005, has a broad host range IncP replicon and encodes tetracycline resistance as the selectable marker. In order to detect successful overexpression and for affinity purification of the gene product it incorporates a StrepII tag at the N-terminus. The *tac* promoter in concert with *lacI^q* facilitates IPTG-regulated expression. The strategy involves the overexpression of each of these genes in *S. meliloti* using this vector, as well as generating reporter gene fusion mutations using a dual reporter plasmid (Cheng, Finan et al., described elsewhere in this volume). The resulting overexpression strains and gene fusion mutants are being characterized on the basis of symbiotic phenotype, competition for nodulation, desiccation tolerance, carbon source utilization, substrate specificity, gene expression, and metabolite analysis.

It is expected that the study of the SDR family in *S. meliloti* will result in a greater understanding of the mechanisms underlying the catabolic diversity of this important model plant symbiont and soil inhabitant. We should also be able to transfer the knowledge of substrate specificity of the *S. meliloti* SDRs to orthologs in other closely related bacteria. Finally, the methods that we are using in this study will be applicable to the study of other genes encoding the synthesis of proteins of unidentified function.

References

Aneja P and Charles TC (1999) J. Bacteriol. 181, 849-857.
Aneja P et al. (2004) FEMS Microbiol. Lett. 239, 277-283.
Bateman A et al. (2004) Nucl. Acids Res. 32, D138-141.
Kallberg Y et al. (2002) Eur. J. Biochem. 269, 4409-4417.
López-Lara IM and Geiger O. (2001) Mol Plant-Microbe Interact. 14, 349-357.

Acknowledgements

Research Funding from NSERC Canada and Genome Canada is gratefully acknowledged.

PROTEOME ANALYSIS ON BACTEROID DIFFERENTIATION OF *BRADYRHIZOBIUM JAPONICUM* USDA110

[1]Shigeyuki Tajima, [2]Le T-P Hoa, [1]Rie Hamaguchi and [1]Mika Nomura
[1]Dept. Life Science, Kagawa U, Kagawa 761-0795, Japan, [2]Faculty of Biology, Hanoi University of Education, Hanoi, Vietnam

Using the genome sequencing data of *B. japonicum* USDA110 and *Mesorhizobium loti*, we tried proteome analysis using 2D-PAGE for revealing the key proteins corresponding to the morphogenesis or metabolism of the symbiotically differentiated bacteria (bacteroid) in soybean nodules (*Glycine max* – *Bradyrhizobium japonicum* USDA110 system). In addition to preparing the 2D-PAGE database this data would be useful to compare the protein profiles during bacteroid differentiation. In this presentation, we show our 2DE proteome database of *B. japonicum* bacteroids, and some application to survey differentiation of *B. japonicum* bacteroids.

2DE Database: Protein fraction was prepared from cultured bacteria and nodule bacteroids. When protein spots (717 spots in cultured bacteria and 816 spots in bacteroids) in 2DE were compared by matching computer software, only 152 spots were common between two tissues. When bacteroid fraction was compared to culture bacteria, 22 spots were up-regulated and 32 spots were down-regulated, suggesting the protein profiles were very different between culture bacteria and bacteroids.

When proteome maps were compared and the characterization of bacteroid-specific proteins was performed by N-terminal amino acid sequencing and matrix-assisted desorption/ionization time of flight mass spectrometry peptide mass fingerprint analysis, putative identity was assigned to 146 bacteroid spots (98 proteins), including many metabolic proteins and ABC transporters as well as nitrogenase proteins like NifH. Although the physiological roles of these proteins are under investigation, NfeC, ACC deaminase, 60 KDa chaperon GroEL3 and AhpC were reported their significant contribution in bacteroids. Enzymes for various metabolisms were also detected as specific proteins[1]. For amino acid metabolism, S-adenoslymetionine synthetase and probable L-asparaginase were detected. Although the biosynthesis of amino acids in *B. japonicum* bacteroids is believed to be down-regulated, various proteins responsible for amino acid transport were detected, and the data is corresponding to the report that amino acid transport is essential for maintaining nitrogen fixation activity in bacteroids of *B. japonicum*. Anti-oxidant protein and alkyl hydroperoxide reductase might help to scavenge peroxide and maintain reductive conditions in bacteroids.

Usefulness of this proteome database was also shown in tracing bacteroid protein expression profiles during nodule development. We tried to compare 2DE profiles among cultured bacteria, young bacteroids from 10 day-after-infection (DAI) nodules (diameter 1-2 mm) and matured bacteroids from 28 DAI nodules (diameter 5-8 mm). Interestingly, nitrogenase proteins were detected very early in 10 DAI and kept high value in 28 DAI. Poly-beta-hydroxybutyrate dehydrogenase, a marker enzyme of bacteroids, was detected in very low value in 10 DAI and cultured bacteria. The content increased significantly in 28 DAI, suggesting the protein content is corresponding to aging of bacteroids and not to nitrogen fixation. Such continuous survey of protein expression during bacteroid development was performed in various proteins and the data suggested the nodule bacteroids were metabolically different due to the ageing.

Due to high sensitivity of peptide mass finger printing analysis in 2DE, we believe that the proteome analysis is a useful tool to compare the metabolism differentiations in various plant organelle[2] and symbionts. The data can be compared with transcriptome data to discuss the metabolic profiles[3]. Such possibilities are now under investigation in our group to analyze bacteroid differentiation in nodules of Fix⁻ mutant *L. japonicus* plant.

References
1) Hoa LTP et al. (2004) Microbes and Environment, 19, 71-75
2) Hoa LTP et al. (2004) Plant Cell Physiol., 45, 300-308
3) Kouchi H et al. (2004) DNA Research 11, 263-274

GLOBAL CHANGES IN GENE EXPRESSION OF *S. meliloti* 1021 NODULE BACTERIA IN *nifA* AND *nifH* MUTANT BACKGROUND

Zhe-Xian Tian[1*], Hua-Song Zou[2*], Jian Li[1], Yuan-Tao Zhang[1], Guan-Qiao Yu[2], Silvia Rüberg[3], Anke Becker[3] and Yi-Ping Wang[1&]
[1] NLPEPGE, College of Life Sciences, Peking Uni., Beijing, P. R. China;
[2] NLPMG, Shanghai Inst. Plant Physiology & Ecology, CAS, Shanghai, P. R. China;
[3] Lehrstuhl für Genetik, Universität Bielefeld, Bielefeld, Germany;
[*] These two authors contributed equally to this work.
[&] Corresponding author. E-mail: wangyp@pku.edu.cn

Sinorhizobium meliloti can fix nitrogen in its symbiotic phase. NifA is a transcriptional activator of many genes involved in nitrogen fixation including *nifHDK* coding for the nitrogenase. As a step towards understanding gene expression linked to *nifA* and *nifH* and its impact on metabolic adaptation in bacteroids, gene expression profiles of nodule bacteria of the wild type and *nifA* and *nifH* mutants were compared using whole genome microarrays. A large scale alteration of gene expression (601 genes) was observed in the *nifA*⁻ background, whereas much less changes (196 genes) was observed in the *nifH* background, when compared with their wild type counterpart (partial results see Table 1).

As expected, *fixABC*, *nifHDK* and associated nitrogen fixation genes were greatly downregulated in the *nifA* mutant nodule bacteria when compared with the wild type. Several nodulation related genes were also downregulated in the *nifA* mutant nodule bacteria compared with the wild type. This may imply that NifA proteins play a regulatory role in the nodulation process as a part of its multi-functions. In the *nifH* mutant nodule bacteria, many nitrogen fixation genes including *fixABC*, *nifHDK* and two nodulation genes (*nodII*, *nodE*) were also observed to be downregulated compared to the wild type. This observation may imply that a global regulatory system exists to reduce the expression level of genes involved in nitrogen fixation in case of a breakdown of a key component of this process.

The translation process was extensively reduced in the wild type nodule bacteria because 10 ribosomal protein-encoding genes, as well as 3 genes required for protein and peptide synthesis (*tufA*, *tufB* and *infB*) were expressed at a lower level compared to both *fix*⁻ nodule bacteria. Many house-keeping genes (for examples, two genes related to cell envelop structure (*murD* and *ddlB*), 3 genes encoding components of the pyruvate dehydrogenase complexes (*pdhAa*, *pdhB* and *lpdA1*) which generate acetyl-CoA, two genes involved in TCA cycles (*acnA* and *sucD*) and 4 genes involved in aerobic respiration (*nuoE1*, *nuoF1*, *nuoL* and *fbcC*)) were downregulated in the wild type when compared with both *fix*⁻ nodule bacteria. In contrast, no house-keeping genes were observed to be upregulated in the wild type when compared with any of two *fix*⁻ mutants. These results indicate that the general metabolism is slower in the wild type than that in the *fix*⁻ *nifA* or *nifH* mutant nodule bacteria. Since the general metabolism appeared slower in nodule than in cultured bacteria (Becker et al. 2004), it is proposed that the lack of effective nitrogen fixation due to the lack of NifA or NifH might have hindered the process of general metabolism slow-down in bacteroids.

It is possible that in nodules inoculated by *fix*⁻ rhizobium strains, the absence of effective nitrogen fixation for the benefit of the host and additional consuming burdens from 'brake-failure' metabolic processes elicit a host defense reaction against the 'pathogen'. This may be partly supported by the microarray results showing that genes related to adaptation to atypical conditions (*csp4* and *clpA*) and chaperoning (*groES1*, *groES2* and *groES3*) were upregulated in *fix*⁻ compared to wild type nodule bacteria. However, direct evidences would be required from the host counterpart to confirm this hypothesis.

Table 1. Partial list of differentially expression genes in the *nifA* and *nifH* mutant compared to the wild type nodule bacteria.

Gene ID	Annotation	$\log_2(\text{wt}/nifA)$	$\log_2(\text{wt}/nifH)$
sma0816	FixX ferredoxin-like protein	2.90	2.89
sma0817	FixC oxidoreductase	4.21	1.91
sma0819	FixB electron transfer flavoprotein a chain	5.43	2.26
sma0822	FixA electron transfer flavoprotein b chain	5.33	3.47
sma0825	NifH nitrogenase Fe protein	4.25	1.67
sma0827	NifD nitrogenase Fe-Mo a chain	3.54	3.05
sma0829	NifK nitrogenase Fe-Mo b chain	3.58	2.34
sma0831	NifX nitrogen fixation protein	4.01	3.02
sma0851	NodH sulfotransferase	2.10	1.08
sma0853	NodE beta ketoacyl ACP synthase	1.41	1.14
sma0762	FixK2 transcription regulator	-2.22	1.61
sma0769	FixP2 cytochrome c oxidase	-2.42	1.16
sma1214	FixQ1 cbb3-type cytochrome oxidase	-1.93	3.23
sma1220	FixN1 Heme b / copper cytochrome c oxidase subunit	-2.38	3.37
sma1225	FixK1 Transcriptional activator	-2.21	2.35
smc03253	putative L-proline 3-hydroxylase	-1.50	3.34

Very unexpectedly, the expressions of *fixK* and *fixNOPQ* operon (both of two copies) which involved in electron transport were downregulated in the wild type compared to the *nifA* mutant. In contrast, these genes were upregulated in the wild type compared to the *nifH* mutant. A gene related to proline synthesis (*smc03253*) appeared to be strongly upregulated in the *nifH* mutant, while it was downregulated in the *nifA* mutant compared to the wild type. Since the expressions of *fixK* and *smc03253* are known to be activated by FixJ (Ferrieres et al. 2004), it is proposed that NifA proteins mediate inhibition of FixJ activation process in nodule bacteria. A quite resembling NifA binding site was predicted on the upstream sequence of *fixK1* by matrix-based statistic analysis. However, its regulatory role needs to be further verified.

References:
Becker A et al. (2004) Mol. Plant Microbe Interact. 17, 292-303.
Ferrieres L et al. (2004) Microbiol. 150, 2335-2345.

Acknowledgements:
This work was supported by grant 2001CB108901 and 2001CB108902 from the Chinese National Key Program for Basic Research (973), grant 031U213D from Bundesministerium für Bildung und Forschung, Germany and grant BIZ 7 from Deutsche Forschungsgemeinschaft.

TOWARD DECIPHERING THE GENOME OF *FRANKIA ALNI* STRAIN ACN14A.

Normand[1] P, Felix[1] S, Alloisio[1] N, Marechal[1] J, Lavire C[1], Berry AM[2], Mullin[3] BC, Tomkins[4] J, Choisne[5] N, Demange[6] N., Truong Cong[6] YC., Coulloux[6] A, Vallenet[6] D, Cruveiller[6] S, Médigue[6] C.
1-Ecologie Microbienne, UCBL, 69622 Villeurbanne Cedex, France; 2-UCDavis, Ca, USA; 3-UTK, Kn, USA, 4-CUGI, Clemson, USA, 5-INRA-UMRGV, INRA-UMRGV, Evry, and 6-Genoscope, Evry, France.
normand@biomserv.univ-lyon1.fr

The sequencing of the genome of *Frankia alni* strain ACN14a is being completed in collaboration with Genoscope (www.genoscope.fr/externe/English/Projets/ Projet _HF/HF.html). This strain was isolated from an *Alnus viridis* subsp. *crispa* growing in Tadoussac, PQ, Canada in 1982 (Normand and Lalonde, 1982). 140871 reads from 3 banks (94407 reads from the 3kb bank, 30350 reads from the 10kb bank and 16114 reads from the BACs) were used for a 12X coverage. Assembly of a temporary scaffold has yielded 11 ordered contigs. Gap closure is under way using a walking approach. This scaffold covers more than 7.5Mb and contains over 6800 CDS. The G+C is 72%, varying from 43 to 79% over 5000nt windows.

Automated annotation has permitted the reconstitution of major biochemical pathways, as well as the identification of a rich secondary metabolism. Based on an automatic annotation procedure with a stringent threshold, more than half of CDSs (> 4000) had no recognizable functions, a proportion that should be modified following manual annotation. At the moment, there have been more than 5000 CDS annotated and this part of the work is expected to be finished by the end of the year. About 47% of *Frankia alni* genes are in synteny with other genomes, the highest proportion being obtained with *Streptomyces avermitilis, Streptomyces coelicolor* and other actinobacteria but also with symbiotic Proteobacteria such as *Bradyrhizobium japonicum*. Several other scientists working on various aspects of the biology of *Frankia* are participating in the annotation part of the project.

The CDSs were translated into a putative protein database against which proteins separated by 2D electrophoresis were compared. Proteins that were up- or down-regulated following growth under different physiological conditions (basal medium (BAP) vs N-free basal medium (BAP-N), basal medium vs complex medium (Qmod), complex medium vs complex medium +benzoic acid) could be identified. This approach demonstrated that a rich array of stress proteins were predominant. However, other proteins involved in various functions were also identified.

References
Normand P & Lalonde M (1982) Can. J. Microbiol. 28, 1133-1142.

GENOME-WIDE VIEW OF *MESORHIZOBIUM* AND *BRADYRHIZOBIUM*, EXPRESSION ISLAND CLUSTARD ON SYMBIOSIS ISLAND AND GENOME DIVERSITY

Toshiki Uchiumi,[1] Takuji Ohwada,[2] Takakazu Kaneko,[3] Tadashi Yokoyama,[4] Kazuhiko Saeki,[5] Makoto Hayashi,[6] Shigeyuki Tajima,[7] and Kiwamu Minamisawa[8]
[1]Dept. Chem. BioSci. Kagoshima U. Kagoshima, Japan; [2]Dept. Agricul. Life Sci., Obihiro U Agricul. Veteri. Med., Obihiro, Japan; [3]Kazusa DNA Res. Insti., Chiba, Japan; [4]Dept. Biol. Produc., Tokyo U Agricul. Tech., Tokyo, Japan; [5]Grad. Sch. Sci., Osaka U, Osaka, Japan; [6]Grad. Sch. Engi., Osaka U, Osaka, Japan; [7]Dept. Life Sci., Kagawa U, Kagawa, Japan; [8]Grad. Sch. Life Sci., Tohoku U, Sendai, Japan

Rhizobia are symbiotic nitrogen-fixing soil bacteria associated with host legumes. The signal exchanges between partners under microaerobiosis are required to establish symbiosis. We have developed macroarray of *Mesorhizobium loti* MAFF303099, a microsymbiont of the model legume *Lotus japonicus*, and monitored the transcriptional dynamics of the bacterium during symbiosis, microaerobiosis, and starvation. Global transcriptional profiling demonstrated that the clusters of genes within the symbiosis island (611 kb) were collectively expressed during symbiosis, whereas genes outside the island were downregulated. This finding implies that the huge symbiosis island functions as clustered expression islands to support symbiotic nitrogen fixation. A highly upregulated gene in bacteroids, *mlr5932* (1-aminocyclopropane-1-carboxylate deaminase), and another upregulated gene in starved cells, *mlr2852* (cardiolipin synthase) were disrupted and these genes were confirmed to be involved in symbiosis and in cell viability under starvation, respectively. These results indicate that the disruption of highly expressed genes is a useful strategy to explore novel gene functions in the environments. By the genomic macroarray of *Bradyrhizobium japonicum* USDA110, the similar transcriptional profiling in bacteroids was obtained. The macroarray of *B. japonicum* USDA110 was also revealed the relationship between the insertion of trn elements and phylogeny of *B. japonicum* strains.

INVESTIGATION OF THE EARLY STAGE SYMBIOTIC INTERACTION OF *SINORHIZOBIUM MELILOTI-MEDICAGO TRUNCATULA* BY FUNCTIONAL GENOMICS

Chunxia Wang[1], Xiaoyan Sheng[1], Chunhong Mao[1], Raymie Equi[1], Trevor C. Charles[2], and Bruno Sobral[1]
[1]Virginia Bioinformatics Institute, VA, USA; [2]Dept. Biol. University of Waterloo, Canada

Symbiotic nitrogen fixation plays an essential role in the global nitrogen cycle. Understanding how symbiotic interactions of legumes-rhizobia establish and develop is important in the development of sustainable and environmentally friendly agriculture. Despite intensive research over several years, we still lack a global view on how the legume-rhizobia symbiosis develops. We are using a combination of proteomics (2-D gel/Mass Spec), DNA microarray, and bioinformatics to investigate the complex developmental processes during the interaction of *S. meliloti-M. truncatula*. The aim of this study is to identify and characterize the genes that are being expressed in both organisms during symbiotic development, with a particular focus on those genes that are expressed during the early stages of nodulation. This is an ideal system for symbiotic study because 1) *S. meliloti*1021 genome has 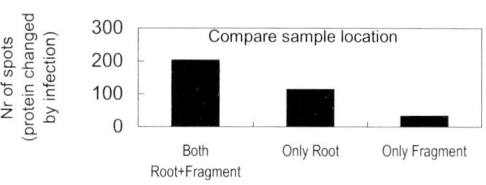 been sequenced, annotated, and the genetic tools are highly developed; 2) *M. truncatula* has a relatively small genome size, a short life cycle, and many genetic and genomic tools are available. We have established biological and genetic approaches to obtain comparative information at the transcriptomic and proteomic levels during a time course of the symbiotic interaction. Nodulation assay in growth pouches showed that the visible nodules always locate at the place where the root tip was at the time of the inoculation. In order to determine the optimal sampling location, we collected samples (4 days post-inoculation) from entire roots, and root fragments (1 cm around root tip at the time of inoculation), respectively. The initial results of protein expression profiles demonstrated that 90.8% of the total changed-spots (compared with non-infected plant) were detected by protein extracted from the entire root samples; while only 67.3% of the total changed-spots were detected by protein extracted from root fragments (see figure). Based on these results, further work will focus on samples of the entire root.

GLOBAL ANALYSIS OF GENOME EXPRESSION IN *SINORHIZOBIUM MELILOTI* STRAIN 1021, USING MICROARRAYS.

Julio Chávez-Zamora, Emmanuel Salazar and Sergio Encarnación.
Centro de Investigación Sobre Fijación de Nitrógeno, Universidad Nacional Autónoma de México, Cuernavaca, Mor., México.

Sinorhizobium meliloti interacts symbiotically with alfalfa (*Medicago sativa*) by inducing the formation of root nodules. Using microarray methodology (Qiagen oligo set), we are analyzing the global genome expression in *S. meliloti* in free living conditions. Recently, DNA microarray technology has revolutionized gene expression studies by providing a powerful tool for parallel measurement of gene expression on a whole genome scale. Using this methodology we compared transcriptional profiles of cells grown in rich medium (PY), and minimal medium (MM) with different carbon sources, in stationary and exponential phase. Our preliminary results showed co-ordinated upregulation and downregulation of specific mRNAs.

As expected, genes of known function, such as *pckA*, were more induced in succinate as carbon source; data was corroborated with proteome results (data not shown). *pgl*, was more induced in cells grown in MM with dextrose as carbon source. Interestingly, genes that codify for the GroEL chaperonins were expressed differentially according to the carbon source, that is, mRNA from *groEL* and *groEL2* accumulated in succinate as carbon source, unlike *groEL5* which has greater expression in dextrose. Also we detected a series of hypothetical proteins that were clearly differentially regulated by the conditions analyzed here. Besides, our microarray experiment revealed that cell growth in dextrose as carbon source decreased flagellum-gene expression. This result correlates with microscopy experiments where we observed reduced motility in this carbon source in comparison with succinate. In contrast to genes related to the formation of flagella genes involved in surface polysaccharide biosynthesis and its regulation were found to be more induced in dextrose; these data indicate that in MM-dextrose *S. meliloti* increased its production of polysaccharides in comparison with cell growth in MM-succinate. Finally, an important group of genes involved in protein synthesis increases its expression in succinate, which correlates with a better growth observed in *S. meliloti* on this carbon source.

Part from this work is supported by CONACYT grant 40046-Z, DGAPA grants IN203003-3 and IX250004.

PROTEOMIC AND TRANSCRIPTOMIC APPROACHES TO STUDY GLOBAL GENOME EXPRESSION IN FREE-LIVING AND SYMBIOTICALLY-ASSOCIATED *RHIZOBIUM ETLI* AND *SINORHIZOBIUM MELILOTI*.

Sergio Encarnación, Emmanuel Salazar, Magdalena Hernández, Gabriel Martínez, Juan García, Humberto Flores, Agustín Reyes, María del Carmen Vargas, Sandra Contreras, Yolanda Mora and Jaime Mora.
Centro de Investigación Sobre Fijación de Nitrógeno, Universidad Nacional Autónoma de México, Cuernavaca, Mor., México.

The proteome of the symbiotic bacteria, *Rhizobium etli* and *Sinorhizobium meliloti* were examined to determine the enzymatic reactions and cell processes that occur in free life (biofilms, aerobic and fermentative metabolism), and symbiosis. The identity of 628 proteins from 2-D gels was determined using peptide mass fingerprinting. Using the databases of the genome sequences from *R. etli* and *S. meliloti* and the software "Pathway tools" we constructed *in silico* the pathways and metabolic reactions (Metabolome), potentionally functional in both bacteria. We called these databases Sinocyc (*S.meliloti*) and Rhizocyc (*R. etli*). 45 pathways involved in many common anabolic and catabolic cellular processes of small molecule metabolism are active in *R.etli*. The proteins identified are involved in: (i) biosynthesis (of amino acids, cofactors, fatty acids, nucleotides); (ii) global functions and intermediary metabolism (gluconeogenesis, pentose pathway, nitrogen fixation); (iii) electron transport; (iv) energy metabolism; (v) carbon metabolism (fermentation, aerobic respiration, TCA cycle, glycolysis) and degradation (amino acids, carbon compounds).

Additionally we constructed a microarray with the complete sequence of the symbiotic plasmid of *R.etli* and performed time-course studies at different time points during nitrogen fixation with *P. vulgaris*. 189 genes were expressed in symbiosis, 37 in free life (Rich medium), and 32 in both conditions. The majority of genes expressed during nitrogen fixation belong to a 200 kb region.

In the rhizosphere, microbes grow as organized biofilms on surfaces. Genes (microarray) and proteins (proteome analysis) involved in adhesion (type 1 fimbriae) and auto-aggregation were highly expressed in the adhered population, consistent with current models of sessile community development. Several novel genes and proteins were induced on biofilm growth, including those expressed under O_2-limiting conditions.
Part of this work was supported by CONACYT grant 40046-Z and DGAPA grants IN203003-3 and IX250004.

NODMUTDB: A COMPREHENSIVE DATABASE FOR PLANT AND BACTERIAL GENES AND MUTANTS INVOLVED IN NODULATION AND NITROGEN FIXATION

Chunhong Mao[1], Jing Qiu[1], Chunxia Wang[1], Trevor Charles[2] and Bruno Sobral[1]
[1]Virginia Bioinformatics Institute, Virginia Polytechnic Institute and State University, Blacksburg, VA 24061, USA; [2]Department of Biology, University of Waterloo, Waterloo, ON N2L 3G1, Canada

Functional genomics activities will generate large amounts of information about the symbiosis-related functions of individual genes. It is important for this information to be placed in the context of the existing published literature. We have developed a database to address this. NodMutDB (No_dulation Mu_tant D_ataba_se) is a database system that archives plant and bacterial genes and mutants that are involved in nodulation, nitrogen fixation and other aspects of symbiosis. NodMutDB is designed to provide a comprehensive resource for depositing and retrieving information on nodulation/nitrogen fixation related genes, mutants and literature. Special emphasis is placed on quality control of data input and on the ability of researchers to use the database to search for general patterns. Initially, we are focusing on two model species for symbiosis research, *Medicago truncatula* and its symbiont *Sinorhizobium meliloti* strain SU47 and its derivatives. Data are collected through literature review and public database searches by a team of curators. The current data curation status is summarized in Table I. The database can be accessed and searched through a user-friendly Web interface (http://nodmutdb.vbi.vt.edu). Currently available information includes 1) lists of genes, mutants and literature that fit user-specified criteria; 2) detailed information about a gene such as its product, function, location, mutant collection, and external links to GenBank, SwissProt and other public databases; 3) detailed information about mutants such as mutagenesis methods, mutation locations, phenotypes at different nodule development stages, genes affected, allelism and related publications. Users can also submit new genes and mutants to the database, which will be validated by our curators before releasing to public.

Table I. NodMutDB data curation status.

	Medicago truncatula	*Sinorhizobium meliloti*
Number of genes	35	227
Number of mutants	35	561
References on mutant study	26	181
Total references collected	228	1656

SECTION IV.

ESTABLISHMENT OF SYMBIOSIS AND NODULE FUNCTION

CELL CYCLE AND SYMBIOSIS

Adam Kondorosi, José Maria Vinardell, Toshiki Uchiumi, Peter Mergaert, Eva Kondorosi
Institut des Sciences du Végétal UPR2355 CNRS, 91198 Gif sur Yvette, France

1. Root nodule development and endoreduplication.
The cell cycle plays a crucial role in plant development, where organogenesis takes place continuously and most cells maintain their ability to re-enter and to regulate the cell cycle in response to a wide range of external signals. The relationship between cell-cycle and development is complex and characterized by mutual dependencies. Phytohormones, particularly auxins and cytokinins are the major regulators of the cell cycle, which act at multiple levels affecting transcription of cell cycle genes or the activity of the cyclin-dependent kinases. Moreover, non-equivalent distribution of auxins with concentration maxima in certain cell types seems to control plant organogenesis and pattern formation. In addition to the hormonal control, endogenous developmental programs as well as many environmental biotic and abiotic signals converge on the cell cycle, leading to its activation or inhibition.

Plant organs originate from existing or *de novo* formed meristems, while proliferation arrest is a prerequisite of differentiation. Though cells undergoing differentiation do not divide, developmental programs often involve specific cell-cycle modes such as the endoreduplication cycles, repeated replication of the genome without mitosis that may occur in most cell-types and result in polyploid genome content. Root nodules are formed in symbiosis of *Rhizobium* soil bacteria with leguminous host plants. This unique plant organ is induced by external mitogen signals of the bacteria, the Nod factors when combined nitrogen is limited in the soil. Nodules, similarly to lateral roots, originate from *de novo* formed meristems arising in front of the protoxylem poles. Auxin triggers lateral root development but it is insufficient for nodule development. On the other hand, altered auxin levels might also be required for nodule organogenesis as Nod factors seem to inhibit the acropetal auxin transport in the root indicating that the "auxin maximum" as a general rule for organ development may also hold for nodules.

Nodules are classified as determinate and indeterminate types depending on the cortical cell type where the first cell division occurs and on the activity of the nodule meristem. The shape of the nodules is also determined by the activity and the position of the meristem that is elongated in the indeterminate types and globular in the determinate ones.

Medicago truncatula nodules formed in symbiosis with the *Sinorhizobium meliloti* is considered as a model for indeterminate nodule development, where cell division starts in the inner cortex and intensive proliferation of these cells leads to the formation of the organ primordium. The nodule primordium, after its outgrowth of the root, differentiates into various nodule cell types resulting in a complex nodule structure (Figure 1). The meristem (zone I) persists in the apical region, from which

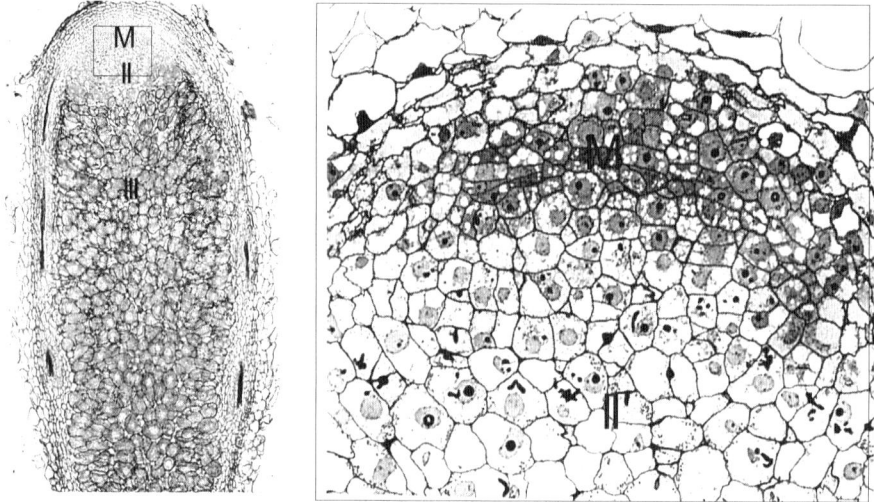

Figure 1. The structure of the indeterminate *Medicago truncatula* nodule. M: meristem, II: invasion zone where endoreduplication cycles and gradual increase of infected cells occur; III: nitrogen-fixing zone.

cells differentiate constantly to various nodule cell types. Infection of nodule cells with bacteria and gradual differentiation of symbiotic cells occurs in 12-15 cell layers of zone II (infection zone). In the submeristematic cells, the bacteria are released from the infection threads in the form of symbiosomes; representing single bacteria encapsulated by a membrane envelop. The symbiosomes first multiply in the infected cells but by maturation of the symbiotic cells in the deeper cell layers they loose their ability to divide; instead they start to grow and differentiate to huge, elongated bacteroides. In zone II, the bacteria continue Nod factor production. These mitogenic signals maintain the cell cycle activities, although the nodule cells do not divide, these cells undergo successive rounds of endoreduplication cycles (Cebolla et al., 1999; Foucher and Kondorosi, 2000). As the nuclear DNA content increases from 2C up to 64C, the cells enlarge proportionally to their genome size as they become older and more distant from the meristem during the longitudinal nodule growth. In zone III, the expression of cell cycle genes is switched off, and the symbiotic cells are terminally differentiated; they are at their final size and are fully packed with extremely enlarged, elongated, nitrogen-fixing bacteroids. Due to the persistence of the meristem, the *Medicago* nodules are of indeterminate type, which are under continuous and dynamic development. Except the *de novo* meristem formation, the different developmental stages can be monitored within a single nodule exhibiting a gradient of differentiation from the distal meristem to the older, proximal zone III region. Therefore, this organ is particularly suited for studying various aspects of developmental processes, such as cell cycle and endoreduplication control or maturation of symbiotic cells.

For determinate nodules, the *Mesorhizobium loti-Lotus japonicus* system is used as a model. In the determinate nodules, division of the outer cortical cells leads to the nodule primordium formation. However, as the activity of the meristem decays rapidly, further nodule growth is due to enlargement of cells, which leads to a homogenous central zone containing the nitrogen-fixing cells. The cell cycle activities and mechanism of cell growth are largely unexplored in the determinate nodules.

Here, we focus on differential regulation of the cell cycle during indeterminate nodule development in *M. truncatula*. In addition, we provide new data on determinate nodules in respect to endoreduplication and cell growth.

2. Cell cycle is controlled by ordered action of cyclin-dependent kinases (CDKs).
The regulatory subunits of cyclin-dependent kinases are the cyclins, classified as G1-types and mitotic A- and B-type cyclins that regulate the G1-S and S to M transitions, respectively. Previous studies showed that Nod factors trigger reactivation of G0-arrested cells (Savoure et al., 1994; Yang et al., 1994). To get an insight in the mechanism of cell cycle activation, we studied the expression of a set of cell cycle marker genes (Foucher and Kondorosi, 2000). Surprisingly, Nod factors induced a mitotic cyclin, *cycA2* within 5 hours that coincided with or even preceded the activation of G1 regulators and many early nodulin genes. This particular behaviour of this cyclin A2, suggested its implication in re-entering of differentiated cells to the cell cycle. In contrast to other mitotic cyclins, *cycA2* expression and protein levels showed weak oscillation in the cell cycle (Roudier et al., 2000). In the primary root, *cycA2* was expressed in the root apical meristem and faintly in the phloem cells. The *cycA2* promoter contains several putative auxin-response-like elements. We showed that these elements were functional and that auxin-treatment resulted not only in the activation of *cycA2* but also in a drastic change in the spatial expression pattern of *cycA2*, inducing *de novo* transcription of *cycA2* in front of the xylem poles, where lateral roots as well as nodules initiate (Roudier et al., 2003). In nodules and lateral roots, *cycA2* expression

was localised in meristems but was repressed in the differentiating cells. These data suggested that the *cycA2* function is required for mitotic cycles, leading to secondary meristems whereas it is dispensable or incompatible with endoreduplication cycles (Roudier et al., 2003).

3. Endoreduplication mediated by the Anaphase-Promoting Complex activator CCS52A is required for symbiotic cell differentiation in nitrogen-fixing nodules.

Symbiotic cells enlarge both in the indeterminate and determinate nodules. Growth of eukaryotic cells is often linked to genome amplification which led to the "nuclear-cytoplasmic ratio" theory, establishing a direct relationship between nuclear DNA content and cell size in the endoreplicative tissues (reviewed in Kondorosi et al., 2000; Kondorosi and Kondorosi, 2004). The occurrence of endoreduplication cycles during the development of symbiotic cells has been confirmed in *M. sativa* and *M. truncatula* (Truchet, 1978; Cebolla et al., 1999). As it was unclear how cell growth occurs in determinate nodules, we measured the DNA content of nuclei in *L. japonicus* nodules by flow cytometry. This measurement clearly demonstrated the existence of polyploid cells also in determinate nodules (2C 6%, 4C 78.3%, 8C 8.9%, 16C 0.9%, 32C 5.8%), indicating that formation of giant symbiotic cells with polyploid genome content is a common feature of nitrogen-fixing nodule development. Polyploid cells are generated via endoreduplication cycles, representing an altered, shortened version of the mitotic cycle composed of the DNA synthesis S-phase and a gap period (Figure 2). It is suggested that endoreduplication requires nothing more than loss of M-phase and oscillations in the activity of S-phase cyclin-dependent kinase.

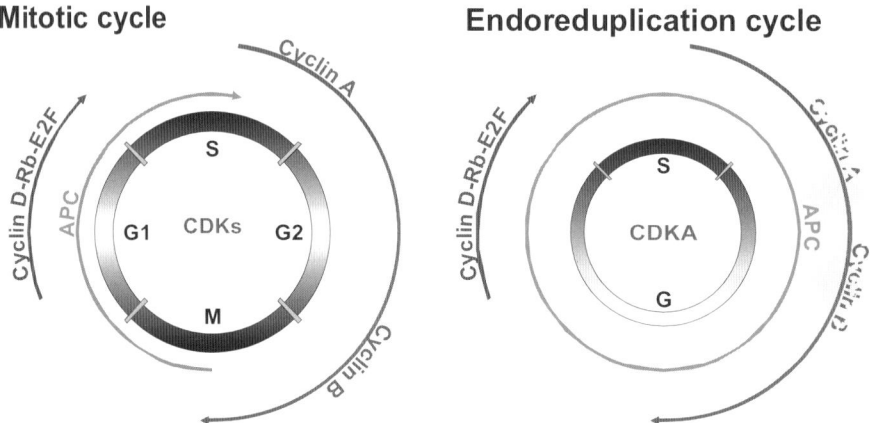

Figure 2. Comparison of the mitotic and the endoreduplication cycles. Premature degradation of the mitotic cyclins by the Ccs52A-APC ubiquitin ligase converts mitotic cycles to endocycles.

Our studies on the organogenesis of *Medicago* nodules, where endocycles persist in a limited region, have led to the identification of the cell cycle switch gene *ccs52* that is a major regulator of the endoreduplication cycles (Cebolla et al., 1999). Ccs52A is a substrate-specific activator of the anaphase-promoting complex (APC), an E3 ubiquitin ligase (Cebolla et al., 1999; Tarayre et al., 2004). We showed that CCS52A provoked degradation of cyclin B, which by inactivation of mitotic CDKs inhibited the mitosis and converted the mitotic cycles to endocycles.

Expression of *ccs52A* was undetectable during the mitotic cycles and nodule primordium formation while it was linked to cell differentiation involving endoreduplication. A positive correlation between the *ccs52A* transcript and the ploidy differentiation supported a *ccs52A* function in the formation of highly polyploid cell types. The importance of somatic endoploidy in nodule development was studied in *ccs52A* antisense plants where the formation of nodule primordia was unaffected while reduction in the *ccs52A* transcript level aborted nodule development. These nodules displayed significantly lower degree of ploidy, smaller cells, which were poorly infected and contained no nitrogen-fixing cells. Our work, thus, demonstrated that *ccs52A* plays a key role in symbiotic cell differentiation and showed that repeated endoreduplication cycles are indispensable for nitrogen-fixing nodule development (Vinardell et al., 2003). The *ccs52A* gene is highly conserved in legumes and its expression both in the indeterminate and determinate nodules indicates that formation of large cells is likely mediated by endoreduplication cycles via Ccs52A in all nodule types.

4. References
Cebolla A et al. (1999) EMBO J. 18, 4476–4484.
Favery B et al. (2002) Mol. Plant-Microbe Interact. 15, 1008-1013.
Foucher F and Kondorosi E (2000) Plant Mol. Biol., 43, 773-786.
Kondorosi E et al. (2000) Curr. Opin. Plant Biol. 3, 488-492.
Kondorosi E and Kondorosi A (2004) FEBS Letters 567, 152-157.
Roudier F et al. (2000) Plant J. 23, 73–83.
Roudier F et al. (2003) Plant Physiol. 131, 1091-1103.
Savoure A et al. (1994) EMBO J. 13, 1093-1102.
Tarayre et al. (2004) Plant Cell 16, 422-434.
Truchet G. (1978) Ann. Sci. Nat. Bot. Paris 19, 3-38.
Vinardell JM et al. (2003) Plant Cell 15, 2093–2105.
Yang WC et al. (1994) Plant Cell 6, 1415-1426.

EXPRESSION PATTERN OF DMI GENES IN *MEDICAGO* NODULES

Rossana Mirabella, Marijke Hartog, Carolien Franken, René Geurts and Ton Bisseling
Department of Molecular Biology, Wageningen University, The Netherlands

Genetic analysis in pea and in the model legumes, *Medicago truncatula* (Medicago) and *Lotus japonicus* (Lotus), have identified sets of symbiosis specific genes that are essential for Nod factor perception and transduction. Recently, most of these Nod factor signaling (NFS) genes have been cloned and they appear to be orthologous in the three above mentioned legumes. Nod factor perception most likely involves transmembrane LysM-domain-containing receptor kinases that are proposed to form heterodimeric Nod factor receptors (Limpens et al., 2003; Madsen et al., 2003; Radutoiu et al., 2003). In Medicago, these include LYK3 and LYK4 (Limpens et al., 2003) and NFP (Amor et al., 2003; C. Gough and R. Geurts, unpublished). Downstream components of the Nod factor signaling pathway in Medicago are the DOES NOT MAKE INFECTION1 (*DMI1*; Ane et al., 2004), the *DMI3* (Levy et al., 2004) and the NODULATION RECEPTOR KINASE (*DMI2*; Endre et al., 2002), encoding for a putative cation channel, a calcium/calmodulin-dependent protein kinase, and a leucine-rich repeat receptor-like protein kinase, respectively. The Lotus orthologue of DMI2 has also been cloned (Stracke et al., 2002). Plants carrying mutations in these genes are blocked at the stage of root hair deformation (Catoira et al., 2000; Wais et al., 2000; Stracke et al., 2002).

Temperate legumes, such as Medicago, form indeterminate nodules. These nodules contain a persistent meristem, also designated as zone I, located at the nodule apex and characterized by small cells rich in cytoplasm that do not contain any bacteria or infection threads. The meristem continuously divides by which it is maintained and also adds new cells to the different nodule tissues. As a consequence, the nodule central tissue can be divided into several zones representing successive developmental stages, from the apex (distal region) to the root attachment point (proximal region). The cell layers adjacent to the meristem form the infection zone, also indicated as zone II. In the distal region of zone II, infection threads branch and penetrate host cells into which bacteria are released. The released bacteria, enclosed in a membrane envelop and now called symbiosomes, proliferate and start to differentiate into the nitrogen-fixing forms. In the nitrogen-fixing zone (zone III), at the base of the nodule, the plant cells are completely filled with thousands of nitrogen-fixing bacteroids.

Nod factors induce several responses in the epidermis and are also sufficient to induce the formation of nodule primordia (Cullimore et al., 2001; Geurts and Bisseling, 2002). However, their involvement in later stages of nodule development is less clear. The best indication that Nod factors play a role at such later stages comes from studies showing that rhizobial *nod* genes are expressed in the distal part of the nodule infection zone (Sharma and Signer, 1990; Schlaman et al., 1991). This suggests that the Nod factors are produced in the nodule in the zone where infection threads grow and bacteria are released. If Nod factors do induce responses in the nodule, this probably requires the presence of Nod factor perception and transduction machinery. To obtain insight into this, we studied by in situ hybridization whether and where *DMI1* and *DMI2* are expressed in Medicago nodules.

We first determined whether in Medicago nodules sharp boundaries could be identified between meristem and infection zone, and infection and fixation zone, respectively, as already described for Pea and alfalfa nodules (Scheres et al., 1990; Vasse et al., 1990; Yang and Bisseling, 1993). For this purpose, we analyzed the expression pattern of the same set of marker genes that were used in those species on longitudinal sections from nodules 14 days after inoculation; namely the early nodulin gene *MtENOD12*, *LEGHEMOGLOBIN* (*LB*) and the rhizobial *nifH* gene.

MtENOD12 mRNA is present at its maximal level in the distal cell layers of the infection zone II adjacent to the meristem but it is not active in this meristem. Therefore, we conclude that also in Medicago a sharp boundary exists between the meristem and the infection zone. *LB* mRNA is present at a maximal level in the first cell layers of the fixation zone (previously named interzone; Vasse et al., 1990), coinciding with the appearance of amyloplasts in the same cell layers. This shows that a sharp boundary is also present between infection and fixation zone. Also, the bacterial *nifH* gene is induced at the transition between these two zones. In conclusion, these data show that, like in pea and alfalfa, the nodule central tissue is characterized by two rapid developmental switches. This results in sharp boundaries between the three successive zones; merisitem, infection zone and fixation zone. Further, within the infection and fixation zone, respectively, a gradual progression of differentiation occurs.

In situ hybridization was carried out on adjacent longitudinal nodule sections using *DMI2* and *DMI3* antisense probes. This showed that *DMI2* and *DMI3* expression patterns coincided. The two genes appeared to be highly expressed in the most distal 2-3 cell layers of zone II, just adjacent to the meristem. This region is markedly smaller than the region where *MtENOD12* is expressed. The highest expression of the *DMI* genes was detected in the first three cell layers of zone II. Previously, it was shown that *Rhizobium nod* genes are active in this region as reported by Sharma and Signer (1990), using reporter constructs based on the *uidA* gene. Further, Schlaman et al. (1991) showed by *in situ* hybridization that expression of various *nod* genes specifically occurs in the more distal cell layers of zone II. Therefore, it is probable that Nod factors are produced by the rhizobia in the cell layers where the *DMI* genes are expressed at the highest level.

In the most distal layers of zone II, in which the expression of the *DMI* genes is relatively high, infection threads branch, penetrate plant cells and bacteria are released into the host cells. The correlation between infection thread growth/bacteria release and the high level of expression of *DMI* genes suggests a role for the Nod factor signaling in these processes.

References
Amor BB et al. (2003) Plant J 34, 495-506.
Ane JM et al. (2004) Science 303,1364-1367.
Catoira R et al. (2000) Plant Cell 12, 1647-1666.
Cullimore JV et al. (2001) Trends Plant Sci 6, 24-30.
Endre G et al. (2002) Nature 417, 962-966.
Geurts R, Bisseling T (2002) Plant Cell 14 Suppl, S239-249.
Levy J et al. (2004) Science 303, 1361-1364.
Limpens E et al. (2003) Science 302, 630-633.
Madsen EB et al. (2003) Nature 425, 637-640.
Radutoiu S et al. (2003). Nature 425, 585-592.
Scheres B et al. (1990) Plant Cell 2, 687-700.
Schlaman HR et al. (1991) J Bacteriol 173, 4277-4287.
Sharma SB, Signer ER (1990) Genes Dev 4, 344-356.
Stracke S et al. (2002) Nature 417, 959-962.
Vasse J et al. (1990) J Bacteriol 172, 4295-4306.
Wais RJ et al. (2000) Proc Natl Acad Sci USA 97, 13407-13412.

ACTINORHIZAL SYMBIOSES

Pawlowski Katharina[1], Zdyb Anna[1], Hause Bettina[2], Göbel Cornelia[1], Feussner Ivo[1], Demchenko Kirill[1,3]
[1]Plant Biochemistry, Göttingen University, Justus-von-Liebig-Weg 11, 37077 Göttingen, Germany, [2]Leibniz Institute for Plant Biochemistry, Weinberg 3, 06120 Halle, Germany, and [3]Komarov Botanical Institute, Russian Academy of Sciences, St.-Petersburg 197376, Russia

Two groups of plants can enter root nodule symbioses with nitrogen-fixing soil bacteria. Gram-negative, unicellular rhizobia induce nodules on legume roots, while Gram-positive mycelial actinomycetes of the genus *Frankia* induce nodules on the roots of plants from eight angiosperm families, mostly woody shrubs, collectively called actinorhizal plants. Molecular phylogenetic studies indicate that legumes and actinorhizal plants belong to the same clade (Soltis et al., 1995), suggesting that the common ancestor aquired the underlaying capacity to develop a root nodule symbiosis originated, probably in a single evolutionary event. To identify this underlaying capacity, we compared different aspects of the symbiosis in three diverse systems, one legume (*Medicago truncatula*) and two actinorhizal plants (*Casuarina glauca* and *Datisca glomerata*).

In order to develop an intracellular symbiosis, plants had to evolve a mechanism to internalize the microsymbionts and to control the nutrient exchange, i.e. to enclose the microsymbionts in a plant-derived membrane. Bacterial microsymbionts (*Frankia* and rhizobia) can enter the root of the host plant either intercellularly (Coriariaceae, Datiscaceae, Elaeagnaceae, Rhamnaceae, Rosaceae) or intracellularly (Betulaceae, Casuarinaceae, Myricaceae, most legumes). During intercellular entry in actinorhizal symbioses, bacteria enter the plant root bewcen epidermal cell and colonize the apoplastic spaces, before they are internalized in primordium cells by the formation of intensely branching infection threads in these cells. During intracellular entry, infection threads are formed in root hairs and grow across cortical cells toward the incipient nodule primordium. Hence, the formation of infection threads that cross cells is an ability of (a subset of) root cells. When the bacteria have reached nodule primordium cells, they can infect them. In actinorhizal nodules, this happens by intense branching of infection threads within the infected cell; in most legume nodules, bacteria are internalized via a complete endocytotic process during which they are enclosed in a peribacteroid membrane derived from the plasma membrane of the plant cell. Infected nodule cells – cells that can stably internalize a bacterial microsymbiont and regulate the nutrient exchange via the perisymbiont membrane - represent a differentiation stage not found in any plant organs other than nodules. Cells that have this ability are not present in roots; they are formed after contact and signal exchange with the microsymbiont.

1) Infected cell-specific transcription factors conserved among actinorhizal plants and legumes
It has been shown previously that there are transcription factors specific for infected cells. A promotor of a soybean leghemoglobin was active in infected cells from *Lotus corniculatus* and *Casuarina glauca* nodules, while the promotor of the symbiotic hemoglobin gene from *C. glauca* was active in the same cell types (Jacobsen-Lyon et al., 1995; Franche et al., 1998), in spite of the fact that the morphological localization of infected cells differs in actinorhizal and legume nodules. Actinorhizal nodules are composed of modified lateral roots with infected cells in the expanded cortex (Pawlowski and Bisseling, 1996), while legume nodules have a stem-like organisation with peripheral vascular system and infected cells in the inner tissue. On the other hand, Svistoonoff et al. (2003, 2004) showed that there are also transcription factors conserved in both systems that are specific to cells containing infection threads – independently whether the cells are passed by infection threads or represent infected cells filled by branching infection threads. The fact that the infection thread-containing cell-specific promoter involved was in the heterologous (legume) system also expressed in dividing nodule primordium cells and in some cells preparing for division, might indicate that infection thread formation signaling in the plant cell is derived from cell cycle signaling.

2) Cytoskeletal changes during infection thread growth
Immunolocalization experiments were performed on sections of actinorhizal and legume nodules using antibodies directed against cytoskeletal compounds. Use of an antibody directed specifically against myosin class VIII (Reichelt et al., 1999) showed that in actinorhizal nodules from *D. glomerata* as well as those from *C. glauca*, myosin VIII accumulates around the tips of infection threads, and to some extent also around the older parts of infection threads. *C. glauca* is infected intracellularly, but *D. glomerata* has an intercellular infection mechanism, i.e. *Frankia* hyphae colonize the apoplast and infection threads are formed directly by the infected cells, not already in root hairs. No myosin VIII accumulation could be detected in legume nodules. We speculate that myosin VIII is required to stabilize the actinorhizal infection threads by connecting them to the actin cytoskeleton. In legume infection thread, the gradual hardening of the infection thread matrix is likely to provide stabilization (Wisniewski et al., 2000), but actinorhizal infection threads do not contain an equivalent of the infection thread matrix and therefore might depend on other means of stabilization.

3) Phytohormones involved in root nodule development

It has already been shown that auxin transport inhibitors and cytokinin can to some extent mimic the effect of rhizobia or *Frankia* on plant development (reviewed by Pawlowski and Sprent, 2004). We were interested in the effects of oxylipin signal factors, jasmonate and its precursor 12-oxo-phytodienoic acid (OPDA) on nodule development. Jasmonates are involved in arbuscular mycorrhizal (AM) symbioses (Hause et al., 2002), and it has been suggested that root nodule symbioses have evolved adapting some mechanism developed for the more ancient AM symbioses (Kistner and Parniske, 2002). In order to understand jasmonate biosynthesis in nodules, we immunolocalized two enzymes of the jasmonate biosynthesis pathway, lipoxygenase (LOX; Feussner et al., 1995) and allene oxide cyclase (AOC; Ziegler et al., 2000) in nodules of *M. truncatula, C. glauca* and *D. glomerata*. LOX catalyszes the first reaction in the oxylipin pathway (reviewed by Feussner and Wasternack, 2002) and is also producing precursors for other pathways than jasmonate formation, while AOC is specific to the jasmonate pathway. Only cytosolic LOX was found in nodules, and it was restricted to uninfected cells in all three systems. AOC was found in plastids, including amyloplasts, in infected and uninfected cells in all three systems. However, in *C. glauca* and *D. glomerata* only younger infected cells contained AOC, while in *M. truncatula*, AOC was present in plastids of infected cells throughout the nitrogen-fixing zone. Hence, like in root cells containing arbuscules (Hause et al., 2002), jasmonate is produced in stably infected cells of root nodules. Yet it is likely that jasmonates fulfil different functions in both systems, since infected cells of different developmental stages contain AOC in legume versus actinorhizal symbioses.

Conclusions

Actinorhizal and legume symbioses go back to a common ancestor, and this is confirmed by conservation of transcription factors specific to symbiotic plant cell types. However, in this study, we found differences between both symbioses with regard to cytoskeletal support of infection thread growth and the stage of development in which infected cells produce jasmonate. Hence, cell infection mechanisms seem to differ dramatically in both symbioses, implying that plant cells might have several ways to internalize a bacterial microsymbiont, or to support tip growth.

References

Feussner I et al. (1995) Plant J. 7, 949-957.
Feussner I, Wasternack C (2002) Annu. Rev. Plant Biol. 53, 275-297.
Franche C et al. (1998) Mol. Plant-Microbe Interact. 11, 887–894.
Hause B et al. (2002) Plant Physiol. 130, 1213-1220.
Jacobsen-Lyon K (1995) Plant Cell 7, 213-222.
Kistner C, Parniske M (2002) Trends Plant Sci. 7, 511–518.
Pawlowski K, Bisseling T (1996) Plant Cell 8, 1899-1913
Pawlowski K, Sprent J (2004) Actinorhizal Symbioses. Kluwer Academic Publishers, in press.
Reichelt et al. (1999) Plant J. 19, 555-569.

Soltis DE et al. (1995) Proc. Natl. Acad. Sci. USA 92, 2647-2651.
Sequerra J et al. (1994) Can. J. Bot. 72, 955-962.
Sequerra J et al. (1995) New Phytol. 130, 545-555.
Svistoonoff S et al. (2003) Mol. Plant-Microbe Interact. 16, 600-607.
Svistoonoff S et al. (2004) Plant Physiol. 136, 3191-3197.
Wisniewski et al. (2000) Mol. Plant-Microbe Interact. 13, 413-420.
Ziegler J et al. (2000) J. Biol. Chem. 275, 19132–19138.

SIGNALING FOR NODULATION IN A WATER-TOLERANT LEGUME

Marcelle Holsters, Ward Capoen, Jeroen den Herder, and
Sofie Goormachtig
Department of Plant Systems Biology, Flanders Interuniversity
Institute for Biotechnology, Ghent University, Technologiepark
927, B-9052 Gent, Belgium

Legume nodulation takes place in the legume roots, in the zone of developing root hairs. Rhizobia invade the host via root hair curling (RHC), a process that is triggered by bacterial nodulation factors (Nod factors or NFs). Water tolerance of legume nodulation also depends on NFs but uses an alternative rhizobial invasion strategy with decreased signaling demands while skipping epidermal responses.

A number of legume plants belonging to the Papilionoideae and the Mimosoideae, the two largest of the three legume subfamilies, have developed an adaptation to grow and nodulate in (temporarily) submerged environments (Sprent, 2002). The best studied example of such water-adapted legume nodulation is that of *Sesbania rostrata* (higher Papilionoideae). *S. rostrata* is a (semi)tropical legume from West Africa where it grows in temporarily flooded habitats of the Sahel region. The plant carries numerous preformed adventitious rootlets on the stem. Upon submergence, these rootlets grow out to form adventitious roots and, upon inoculation with the microsymbiont *Azorhizobium caulinodans*, they develop into stem-located nodules.

Under hydroponic conditions, roots of *S. rostrata* are nodulated extremely efficiently at positions of lateral root bases (LRBs). This LRB nodulation, as well as stem nodulation, is an atypical nodulation process because nodule initiation usually takes places at positions of developing root hairs in a zone behind the root apical meristem. Adventitious and LRB nodulation are similar processes, which do not involve root hairs but which do strictly depend on Nod factors for signaling for infection and cell division. Bacterial invasion depends on local cell death for the creation of intercellular infection pockets (D'Haeze et al., 2003) in a process that resembles aerenchyma formation (Drew et al., 2000).

In contrast to RHC nodulation, LRB nodulation requires the action of the plant stress hormone ethylene, as shown in pharmacological experiments in which the addition of either Ag^+ ions or L-α-(2-aminoethoxyvinyl)-glycine (AVG), prior to or simultaneously with azorhizobia, prevented nodule development on hydroponic roots and bacterial colonization of the cortex. In LRB nodulation, ethylene acts downstream of NFs to mediate division and death. Application of purified Nod factors to *S. rostrata* roots in water, causes cell division, cell death, and formation of cavities at positions of LRBs. All these Nod factor effects are abolished after ethylene inhibitors are added. Moreover, cell death and cavities were also triggered by exogenous ethylene (D'Haeze et al., 2003). Hence, for LRB nodulation, *S. rostrata* has coupled ethylene responses to Nod factor perception as a strategy to allow colonization of the root cortex independently of the epidermis. Infection pocket formation is reminiscent of aerenchyma formation induced by ethylene or hypoxia in roots of maize (Drew et al., 2000). This special adaptation is consistent with the notion that ethylene plays a central role in submergence-tolerant plant species.

Interestingly, besides this unorthodox nodulation pathway, *S. rostrata* can also develop nodules via the regular RHC process. When roots are growing under non-flooded conditions, in vermiculite in Leonard Jars, nodules form in zone 1 of developing root hairs (Goormachtig et al., 2004a). Root hair curling is observed and green fluorescent protein-marked bacteria can be followed in intracellular infection threads that guide the bacteria to the nodule primordia. Also, in another water-adapted legume, *Neptunia plena*, nodules can form either at LRB, under flooded conditions, or via RHC under non-flooded conditions (Goormachtig et al., 2004a).

Why does *S. rostrata* offer an alternative invasion route to its microsymbiont? Why did this semi-aquatic plant develop versatile nodulation features? An answer to these questions can be derived from the study of the role of ethylene in root growth and nodule initiation. Indeed, flooding causes accumulation of ethylene and ethylene interferes with RHC nodulation and with initiation of root hair growth. First, exogenous ethylene inhibits RHC nodulation at positions of susceptible root hairs on roots grown in Leonard jars. When ethylene is added to such roots, even up to 1 hour prior to bacterial inoculation, RHC nodulation is inhibited. On the contrary, addition of AVG enhances the number of RHC nodules in the Leonard jar system.

Moreover, ethylene interferes with the initiation of root hairs. Hydroponically grown roots of *S. rostrata* or *N. plena* have no root hairs, whereas roots grown under well-aerated conditions have plenty. However, when seedling roots are grown in water in the presence of AVG, a partial restoration of root hairs in zone 1 can be observed and, indeed, RHC nodulation takes place. Hence, blocking ethylene levels restores root hair formation, and the root hairs are susceptible to nodulation responses in contrast to submerged root hairs of roots grown in Leonard jars.

As RHC nodulation on *S. rostrata* roots is blocked upon flooding, in order to cope with the variable conditions of the habitat, an alternative way had to be recruited that avoids the ethylene-sensitive epidermal step. This atypical invasion track actually depends on ethylene and, in its ethylene dependency for infection pocket formation, resembles aerenchyma formation in maize (Drew et al., 2000).

Both RHC and LRB nodulation on *S. rostrata* are strictly dependent on NFs, but have different stringency demands (D'Haeze et al., 2000; Goormachtig et al., 2004a). Azorhizobial NFs carry different substitutions at both reducing and non-reducing chitin backbone residues (Mergaert et al., 1993). Infection pocket formation during LRB nodulation can still occur – albeit at low frequency – upon inoculation with mutant bacteria that produce undecorated NFs (D'Haeze et al., 2000) and nitrogen-fixing functional nodules can be formed. However, RHC nodulation is more demanding, and the lack of decorations prevents formation of functional nodules. Instead, swollen roots are observed with numerous primordia but without bacterial invasion (Goormachtig et al., 2004a), suggesting a stringent checkpoint for root hair entry.

In summary, *S. rostrata* has developed two alternative ways to allow entry of its microsymbiont. Both ways have in common a strict NF dependency for inducing cortical cell division and preparing for invasion. The major differences reside in the primary infection events, which happen at different positions on the root and either make use of or avoid the epidermis for primary signaling with ethylene sensitivity or dependency as a consequence (Goormachtig et al., 2004b).

As these processes occur on the same host plant, we can compare transcript profiles. Therefore, we used a cDNA-AFLP approach to studying comparable early stages of initiation of RHC or LRB nodulation. A collection of approximately 7000 sequences were analyzed, of which 648 gave a significant hit by either BLASTX against protein databases or BLASTN against expressed sequence tag databases of The Institute for Genome Research. Of these tags, 9% were specific for LRB invasion, 25% for RHC invasion, and 66% were common. A few potentially interesting genes will be further investigated.

Apart from this general approach, we address specific questions concerning the role of different signaling receptors in RHC versus LRB nodulation. Currently, we are cloning the orthologous *S. rostrata* genes corresponding to the LysM-type receptors that were isolated from Medicago and Lotus (Oldroyd and Downie, 2004) and will determine their involvement in epidermal, cortical, or both responses. Furthermore, we will investigate the role of the unique *SymRK* ortholog of *S. rostrata* in LRB nodulation, using RNA interference in transgenic roots to obtain functional knockdown of gene expression. Together, these approaches will allow us to sort out in which processes NF perception and SymRK function are linked and for which processes NF receptors and SymRK work independently.

Acknowledgements
The authors thank Martine De Cock for help with the preparation of the manuscript. This work was supported by grants from Interuniversity Poles of Attraction

Programme-Belgian Science Policy (P5/13) and the Fund for Scientific Research-Flanders ("Krediet aan Navorsers" 1.5.088.99N and 1.5.192.01N). W.C. is indebted to the "Instituut voor de aanmoediging van Innovatie door Wetenschap en Technologie in Vlaanderen" for a predoctoral fellowship and J.d.H. is a Research Fellow of the Fund for Scientific Research-Flanders.

References
D'Haeze W et al. (2000) J. Biol. Chem. 275, 15676-15684.
D'Haeze W et al. (2003) Proc. Natl. Acad. Sci. USA 100, 11789-11794.
Drew MC et al. (2000) Trends Plant Science 5, 123-127.
Goormachtig S et al. (2004a) Proc. Natl. Acad. Sci. USA 101, 6303-6308.
Goormachtig S et al. (2004b) Trends Plant Sci., in press.
Mergaert P et al. (1993) Proc. Natl. Acad. Sci. USA 90, 1551-1555.
Oldroyd GED and Downie JA (2004) Nat. Rev. Mol. Cell Biol. 5, 566-576.
Sprent JI (2002) *Nodulation in Legumes*. Kew, Royal Botanical Gardens.

GENETIC AND MOLECULAR ANALYSIS OF NOD FACTOR SIGNALLING IN
MEDICAGO TRUNCATULA

F. Debellé[1], C. Bres[1], J. Lévy[1], B. Ben Amor[1], JF. Arrighi[1], F. Maillet[1], JM. Ane[1], C. Rosenberg[1], J. Dénarié[1], S. Shaw[2], G. Oldroyd[2], S. Long[2], R. Penmetsa[3], D. Cook[3], R. Geurts[4], T. Bisseling[4], G. Duc[5], and C. Gough[1]
[1]Laboratoire des Interactions Plantes-Microorganismes, INRA-CNRS UMR 441/2564, 31326 Castanet-Tolosan, France. [2]Department of Biological Sciences, Stanford University, CA 94395-5020, USA. [3]Department of Plant Pathology, University of California, Davis, CA95616, USA. [4]Laboratory of Molecular Biology, Wageningen University, 6703HA Wageningen, Netherlands. [5]Unité de Recherches de Génétique et Ecophysiologie des Légumineuses INRA, 21065 Dijon, France.

1. Introduction

Establishment of the *Rhizobium*-legume symbiosis requires the production by rhizobia of lipochito-oligosaccharidic molecules, called Nod factors, which act as signals in the molecular dialogue between symbiotic partners. The nature of the Nod factor N-acyl chain and other substitutions are important determinants of rhizobial host specificity. Pure Nod factors, at very low concentrations, can induce diverse symbiotic responses in host roots, suggesting the existence of high affinity receptors and complex signalling mechanisms (Geurts and Bisseling, 2002). Genetic dissection of Nod factor signalling in *Medicago truncatula* has enabled us to identify key components that control the early steps of Nod factor perception and signal transduction. Here, we summarise these results and describe the recent cloning of two *M. truncatula* genes controlling Nod factor signalling, *NFP* and *DMI3*. Together with *DMI1* and *DMI2*, which have also been cloned (Ané et al., 2004; Endre et al., 2002), the molecular identities of four of the earliest components are now known and rapid progress towards understanding the detailed mechanisms by which Nod factors induce symbiotic responses and nodulation should be possible.

2. Identification of six components of a Nod factor-activated signal transduction pathway, three of which also control the establishment of arbuscular mycorrhizal symbioses

Among a large collection of nodulation-deficient (Nod⁻) mutants of *M. truncatula*, we screened for those deficient for infection by *Sinorhizobium meliloti*. Such mutants were subsequently tested to determine whether they could still respond to Nod factors by root hair deformation, calcium spiking, early nodulin gene induction (*rip1*, *MtENOD11* and

MtENOD40) and cell division. The genetic analysis of mutants, which were pleiotropically affected in these responses, led to the identification of six genes; these were *NFP* (Nod Factor Perception), *DMI1*, *DMI2*, *DMI3* (Doesn't Make Infections), *NSP1* and *NSP2* (Nodulation Signalling Pathway) (Catoira et al., 2000; Wais et al., 2000; Ben Amor et al., 2003). According to the precise phenotypes of the corresponding mutants, these genes were ordered on a signal transduction pathway that is predicted to be activated by Nod factors and to lead to the induction of symbiotic responses and nodulation (Figure 1). Mutations in the three *DMI* genes result not only in a Nod⁻ phenotype, but also block the establishment of symbiosis with arbuscular mycorrhizal fungi. Such "common symbiotic" genes apparently define an ancient signalling pathway that appears to have been recruited by legumes for transducing rhizobial Nod factors (Figure 1).

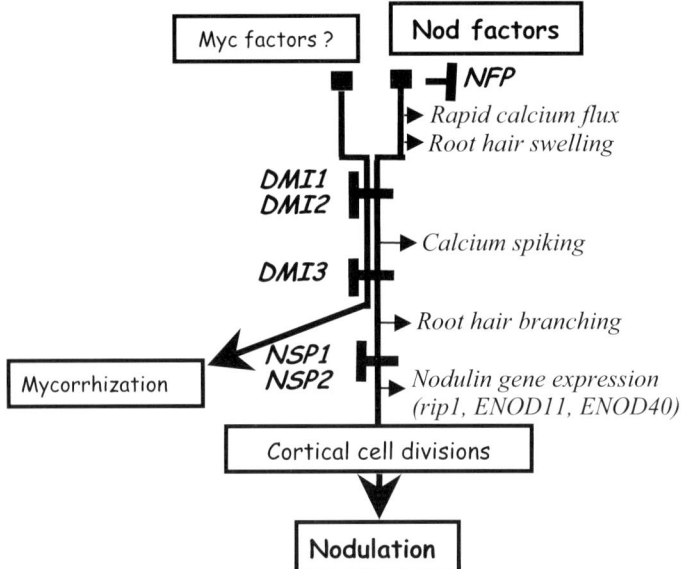

Figure 1. Model for a Nod factor signalling pathway in *Medicago truncatula*. The part of the pathway common to mycorrhization might be activated by putative Myc factors (Ben Amor et al., 2003).

3. NFP controls a very early step of Nod factor signal transduction and probably constitutes part of a Nod factor receptor

An *nfp* mutant of *M. truncatula* does not respond to Nod factors by any of the responses tested; there was no rapid calcium flux, no root hair deformation, no calcium spiking, no early nodulin gene expression, and no cell division (Ben Amor et al., 2003). On the other hand, this mutant shows a wild-type mycorrhizal phenotype, indicating that *NFP* controls a very early step of Nod factor signalling specific to nodulation (Ben Amor et al., 2003) (Figure 1). Mutants with the same phenotype exist in *Pisum sativum* (*sym10*) and *Lotus japonicus* (*nfr1* and *nfr5*), where the corresponding genes have been cloned (Radutoiu et al., 2003; Madsen et al., 2003). Based on synteny studies showing that *NFP* is the likely orthologue of *SYM10*, a candidate gene for *NFP* was identified, cloned and confirmed. *NFP*, like *SYM10*, *NFR1* and *NFR5*, is predicted to encode a transmembrane receptor-like kinase containing extracellular LysM domains (LysM-RLK) (Figure 2). The LysM domain is widespread in bacterial cell wall-degrading enzymes, where it is thought to have a peptidoglycan-binding function. LysM domains can also be found in eukaryotic proteins, for example, in association with chitinase domains. The structural similarity between the β1-4 linked units of N-acetyl-glucosamine that constitute the backbone of Nod factors and potential ligands of LysM domains, together with the phenotypes of *sym10*, *nfr1*, *nfr5* and *nfp* mutants, are good arguments in favour of these LysM-RLKs being components of Nod factor receptors (Gough, 2003; Cullimore and Dénarié, 2003). In *M. truncatula*, NFP could be part of a receptor complex responsible for activating the *DMI-NSP* signalling pathway.

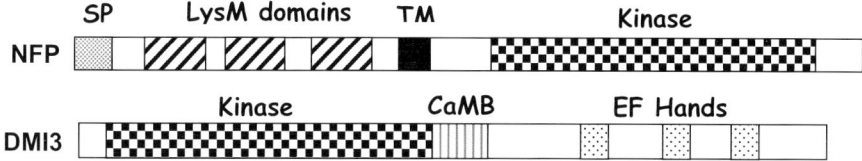

Figure 2. Predicted domain structure of the *Medicago truncatula* NFP and DMI3 proteins. SP: signal peptide; TM: transmembrane domain; CaMB: calmodulin-binding domain; EF Hands: calcium-binding EF hands.

Two other symbiotic LysM-RLKs, LYK3 and LYK4, have been identified in *M. truncatula* (Limpens et al., 2003). They are thought to control bacterial entry into root hairs and infection thread formation in a manner dependent on Nod factor structure. Nod factor perception in *M. truncatula* could thus intervene at least twice during the establishment of the symbiotic interaction; at the very earliest stage via a receptor involving NFP, then, following root hair curling, via a second receptor involving LYK3 and LYK4 (Gough, 2003; Cullimore and Dénarié, 2003).

The next challenges include understanding the potential roles of LysM domains in discriminating Nod factor structures for specificity. NFP could act either alone or, by analogy with the model proposed in *L. japonicus* (Radutoiu et al., 2003), as a heterodimer with another LysM-RLK, to activate the *DMI-NSP* Nod factor signalling pathway. To better understand these mechanisms, we have identified other LysM-RLKs of *M. truncatula*. Their potential roles in symbiosis are being investigated by expression studies and RNA-interference.

3. *DMI3* encodes a putative Ca^{2+} and Calmodulin-Dependent Protein Kinase

Contrary to mutants in *NFP*, *DMI1* and *DMI2*, a mutant in *DMI3* still responds to Nod factor application by calcium spiking (Wais et al., 2000). Therefore, *DMI3* probably acts downstream of *NFP*, *DMI1*, *DMI2*, and calcium spiking in the Nod factor signalling pathway (Figure 1). By map-based cloning, we identified DMI3 as a putative Ca^{2+} and calmodulin-dependent protein kinase, highly similar to plant proteins previously well characterized at a biochemical level, but for which no biological role was known (Lévy et al., 2004). We also characterised the *P. sativum* orthologue of *DMI3*, *SYM9*, in which several mutations had been described (Lévy et al., 2004). The predicted domain structure of DMI3 (Figure 2), together with biochemical analysis performed on similar proteins, suggest that it might sense calcium both directly through three calcium-binding EF hands and indirectly by binding a calcium-activated calmodulin (Lévy et al., 2004; Mitra et al., 2004). DMI3 could thus decode complex calcium signatures, such as calcium spiking in response to Nod factors. This theory is compatible with the position of *DMI3* in the Nod factor signalling pathway.

The identification of DMI3 as a calcium sensor clearly establishes calcium as a central component of the Nod factor signalling mechanism and indicates that calcium plays a major role in the establishment of mycorrhizal symbioses. An interesting question concerns the origin of the calcium signals. Based on the recent finding that *DMI1*, a gene located upstream of *DMI3* in the Nod factor signalling pathway, might encode a cation channel (Ané et al., 2004), DMI1 might, directly or indirectly, be involved in generating the observed changes in calcium concentration. In the case of the mycorrhizal symbiosis, the nature of the calcium signal involved is not known. Future work should focus on both characterising this signal and identifying the targets of DMI3 kinase activity in both rhizobial and mycorrhizal symbioses. Since DMI3 is the last known step common to rhizobial and mycorrhizal symbioses in *M. truncatula*, it could act as a switch by recognising different calcium signatures and activating accordingly the appropriate transduction pathway for nodulation or mycorrhization.

4. References

Ané JM et al. (2004) Science. 303, 1364-1367.
Ben Amor B et al. (2003) The Plant Journal 34, 495-506.
Catoira R et al. (2000) Plant Cell 12, 1647-1666.
Cullimore JV and Dénarié J (2003) Science 302, 575-578.
Endre G et al. (2002) Nature 417, 962-966.
Geurts R and Bisseling T (2002) Plant Cell S239-S249.
Gough C (2003) Current Biology 13, R973-R975.
Lévy J et al. (2004) Science, 303:1361-1364.
Limpens E et al. (2003) Science 302, 630-633.
Madsen EB et al. (2003) Nature 425, 637-640.
Mitra RM et al. (2004) Proc. Natl. Acad. Sci. USA 101, 4701-4705.
Radutoiu S et al. (2003) Nature 425, 585-592.
Wais R et al. (2000) Proc. Natl. Acad. Sci. USA. 97, 13407-13412.

ACTIVATION AND PERCEPTION OF CALCIUM OSCILLATIONS DURING NOD FACTOR SIGNALLING

Cynthia Gleason[1], Raka Mitra[2], Péter Kaló[1], Christine Galera[3], Clare Gough[3], Jean Dénarié[3], Sharon R. Long[2] and Giles E. D. Oldroyd[1]
[1] Department of Disease and Stress Biology, John Innes Centre, Norwich Research Park, Norwich, UK. [2] Department of Biological Sciences, Stanford University, Stanford, CA 94305, USA. [3] Laboratoire des Interactions Plantes-Microorganismes, INRA-CNRS UMR 441/2564, 31326 Castanet-Tolosan, France.

Calcium is a critical secondary messenger in diverse plant signalling pathways. A conundrum for such a ubiquitous secondary messenger is how specificity is maintained in individual signalling pathways. One possible mechanism is the activation of different modes of calcium responses, termed calcium signatures, which can have differing effects on downstream components. One of the most complex calcium signatures is calcium spiking, a signature common to a number of animal signalling pathways, but so far unique in plants to the legume/rhizobial symbiosis(1). Calcium spiking is defined as rapid and repetitive oscillations in cytosolic calcium levels. Repetitive oscillations in calcium are also observed in guard cells in response to abscisic acid, but this response lacks the rapidity common to calcium spiking(2).

Calcium spiking is activated in legume root hair cells in response to the rhizobial signalling molecule Nod factor. Since Nod factor-induced calcium spiking is not inhibited by the presence of EGTA in the external medium, the origin of the calcium is an internal store. The calcium changes are mostly restricted to the cytosol associated with the nucleus. This could indicate a role for the nucleus itself as an internal store for calcium, however, more likely it is the endoplasmic reticulum that is mostly restricted to nuclear regions. Such an ER source for calcium is consistent with animal systems, where the IP3 receptor located on the ER acts as the calcium-mobilising channel in most calcium-spiking responses. Currently, we have very little understanding of how Nod factor perception leads to activation of calcium spiking, however, the tools available in the model legume *Medicago truncatula* provide an unprecedented opportunity to use genetics to dissect the activation and perception of this complex cellular response.

1. Nod factor activation of calcium spiking
The genetic dissection of nodulation in *M. truncatula* has revealed six genes that are essential for Nod factor perception and signal transduction(3-5). These genes can be

placed in three major classes according to their level of macroscopic responses to Nod factor: *nfp* shows no responses to Nod factor; the *dmi* mutants (*dmi1*, *dmi2* and *dmi3*) are blocked for all Nod factor responses, but do show swelling of the root hair tip; and the *nsp* mutants (*nsp1* and *nsp2*) and also blocked for all Nod factor responses, but show root hair deformation (Figure 1). The characterisation of calcium spiking in these mutants revealed that *nfp*, *dmi1* and *dmi2* are unable to mount Nod factor-induced calcium spiking, whereas *dmi3*, *nsp1* and *nsp2* show a wild-type calcium-spiking response(6). This work places calcium spiking downstream of NFP, DMI1 and DMI2.

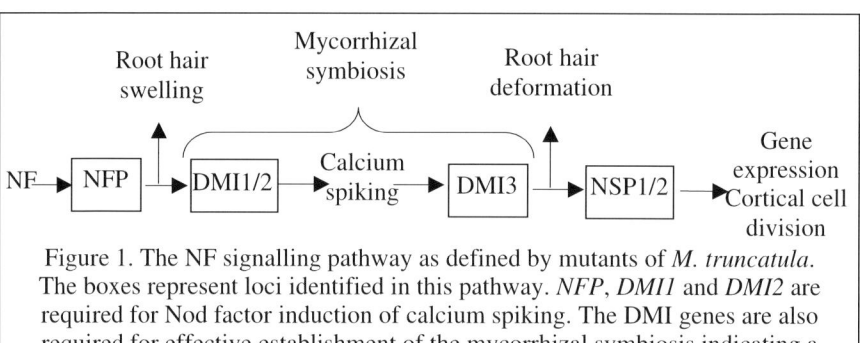

Figure 1. The NF signalling pathway as defined by mutants of *M. truncatula*. The boxes represent loci identified in this pathway. *NFP*, *DMI1* and *DMI2* are required for Nod factor induction of calcium spiking. The DMI genes are also required for effective establishment of the mycorrhizal symbiosis indicating a possible role for calcium spiking in this interaction.

The genetic identity of *NFP*, *DMI1* and *DMI2* have now been defined. *NFP*, like *NFR1* and *NFR5* of *Lotus japonicus*, encodes a receptor-like kinase with LysM domains in the extracellular region (7, 8; C. Gough, per. com.). LysM domains have previously been shown to bind polysaccharides, including *N*-acetyl glucosamines, analogous to the backbone of Nod factor. Therefore, NFP, NFR1 and NFR5 are likely candidates for the Nod factor receptor. *DMI2* encodes another receptor-like kinase, but with leucine-rich repeat domains in the extracellular region (9). The mechanism that this protein plays is unclear at this stage. *DMI1* encodes a protein with weak, but broad homology to bacterial cation channels (10), implicating this protein in calcium signalling. However, in addition to calcium spiking, there are early calcium and potassium fluxes across the plasma membrane that are activated by Nod factor, and DMI1 may be playing a role in these earlier cationic changes. In order to define the role that DMI1 is playing, it is vital to understand what, if any, cations it can transport and the localisation of this protein in the cell. Such positional information will shed light on its possible role in Nod factor signalling, since the diverse ionic changes induced by Nod factor show specific cellular locations. We are now in a position to dissect the role these proteins play in Nod factor activation of calcium spiking. Clearly, a kinase cascade is involved, since at least 2 receptor-like kinases are present in this pathway. Their targets of phosphorylation and the role of additional secondary messengers, which are a common feature of other calcium-signalling pathways, is a major challenge for the next few years.

2. Perception of Nod factor-induced calcium spiking

Because DMI3, NSP1 and NSP2 lie downstream of calcium spiking, they are likely candidates for perceiving and transducing the signal from this calcium response. Although a number of calcium-activated kinases and calmodulins are present in plants, their role in decoding calcium signals is still unclear and this genetic dissection of Nod factor signalling provides an opportunity of defining the proteins essential for decoding this calcium-spiking response.

A number of mutants in the Nod factor-signalling pathway, including *dmi3-1*, were generated with mutagens that cause deletions. We hypothesised that such deletions or RNA instability generated by the introduction of premature translational termination may be detectable on gene microarrays. We tested this using *dmi2-1*, a frameshift mutation that introduces premature translational termination. From a gene microarray of 9,935 root expressed *M. truncatula* genes, we identified 4 tentative consensuses (TCs; contigs of homologous ESTs) that were significantly down-regulated in untreated roots of the *dmi2-1* mutant (Table 1). Two of these were non-overlapping TCs of the *DMI2* gene (11). This work indicated that such an approach can be used to identify mutated genes.

Table 1. Identification of down-regulated transcripts reveals *DMI2* in the *dmi2-1* mutant and a calcium/calmodulin dependent protein kinase in the *dmi3-1* mutant.

TC number	Sequence homology	$\log_2$ (fold change) wild-type vs mutant	
		dmi2-1	*dmi3-1*
37248	FRO2-line	-1.5	-1.4
39727	Unknown	-1.2	0.0
29359	DMI2	-2,2	-0.1
38936	DMI2	-2.2	0.0
30957	Unknown	-1.1	-1.6
36850	Nitrate transporter	-1.2	-1.6
33626	CCaMK	0.1	12.0
41561	CCaMK	0.1	-2.1

We undertook a similar process with the *dmi3-1* mutant. Again, only 4 TCs were significantly down-regulated in the *dmi3-1* mutant (Table 1). Two TCs were non-overlapping TCs of the same calcium/calmodulin dependent protein kinase. Analysis of this gene in the *dmi3-1* mutant revealed a 14bp deletion that causes a frameshift leading to premature translational termination. Furthermore, analysis of the same gene in 5 mutant alleles of pea *SYM9* revealed point mutations in all 5 alleles (11). This indicates that we have successfully identified the *DMI3* gene. To validate this hypothesis, we undertook complementation of the *dmi3-1* mutant with the *DMI3* cDNA driven by the 35S promoter. *Agrobacterium rhizogenes*-mediated transformation of this construct into *dmi3-1* led to complementation of the mutant, indicating that the calcium/calmodulin-dependent protein kinase is indeed the genetic identity of *DMI3*.

The genetic data positions *DMI3* immediately downstream of calcium spiking and therefore, suggests it is a candidate for decoding the calcium signal. The genetic identity of *DMI3* makes it a strong contender for fulfilling this role. However, clearly this still needs to be proved and, more interestingly, the mechanisms by which this protein can achieve it need to be defined. A good framework for such studies is mammalian calmodulin-dependent protein kinase II (CaMKII), which has the capacity to activate its kinase activity in response to oscillatory calcium signals. This is achieved through a subtle play of affinities for the calmodulin molecule. It is possible that CCaMK uses a similar mechanism for decoding calcium spiking, however, unlike CaMKII, that only has the capacity to bind calcium in a complex with calmodulin, CCaMK can bind calcium directly through the EF-hand domains located on the protein and indirectly through calmodulin association. Therefore, CCaMK appears to have a more complex calcium-binding capacity, and therefore response, than CaMKII. Clearly, defining the mechanism of CCaMK activity will provide significant insights into calcium signalling in plants.

We have localised DMI3, using a GFP fusion, to the nucleus despite the fact that DMI3 lacks obvious NLS sequences. Maybe DMI3 requires other members of the Nod factor-signalling pathway for nuclear localisation. However, DMI3 shows the same nuclear localisation in both *nsp1* and *nsp2* mutants. If DMI3 does decode calcium spiking, then the importance of the nucleus in this calcium response is highlighted and indicates that calcium changes inside the nucleus as opposed to in the cytosol are most significant. We have used the calcium reporter, cameleon, to analyse the calcium changes in the nucleus and find that calcium spiking is occurring there and synchronously with the cytosol. However, these techniques may not be sensitive enough to detect minor differences in the timing of calcium spikes in these two compartments.

References
1. Ehrhardt DW et al. (1996) Cell 85, 673-81.
2. McAinsh MR et al. (1995) Plant Cell 7, 1207-1219.
3. Amor BB et al. (2003) The Plant Journal 34, 495-506.
4. Catoira R et al. (2000) Plant Cell 12, 1647-1665.
5. Oldroyd GE, Long SR (2003) Plant Physiol 131, 1027-32.
6. Wais RJ et al. (2000) Proc Natl Acad Sci USA 97, 13407-13412.
7. Madsen EB et al. (2003) Nature 425, 637-40.
8. Radutoiu S et al. (2003) Nature 425, 585-92.
9. Endre G et al. (2002) Nature 417, 962-6.
10. Ane JM et al. (2004) Science 303, 1364-7.
11. Mitra RM et al. (2004) Proc Natl Acad Sci USA 101, 4701-5.

FUNCTIONAL GENOMICS OF THE REGULATION OF NODULE NUMBER IN LEGUMES

Peter M. Gresshoff[1,2], Gustavo Gualtieri[1,2], Titeki Laniya[1,2], Arief Indrasumunar[1,2], Akira Miyahara[1,2], Sureeporn Nontachaiyapoom[1,2], Tim Wells[1,2], Bandana Biswas[1,2], Pick Kuen Chan[1,2], Paul Scott[1,2], M. Kinkema[1,2], M. Djordjevic[5], Dana Hoffmann[1,2], Lisette Pregelj[1,2], Diana M. Buzas[1,3], Dong Xi Li[1,2], Artem Men[4], Qunyi Jiang[1,2], Cheol-Ho Hwang[1,2] and Bernard J. Carroll[1,3]

[1]ARC Centre of Excellence for Integrative Legume Research; [2]School of Life Sciences, and [3]School of Molecular and Microbial Sciences and LAFS, The University of Queensland, St. Lucia, Brisbane QLD 4072, [4]AGRF; [5]Genome Interaction Group, RSBS, ANU, Canberra, ACT, Australia.

The nodule number in legumes is regulated by numerous factors including the number and efficiency of the interacting *Rhizobium* bacteria, environmental factors such as heat, drought, salinity, nutritional aspects such as nitrate and phosphate, environmental toxins such as cadmium, as well as endogenous plant factors including phytohormones, plant nodulation reception systems and autoregulation of nodulation (AON; based on original thoughts of Wolfgang Bauer). The cloning of the AON-control gene *GmNARK/LjHAR1* and the recognition of its function as a LRR receptor kinase now facilitates the mechanistic analysis of the process in terms of ligands, interacting proteins and signal transduction.

Not all forms of increased nodulation should be viewed as alterations of AON. For example, delayed inoculation coupled with limited nitrogen fertilization leads to plants with abundant nodules. Absence of AON leads to the supernodulation or hypernodulation phenotype. This term presently is used interchangeably, although Carroll et al (1985), who first discovered the trait, coined and then distinguished the terms to reveal a difference in the degree of nodulation. Unfortunately, subsequent researchers used the terms interchangeably without reference to the original literature!

Several properties are essential features of the absence of AON. First, nodule number per plant must be increased significantly; second, the nodulation interval (NI), being the proportion of the root covered by nodules must be larger; (N.B., thus the sickle mutant of *M. truncatula* should not be termed "hypernodulation mutant", because its NI is relatively small); third, nodule mass per plant must be increased. This parameter relative to nodule number is very plastic. Already Philip Nutman pointed out the inverse relationship between these parameters; fewer nodules usually are larger, but the total mass of nodules per plant would stay constant. Fourth, loss of AON appears to lead to increased nodule number (but not necessarily nodule mass) in the presence of otherwise inhibitory levels of nitrate, suggesting that the external regulation of nodulation by nitrate and AON are interactive. However, the processes may be different, as nitrate inhibition of nodulation acts locally (as seen in split root experiments (Carroll and Gresshoff, 1983) and interestingly in *Sesbania* stem nodulation in plants exposed to nitrate on the roots. Fifth, absence of AON leads to increased nodulation that affects negatively the development of the lateral root system (Men et al. (2002). At times associated effects on root morphology are even seen in uninoculated plants with absence of AON. A clear example is *Lotus japonicus* mutant *har1-1* (Wopereis et al., 2000) where the inoculated plant literally has a crippled root system with abundant nodules, and the uninoculated, nitrate grown plant has abundant lateral branching and reduced tap root growth. This severe root morphology effect is not seen in *nts1007* and *nts382* mutants of soybean, where the roots are decreased under inoculated (supernodulating) conditions, but nitrate–grown plants show normal morphology. The two characteristics co-segregate in *L. japonicus*. Although the altered root trait of *har1-1* is claimed to be shoot-controlled by reciprocal grafting experiments (Krusell et al., 2002), no evidence was given for it. This issue seems to require some further resolution.

Plant mutants lacking AON generally exhibit tolerance to otherwise inhibitory levels of nitrate. Many bacterial mutants and naturally occurring strains cause enhanced nodulation, often associated with an ineffective nitrogen fixing symbiosis. AON works systemically and involves the leaf. Nodulation steps are blocked either early (infection level as seen in alfalfa) or latter at the primordium advancement stage (in soybean). Only one root-controlled AON mutant was isolated, namely *nod3* in *Pisum sativum*. Other plant mutants causing increased nodulation (but not necessarily caused by the absence of AON) are: *ASTRAY* and *DISTANS* in *Lotus japonicus (*M. Kawaguchi, Japan) as well as the ethylene-insensitive mutant *sickle* (mutated in *EIN2*) in *Medicago truncatula* (Penmetsa et al. 2003).

Autoregulation mutants of legumes
Forward genetics have been productive for the isolation of mutants that have helped to dissect the nodule ontogeny of several legume species. Model legumes such as *Lotus japonicus* and *Medicago truncatula* further accelerated the progress with their experimental advantages including small seed, plant and genome size (about 460 megabases for either), short generation time with abundant progeny seed, self-fertility, and the ease of genetic analysis and transformation, especially in *Lotus*. High throughput transformation protocols as used in *Arabidopsis* so far have been unsuccessful or irreproducible in the two model legumes, demanding continued use of *Agrobacterium tumefaciens* tissue culture approaches or chimeric hairy root systems (Stiller et al.1997; Martirani et al. 1999).

Induced mutagenesis is the method of choice when looking for AON mutants. Usually EMS is used although Men et al (Men et al. 2002) described and used in positional cloning a fast neutron induced mutant of *G. soja* (*FN37*). The first AON mutants were isolated not from the model legumes but from soybean (Carroll et al. 1985) and pea (Jacobsen and Feenstra 1984). Subsequently they were isolated in *Phaseolus* bean, other soybeans, *Lotus* and *Medicago truncatula*.

With the advent of molecular genetics and genomics, the isolation of mutant loci became feasible. The last 3 years have seen an acceleration of gene discovery for many nodulation related events. The procedure of choice is map-based cloning in which segregating populations of the mutant locus are surveyed for close association of molecular markers, which then serve to isolate BAC clones, which in turn are interrogated for candidate genes by whole scale sequencing, and candidate gene testing.

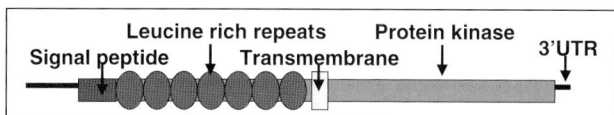

Figure 1: A generalized structure of the NARK-type LRR receptor kinase. The LRR is divided into two domains. A single intron is conserved in the kinase domain but varies in length from 76-1576 bp.

This was successfully achieved for soybean by us (*SciencExpress* publication was October 31, 2002), which represented the first cloning of an autoregulation gene in legumes and the first positional cloning of any gene in soybean (Searle et al 2003). Concurrently the *Lotus har1* gene was cloned independently (Krusell et al. 2002; Nishimura et al. 2002). Soybean contains a close relative (90% identical) to *GmNARK*, namely *GmCLV1A*. This may be the true *CLAVATA1* homologue, a possibility that we are presently testing through TILLING and RNAi in soybean. *Lotus* surprisingly does not have a similar partner to *LjHAR1*, suggesting perhaps that the severe root architecture effect in mutants may be caused by the reliance for AON, lateral root emergence and tap root growth on one gene, namely *HAR1*.

The *GmNARK* gene encodes a single transmembrane spanning, leucine-rich repeat (LRR) receptor kinase with 19 extracellular LRR spans with structural similarities to the *CLAVATA1* of *Arabidopsis* (Flechter 2002). CLV1 is part of a complex functioning in the shoot apical meristem controlling cell proliferation and fate. *Arabidopsis*, and thus presumably other plants as well, harbor about 500 receptor kinases, reflecting the complex need to sense the external environment and to make appropriate developmental and biochemical decisions relative to it. [N.B., mammals in contrast possess no more that\n 50 receptor kinases illustrating the "hard-wired" nature of their development and lack of plasticity and adaptation ability as compared to plants]. Mutation results in abundant nodulation caused by the loss of a yet-undefined negative nodulation repressor system.

The fact that the shoot is involved stems from two types of experiments. In the first one, root systems were split to allow time-delayed inoculation. In wild type soybean plants (Olssen et al. 1993), a delay of 7 days results in 100% suppression of nodulation on the delayed inoculated side. For smaller legumes such as clover and alfalfa that time span is shorter. This systemic suppression of nodulation, first demonstrated by Kosslak and Bohlool (1984), is reduced in *GmNARK*-deficient mutants. The split root results demands a root to shoot signal. It also suggests a shoot-derived inhibitor, which we propose to emanate from the leaf.

The importance of the shoot/leaf was further illustrated by reciprocal grafts in the hypocotyl or epicotyl. The genotype of the shoot (scion) controlled the nodule number in the root (the stock; Men et al. 2002; Krusell et al. 2002; Delves et al. 1986; Jiang and Gresshoff 2002). That the leaf seems the major source of autoregulation came from the following experiments. Delves et al (1986) removed all growing parts form wild type soybeans and showed that AON was still functional. Likewise trifoliate leaves were rooted in calcium chloride and nodulated to reveal a nodule number status reflecting the genotype of the isolated and meristem deficient leaf. We also intergrafted two or three internode segments (4-6 cm long) between *nts1007* shoot and roots. The resulting plant maintained the supernodulation ability, showing that a) wild type stem does not block the transit of some factor, and b) stem does not produce the shoot derived inhibitor.

Gene expression analysis revealed a paradox. Roots and leaves of soybean contained similarly high levels of *GmNARK* mRNA, yet the root cannot activate AON. Shoot apical meristem (SAM) showed low *GmNARK* RNA with nodules being intermediate. Young leaves increase *GmNARK* RNA levels as the fully expand. Expression of *NARK*, just like *AtCLV1*, in roots presents a biological paradox.

As diagramed in Figure 2, Nod factor (LCO or NF) leads to an activated state in the root epidermis, which then signals neighboring cells to divide. Cytokinin, auxin, and ethylene interplay to permit nodule primordium formation in the pericycle and the cortex. Note, cortex and pericycle divide before the infection thread delivers NF to the deeper regions. The 'activated state' results in the production of a translocatable root signal that is perceived directly or indirectly by leaf-located GmNARK. This releases factor after direct or indirect interaction with an "exporter of SDI). SDI (or X) targets the common nodulation induction pathway and lowers the susceptibility of the tissue (the cortex or the epidermis) to NF, thereby initiating AON. Different legumes target different steps; alfalfa for example, achieves AON through prevention of early infections as no arrested nodule primordia are found. Soybean in contrast blocks stage III to IV transition (Mathews et al.

1989). Since NF is required for a prolonged period to achieve nodule initiation and then nodule proliferation maintenance, suppression of NF reception facilitates AON.

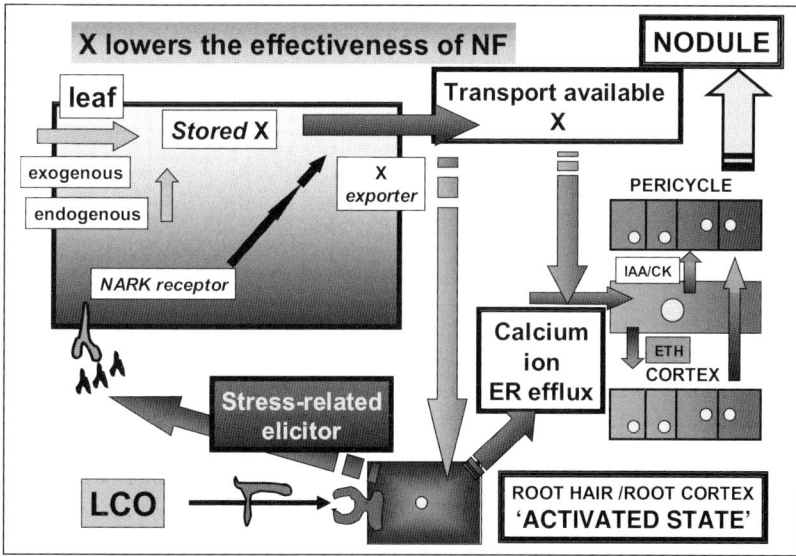

Figure 2: Autoregulation of nodulation (AON) model:

We now believe, based on approach grafting of mutant plants to wild-type plants and using time delayed nodulation suppression that AON is triggered by event after the SYMRK but before the infection thread (*NIN/LYK3/4*) stage.

Both kinase and LRR domain of GmNARK were expressed in *E. coli* and purified. Antibody against *GmNARK* presently is being prepared to facilitate immuno-localization of the protein in the leaf, to ascertain a correlation between RNA and protein levels, and to "pull-down" the entire complex, thus permitting analysis of the ligand and interactors. The cloning of *GmKAPP-2* (kinase associated protein phosphatase) will lead to more detailed protein-protein interaction studies. Site directed mutagenesis of either interacting partners will define structural domains, allowing progressive functional analysis of this fascinating biological process (Gresshoff 2003).

Acknowledgments
We thank the ARC for a Centre of Excellence grant to establish the Centre for Integrative Legume Research, AUSAID, UQ, the Queensland Government Smart State Fund, and the Thai Government.

References

Carroll, B.J. and Gresshoff, P.M. (1983) Zt. Pflanzenphysiol. 110:77-88.
Carroll, B.J. et al (1985). Proc. Natl. Acad. Sci. USA 82:4162-4166.
Delves, A.C, et al (1986) Plant Physiol. 82:588-590.
Delves, A.C. et al (1992) Plant Cell Physiol. 15: 249-254.
Fletcher, J.C. (2002) Annu. Rev. Plant Biol. 53:45-66.
Gresshoff, P.M. (2003) Genome Biology 1, 201.
Jacobsen, E. and Feenstra, W. (1984) Plant Sci.
Jiang, Q. and Gresshoff, P.M. (2002) Functional Plant Biol. 29:1371-1376.
Kosslak, R. and Bohlool, B.B. (1984) Plant Physiol.
Krusell, L. et al (2002). Nature 420:422-426.
Martirani, M. et al. (1999) Mol. Plant Microbe Interact. 12: 275-284.
Mathews, A. et al (1989) Protoplasma 150: 40-47.
Men, A.E. et al (2002) Genome Letters 1:147-155.
Nishimura, R. et al (2002) Nature 420:426-429.
Olssen, J.E. et al (1989) Plant Physiol. 90: 1347-1352.
Penmetsa, R. V. et al (2003) Plant Physiol. 131:1-11.
Searle, I.R., et al (2003) Science 299:108-112 (on line)
Stiller, J. et al (1997) J. Exp. Bot. 48:1357-1365.
Wopereis, J. et al (2000) Plant Journal 23:97-114.

"ACTIVATOR" AND "INHIBITOR" LEADING TO GENERATION AND STABILIZATION OF SYMBIOTIC ORGAN DEVELOPMENT IN LEGUME

Masayoshi Kawaguchi
Department of Biological Sciences, Graduate School of Science, University of Tokyo, Hongo, Tokyo 113-0033, Japan

Nodule is a symbiotic nitrogen fixation organ generated by biological interactions. The symbiotic organ is induced by Nod factor (lipochitin-oligosaccharides) derived from rhizobia and the excessive nodulation is suppressed by a long-distance signal (autoregulation signal) produced in shoots, on receiving the root-derived factor. These positive and negative regulators significantly contribute to generation and stabilization of the symbiotic organ development. Here I point out that Nod factor and HAR1-mediated long-distance signal may correspond to "Activator" and "Inhibitor" in reaction diffusion (RD) system respectively that has been proposed by Alan Turing in 1952. The RD system consists of short distance "Activator" and long range "Inhibitor", and pattern formation starting from initially homogenous conditions is generated by local self-enhancement of "Activator", coupled with a long-range antagonistic effect by "Inhibitor". Trough pointing out the similarity between RD system and systemic regulation of symbiotic organ development, I will focus on the structural originality of Nod factor as a presumptive "Activator".

1. Positive feedback aspect of Nod factor
Besides spontaneous nodules, nodule development requires Nod factor production by rhizobia. The biosynthesis of Nod factor is triggered by flavonoids secreted from plant roots. In 1991, Recourt et al. demonstrated that inoculation of *Vicia* root with rhizobia promotes the new production of Nod inducer (possibly flavonoids). Considering that some flavonoids induced by Nod factor promotes further biosynthesis of Nod factor, amounts of Nod factor is expected to increase with time based on the legume and rhizobia communications. Nod factor also creates cytoplasmic space called preinfection thread in the dividing root cortex where infection thread network is formed (van Brussel et al. 1991). The preinfection thread is needed for bacterial entry via

infection thread formation, and eventually supports proliferation of bacteria that biosynthesize Nod factors. It is also expected that amounts of Nod factor would increase during early infection process. Actually Paulina et al. (2004) observed a strong increase in the amount of Nod factor prior to nodule primordium formation during *Vicia* and *Rhizobium leguminosarum* interactions. These reaction schemes by Nod factor represent its positive feedback aspect, namely autocatalytic aspect of Nod factor based on the biological communications (Fig. 1).

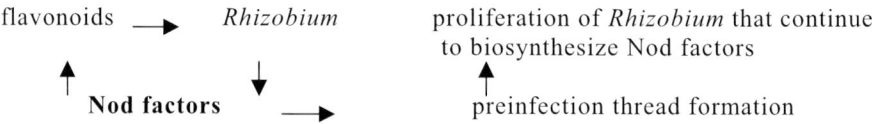

Fig. 1 Autocatalitic aspect of Nod factor

2. HAR1 and DISTANCE-mediated long distance signaling
Although symbiotic root nodules are beneficial to leguminous hosts, excessive nodulation damages the host because it consumes a lot of photosynthates for nitrogen fixation, as well as causing hyper-infection with root symbionts. In order to stabilize symbiotic interaction, leguminous plants program negative regulatory systems into their genome. There is a systemic feedback regulatory system controlling nodule development, termed autoregulation, in which developed nodules or nodule primordia suppress the emergence of further nodules (Nutman 1952). Nodulation control by systemic feedback is of great interest in relation to the flexibility of plant organogenesis. In order to understand the feedback system at molecular level, I and associates conducted EMS and ion beam mutagenesis using *L. japonicus*. As a result, we have identified 9 loci responsible for Nod- phenotype, 5 loci for Hist- (bump), at least 4 loci Fix-, 3 loci for Nod++ (Kawaguchi et al. 2002 and unpublished 10 loci). Among them, *har1* (*Ljsym78*), and *distance* (*Ljsym90*) are typical hypernodulating mutants. Reciprocal and self-grafting studies using *har1* mutants revealed that the shoot genotype of *har1* mutants is involved in the negative regulation of nodule development. A map-based cloning strategy showed that the *HAR1* gene encodes a receptor-like protein containing 21 leucine-rich repeats, a single transmembrane domain and serine-threonine kinase domains (Krusell et al. 2002; Nishimura et al 2002). *HAR1* showed the highest level of identity with *CLAVATA1* (*CLV1*) of all *Arabidopsis* receptor-like kinases. CLV1 is known to negatively regulate formation of the shoot apical meristem via cell-cell communication recognizing the CLV3 peptide. Recently Tomomi Nakagawa in my laboratory found that application of methyl jasmonate to the wild-type and *har1* shoots significantly suppressed nodule development, but not in the case of salicylic acid. The finding indicates that endogenous jasmonate

may work in the downstream of HAR1 and act as autoregulation singal. On the other hand *distance* is a novel hypernodulating mutant with very late flowering which was isolated by He2+ beam mutagenesis. Grafting experiments of *distance* revealed that shoot genotype seems to determine the root phenotype like *har1* and soybean *nts1* hypernodulating mutants. These results indicate that *distance* is unable to produce autoregulation signal from the shoots.

3. Schemes of systemic regulation of symbiotic organ development and Reaction Diffusion (RD) system

What drives the systemic regulation of symbiotic organ development? Recently van Brussel et al. (2002) clearly showed that feedback regulation of nodulation is elicited systemically in the *Vicia* split root system by the application of Nod factors. His finding indicates that Nod factor acts as possitive and simultanenously negative regulators of nodule development.

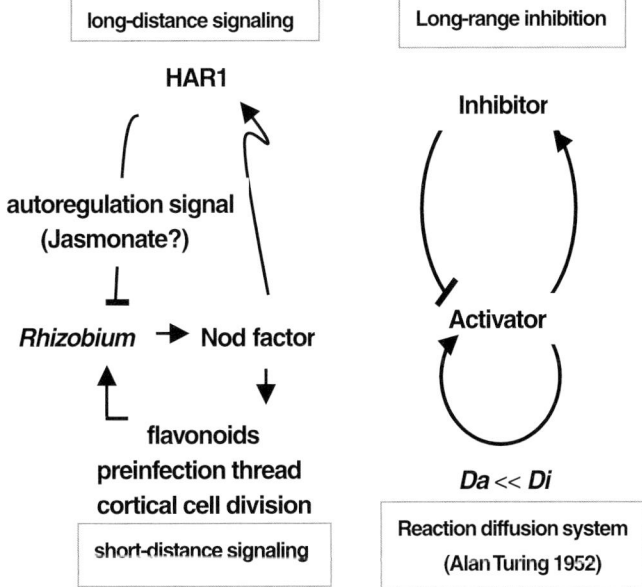

Fig. 2 Reaction schemes; Nod factor and systemic regulation of nodule development (a), RD system (b).

Putting previous and current findings related to nod factor, flavonoids and autoregulation together, a reaction scheme of the systemic regulation of nodulation could be drawn as Fig. 2a. Here I notice that this reaction scheme of symbiosis is similar to that of RD system that was proposed by Alan Turing in 1952. In RD system, a short-range substance "Activator" promotes its own production (autocatalysis) as well as that of its rapid diffusing antagonist information and pattern formation is inevitably generated and stabilized from initially homogenous field. In my view, the theoretical RD system leading to generation and stabilization of pattern may be actualized in the symbiotic systems, especially systemic regulation of nodule development in legume .

4. Structural originarity of Nod factor

In RD system, "Inhibitor" must diffuse much faster than "Activator" to generate pattern in a uniform field. To put it differently, pattern could be generate if the diffusion rate of "Activator" would be very small as compared with that of "Inhibitor" bacause relative difference of diffusion rate of two molecules is crucial. Viewed in this light, chemical structure of Nod factor especially fatty acid moiety is worthy of notice. Non-reducing end of decorated chitin oligomers is acylated with fatty acid by rhizobial NodA and NodB genes (Atkinson et al. 1994), resulting in creation of the amphiphilic molecules. Here it is expected that addition of fatty acid to chitin oligomers may contribute to significantly reduce the diffusion rate of Nod factor by trapping in cell wall and plasma membrane. Actually Goedhart et al. (2000) demonstrated that fluorescently labelled Nod factor shows very low lateral diffusion. Fatty acid moiety of Nod factor is thought to be involved in host specificity. Besides that, judging from theoritical point of view on the essential condition in RD system, Nod factor might have become presumptive "Activator" of symbiotic organ development, by means of the significant reguction in diffusion of chitin oligomers by adding fatty acid to them. To verify the idea, there is a clear need to unvail a reaction scheme of nod factor and HAR1-, DISTANCE-mediated long distance signaling at molecular level.

5. References

Atkinson EM et al. (1994) Proc. Natl. Acad. Sci. USA 91, 8418-8422.
Goedhart J et al. (2000) Plant J. 21, 109-119.
Kawaguchi et al. (2002) Mol. Plant Microbe Interac. 15,
Krusell L et al. (2002) Nature 420, 422-426.
Nishimura et al. (2002) Nature 420, 426-429.
Nutman PS (1952) Ann. Bot. 16, 79-101.
Paulina TT et al. (2004) Mol. Plant Microbe Interac. 17, 816-823.
Recoult e al. (1991) Mol. Plant Biol. 16, 841-852.
Turing A (1952) Phil. Trans. R. Soc. B237, 37-72.
van Brussel et al. (1991) Sience 257, 70-72.
van Brussel et al. (2002) Mol. Plant Microbe Interac. 15,341-349.

REGULATORY MECHANISMS OF SYMRK KINASE ACTIVITY

Satoko Yoshida and Martin Parniske
The Sainsbury Laboratory, John Innes Centre, Colney Lane, Norwich,
NR4 7UH, UK; Present address: Ludwig-Maxmilians-Universität München,
Maria-Ward-Str. 1a, München 80638, Germany.

Most land plants develop a symbiosis with arbuscular mycorrhizal fungi to benefit from phosphate uptake, whereas only members of four plant orders, e.g., legumes, interact with rhizobia to form a specialized nitrogen-fixing organ, the root nodule. Genetic analyses revealed that these two different types of symbiosis have common components and mutation of "common symbiosis (*SYM*) genes" cause defects in both symbiotic processes (Kistner and Parniske, 2002). Several common *SYM* genes have been identified in the model legumes, *Lotus japonicus* and *Medicago truncatula*.

SYMRK (symbiosis receptor-like kinase) is a common *SYM* gene from *Lotus japonicus*, which is required for an early step of symbiosis. The *SYMRK* gene encodes an LRR-type receptor-like kinase, featuring a signal peptide, three LRR motifs, a transmembrane domain and a Ser/Thr kinase domain (Stracke et al., 2002). SYMRK orthologues were identified from *Medicago truncatula* (DMI2), *Medicago sativa* (NORK) and *Pisum sativum* (SYM19), indicating that SYMRK is evolutionarily conserved among legume species (Stracke et al., 2002; Endre et al., 2002).

To understand how SYMRK regulates symbiosis signal transduction, we analysed the mutation points of plant *SYMRK* mutants. Several mutations were found in the conserved kinase domain, and assigned to conserved crucial residues for kinase activity. Therefore, we assumed that kinase activity is required for SYMRK function. To test this hypothesis, we introduced the same mutation into *E. coli* expression clones and analysed kinase activity *in vitro*. As expected, we observed a correlation between kinase activity and symbiosis phenotype, suggesting the kinase activity is important for SYMRK symbiosis function.

SYMRK kinase domain autophosphorylates in an intermolecular manner. Intermolecular phosphorylation is important to activate animal receptor tyrosine kinases, therefore, the intermolecular autophosphorylation of SYMRK may be a first step for activation of

SYMRK. As His-tagged SYMRK kinase domain (His-SYMRK) expressed in *E. coli* shows multiple bands on a SDS-PAGE, we tested whether these low mobility bands represent phosphorylated His-SYMRK. After phosphatase (CIP) treatment, all lower mobility bands disappear and only the smallest molecular weight band was observed (Figure 1A). This result indicates His-SYMRK is phosphorylated in *E. coli*. To test whether the phosphorylation status of SYMRK affects its kinase activity, we performed *in gel* and *in vitro* kinase assay. We found that phosphorylated SYMRK is more active than non-phosphorylated SYMRK both *in gel* and *in vitro* kinase assays (Figure 1B), indicating SYMRK kinase activity is regulated by its phosphorylation status.

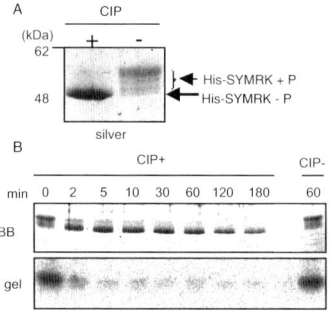

Figure 1. Kinase activity of phosphatase-treated SYMRK.
Purified His-tagged SYMRK protein was incubated with or without CIP for 1 h at 37°C and separated by SDS-PAGE. Multiple bands seen in the CIP- sample were shifted to the lowest size bands after incubation with CIP. The upper bands, therefore, represents phosphorylated His-SYMRK (His-SYMRK+P) and the lowest band non-phosphorylated SYMRK (His-SYMRK-P). (B) *In gel* kinase assay of CIP treated SYMRK. His-SYMRK protein was incubated with CIP at 37°C and samples were taken at the indicated time points. The samples were subjected to SDS-PAGE analysis followed by Coomassie staining (CBB, upper panel) and *in gel* kinase assay (in gel, lower panel). As a control, the protein was incubated without CIP for 60 min at 37°C (CIP-).

To identify which phosphorylation sites regulate to SYMRK kinase activity, we used a mutagenesis approach. Among 25 potential phosphorylation sites in the SYMRK intracellular domain, we selected conserved residues between legume and non-legume SYMRK orthologues as possible mutagenesis targets. We selected 2 Ser (S751 and S754) and 1 Thr (T760) residues from the activation segment, 1 Ser (S580) and 1 Thr (T593) from the juxtamembrane region, and 2 Ser (S870 and S888) from the C-terminal region. These Ser/Thr residues were substituted by Ala in the GST-tagged SYMRK construct, and mutated proteins were analysed *in vitro* and *in gel* kinase assays. Mutations S760A and T593A reduced kinase activity significantly, although these mutant proteins still showed basal *in vitro* kinase activity. The activity of these mutant proteins could not be detected by *in gel* kinase assay. The S754A mutation slightly reduced auto- and substrate-phosphorylation activity *in vitro*, whereas no activity was detected *in gel* kinase assay. Other mutations had no effect on the kinase activity.

To check if these 3 sites are phosphorylated, we analysed *E. coli* expressed SYMRK protein by Q-TOF. Q-TOF identified phosphorylation on S754 and T760 but not on T593. These three sites are largely conserved in plant receptor-like kinases, suggesting their regulatory mechanism is conserved.

In conclusion, we show that SYMRK kinase activity is important for its symbiosis function. We found the SYMRK kinase activity is regulated by its phosphorylation status, and identified 3 crucial residues for activation. We propose that SYMRK is activated by autophosphorylation, probably induced by an external signaling component(s) (Figure 2).

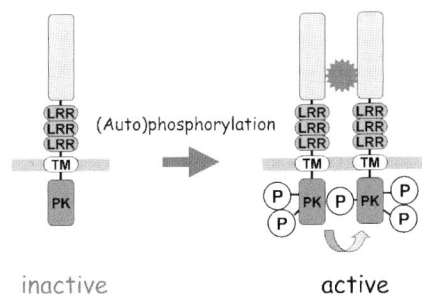

Figure 2. A proposed model for SYMRK activation. Either ligand binding or external signal leads to SYMRK autophosphorylation and kinase activation.

References
Kistner and Parniske (2002) *Trends. Plant Sci.* 7, 511-518.
Stracke et al (2002) *Nature*, 415, 959-962.
Endre et al. (2002) *Nature* 415, 962-966.

A METHOD TO ENABLE CONTINUOUS LIQUID INTRODUCTION INTO AN APOPLAST OF A SOYBEAN-PLANTLET LEAF AND ITS APPLICATION TO ELUCIDATING THE SUPERNODULATION TRAIT OF NOD1-3

Yasuhiro Arima, Hiroko Yamaya and Tadashi Yokoyama
Department of Bioproduction, Tokyo University of Agriculture and Technology, Fuchu, Tokyo 183-8509, Japan

Results of grafting experiments between super- (or hyper-) nodulation mutants of some leguminous species and their parents had revealed that supernodulation is a shoot-controlled phenomenon (1-5). This fact strongly suggests that this nodulation trait of leguminous plants is under the control of a systemic regulation mechanism and some shoot-originated nodulation-controlling substance (SNS) is translocated to the roots. Some interspecies grafting experiments showed that the SNS effect occurred to some extent over a range of leguminous species and genera (2, 5). Thus, identification of this SNS is a key subject in attempts to clarify the systemic mechanisms regulating root nodulation, and a method of SNS detection is essential for its identification.

To search for the SNS involved in the supernodulation trait of NOD1-3, a mutant of soybean (*Glycine max* [L.] Merr.) cv. Williams, we tried to establish a bioassay-system with cuttings of soybean plantlets that enables us to introduce liquid substances continuously into leaves and to assess their effects on root nodulation (6). The first trifoliate leaves of non-nodulated 3-week-old Williams82 and NOD1-3 were collected as cutting material, and planted into sterilized vermiculite beds containing N-free plant culture medium. The plantlet cuttings were kept under saturated humidity and fluorescent light at 25℃ for 1 week, and then placed under conditions of 60% of relative humidity and light ($170 \mu Fs^{-1}m^{-2}$) - dark cycle of 14-10 hr after confirming root genesis. Just before introduction of the assay solution, the central foliolate of the plantlet was cut off at its petiolule, and then the cut-surface was brought into contact with distilled water followed by the assay solution. The introduction of the liquid by self-suction continued for at least 7 days, although the introduction rate declined gradually.

When ^{15}N-labeled urea was introduced, both the emergence and development of root-nodule meristem were significantly repressed, especially in the NOD1-3 plantlets. ^{15}N-analyses revealed that a significant amount of fixed-nitrogen originating from introduced urea was translocated to the roots by the fifth day after initiation of ^{15}N-feeding. In contrast to urea, sucrose introduction enhanced both emergence and development of the root-nodule meristem, irrespective of the plantlet genotype and light conditions. The obvious differences in root-nodule formation between plantlets of NOD1-3 and Williams82 did not disappear under conditions of a normal or plentiful supply of photosynthate to plantlet roots or when it was exhausted.

By introducing a leaf extract of non-infected Williams82, formation of root-nodule meristem in NOD1-3 plantlets tended to be repressed at the 5th day after rhizobial inoculation, and the number of visible nodules at the 7th day decreased to the level of Williams82 plantlets. In contrast to the nodulation of NOD1-3 plantlets, nodulation of Williams82 plantlets did not suffer any repression by the leaf extract of Williams82. Introduction of NOD1-3 leaf extract resulted in no significant change of the nodulation in either NOD1-3 or Williams82 plantlets. The intensity of the repressive effect of Williams82 leaf extract was not changed by rhizobium inoculation of the source plants.

References
1. Gresshoff PM and Delves AC (1986) in A. D. Blaustein and P. J. King (eds.) Plant Gene Research, vol 3, 159-206, Springer Verlag, New York.
2. Delves AC et al. (1987) J. Plant Physiol. 128, 473-478.
3. Francisco Jr. PB and Akao S (1993) J. Exp. Bot. 44, 547-553.
4. Francisco Jr. PB and Harper JE (1995) Plant Science 107, 167-176.
5. Harper JE et al (1997) Crop Sci. 37, 1242-1246.
6. Yamaya H and Arima Y (2004) Jpn. J. Soil Sci. Plant Nutr. 75, in press.

AMINO ACID CYCLING BY *RHIZOBIUM LEGUMINOSARUM* IN PEA NODULES

James White, Alex Bourdes, Arthur Hosie, Seonag Kinghorn, Philip Poole
School of AMS University of Reading UK

We have shown that *R. leguminosarum* has two general amino acid permeases, Aap and Bra, which transport a broad range of amino acids. The *aap* operon consists of four genes *aapJQMP* that code for an ABC (ATP binding cassette) transporter which transports a wide range of L-amino acids, including glutamate, aspartate, histidine, leucine and alanine. The Aap actively accumulates amino acids to high intracellular concentrations leading to a several thousand fold-excess over the extracellular concentration. However, it also promotes the exchange and efflux of amino acids indicating that the system is also bidirectional (Hosie et al., 2002; Walshaw and Poole, 1996). Mutations in the *aap* prevent growth on glutamate as a sole carbon source on agar plates but still allow growth in liquid culture. This led to the discovery of the Bra ABC transport system, which is a second broad range amino acid permease (Hosie et al., 2002). As its name suggests, the Bra system has highest identity to the branched chain amino acid transport systems (also known as LIV) but actually transports a very wide range of amino acids including leucine, alanine, glutamate, histidine and most unusually γ-aminobutyric acid (GABA). All other known GABA transporters are very specific for GABA and belong to the APC family of amino acid transporters. Single mutations in either the *aap* or *bra* reduce the uptake of a range of amino acids, including glutamate, by around 50%. A double mutation in the *aap* and *bra* severely reduces the uptake of a broad range of amino acids and prevents growth on glutamate on agar and in liquid culture. However, the double mutant grows on minimal medium, demonstrating that it can synthesise amino acids.

Peas nodulated by either *aap* or *bra* single mutants were indistinguishable from plants nodulated by the wild type A34, whereas peas nodulated by the double mutant progressively yellowed (Lodwig et al., 2003). There were also dramatic reductions in shoot dry weight and shoot nitrogen content of plants nodulated by double mutant, as well as increased nodule number and mass, typical of plants unable to fix nitrogen. Thus while single mutations in either system have no affect on symbiotic nitrogen fixation the double mutation prevents effective nitrogen assimilation. The phenotype is very unusual, since bacteroids can still reduce acetylene or $^{15}N_2$, up to 60% and 30% of wildtype rates, respectively. In addition the root nodules were pink, unlike the white nodules usually induced by classical non-fixing mutants, or the deep red of wild type pea nodules. Light and electron micrographs showed that the bacteroids were fully formed although most unusually for bacteroids from indeterminate nodules the double mutant accumulated polyhydroxybutyrate. As is often seen in fix minus nodules excess starch had accumulated in the plant. Two related questions could be posed from these results: one, how can the bacteroid retain active nitrogenase and the plant remain so nitrogen starved; two, why should preventing amino acid movement cause this nitrogen starvation. In addition an aspartate aminotransferase (*aatA*) mutant was unable to fix N_2. This suggested that amino acid cycling and transamination is essential for productive N_2 fixation.

We therefore proposed a model, where in addition to dicarboxylates, glutamate, or a precursor/derivative of it, is transported into bacteroids (Fig. 1).

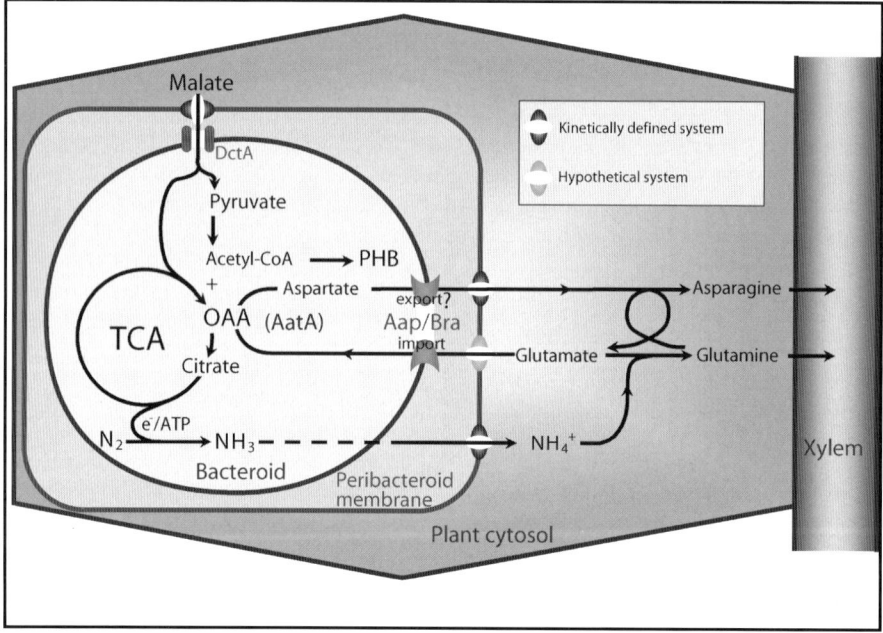

The glutamate, which would enter via Aap/Bra, would act as a transamination donor to produce aspartate or possibly an amino acid such as alanine. In our original model we suggested that the secreted aspartate would be used by the plant for asparagine synthesis. While this may be true, the effect on the plant may not be so direct, such that preventing cycling of amino acids between the bacteroid and plant might have a detrimental effect on amino acid pools generally or on regulation of their synthesis. Glutamate was proposed as the most probable donor amino acid because it is highly abundant in nodules and is known to stimulate transamination of oxaloacetate and pyruvate in isolated pea peribacteroid units, leading to aspartate and alanine secretion (Appels and Haaker, 1991; Rosendahl et al., 1992). Since Aap and Bra have a broad solute specificity, amino acids other than glutamate may be taken up from the plant. However, most bacterial amino acid metabolism is channelled via glutamate, making little difference to the model between direct uptake of glutamate or a precursor/derivative. Aspartate is the most likely secretion product, because glutamate stimulates its synthesis in isolated peribacteroid units and a blockage in its secretion would collapse asparagine synthesis in the plant. However, the secretion of alanine by bacteroids could have a similar affect because alanine, glutamate and aspartate can be interconverted in the plant by glutamate pyruvate transaminase and aspartate aminotransferase. As mentioned above, apart from their role as uptake transporters, Aap and Bra also act as high rate but low affinity exporters and therefore, may also mediate the export of aspartate or other amino acids. Blocking amino acid import, export or both will have similar affects in our model.

It is apparent that to be able to test this model rigorously it is essential to determine which amino acids must move via the Aap and Bra and whether the amino acids are transported into or out of the bacteroid. The difficulty is that the Aap and Bra are both broad specificity amino acid transport systems. We therefore started with a mutant which has the Aap deleted, thereby allowing us to manipulate the Bra system to change its solute specificity. The Bra system consists of BraDEFGC, with BraC the periplasmic solute binding protein. Mutations in BraDEFG completely inactivated transport via the Bra but mutations in BraC knocked out all transport except that of the aliphatic amino acids (leucine and alanine were tested). This indicates that there is a second solute binding protein, BraC3, that interacts with the BraDEFG membrane complex that has a narrower solute specificity than BraC. A likely candidate for BraC3 was identified by bioinformatic analysis of the *R. leguminosarum* 3841 genome sequence and this was mutated. When *braC3* was mutated the aliphatic specific transport via the Bra was lost and when *braC* and *braC3* were both mutated all transport via the Bra was lost. Critically for this study, a strain in which the Aap is deleted and *braC* is mutated will only transport aliphatic amino acids. When this strain was inoculated onto peas it fixed and assimilated nitrogen at wildtype rates. This shows that the only amino acid that must move via the Aap and Bra to enable productive nitrogen fixation is an aliphatic amino acid, which almost certainly means alanine. Alanine is already known to be synthesised and secreted by pea and soybean bacteroids, so this makes it highly likely that the role of the Aap and Bra is to secrete alanine to the plant (Allaway et al., 2000; Appels and Haaker, 1991; Rosendahl et al., 1992; Waters et al., 1998). This work also shows that glutamate uptake via the Aap or Bra is not needed for productive nitrogen fixation. Therefore, alanine synthesis must be derived either from *de novo* synthesis or transamination from an amino acid that enters via a transport system other than the

Aap/Bra. It has already been shown that the only active system for *de novo* alanine synthesis in pea bacteroids (alanine dehydrogenase) can be mutated without major effect on productive nitrogen fixation (Allaway et al., 2000), so this suggests that another amino acid may be transported into bacteroids to drive a transamination cycle. We are currently investigating which other amino acids might be imported by bacteroids. One intriguing possibility is GABA, since we have identified a GABA specific ABC transporter that is not normally expressed in laboratory culture. Transposon mutants have now been isolated in this system and its effects on peas are now being investigated.

References
Allaway D et al. (2000) Mol. Microbiol. 36, 508-515.
Appels MA and Haaker H (1991) Plant Physiol. 95, 740-747.
Hosie AHF et al. (2002) J. Bacteriol. 184, 4071-4080.
Lodwig EM et al. (2003) Nature 422, 722-726.
Rosendahl L et al. (1992) J. Plant Physiol. 139, 635-638.
Walshaw DL and Poole PS (1996) Mol. Microbiol. 21, 1239-1252.
Waters JK et al. (1998) Proc. Natl. Acad. Sci. USA 95, 12038-12042.

ROOT NODULE EXTENSINS IN INFECTION THREAD DEVELOPMENT

Elizabeth A. Rathbun and Nicholas J. Brewin
John Innes Centre, Norwich, NR4 7UH UK

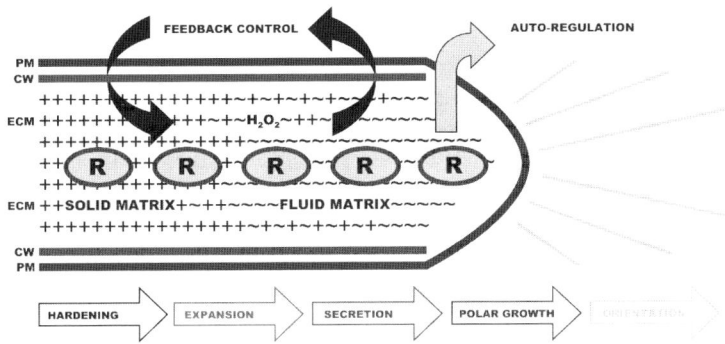

Figure 1. Dynamic model for apical growth of an infection thread.
Rhizobial cells gain access to legume host cells and tissues through the lumen of tubular infection threads, which are inwardly growing cylinders bounded by plasma membrane (PM) and plant cell wall (CW). Rhizobial cells (R) divide only at the apex (15-25 µm).
The model proposes that hydrogen peroxide causes the progressive cross-linking of root-nodule extensins in the infection-thread lumen. Solidification of the extracellular matrix (ECM) prevents further growth/division of bacteria in mature parts of the thread.

Root nodule extensins (RNEs) are highly glycosylated plant glycoproteins localized in the extracellular matrix of legume tissues and in the lumen of infection threads. The deduced 150 amino acid sequence of the secreted polypeptide of PsRNE-1 is typical of this family of (glyco)-proteins. There are 70 (hydroxy)proline residues organized as interspersed motifs characteristic of extensins (contiguous Pro) and arabinogalactan (AGP) proteins (alternating Pro). The combination of extensin and AGP motifs within a single polypeptide appears to be unique to legumes. Examples of RNE glycoprotein have been found as nodule-enhanced gene transcripts in *Medicago, Lotus, Vicia, Sesbania*, and *Glycine* spp. Furthermore, their close physical interaction with rhizobial cells suggests that they may play an important role in the growth of infection threads.

Hérouart et al. (2002) have demonstrated the presence of hydrogen peroxide in the infection-thread lumen. Also, a strain of *Sinorhizobium meliloti*, which overproduces catalase to destroy H_2O_2 in the infection thread, produced grossly distended infection threads (Jamet, Herouart and Puppo, unpublished data), consistent with our model that peroxide-driven protein cross-linking is important for infection thread growth (Fig. 1).

Infection threads are observed to grow at a rate of about 10 μm/hour in *Medicago* and other legumes (Gage, 2002). Bacteria at the tip divide with a generation time of 3-4 hours, but this tapers off rapidly along the maturing infection thread, such that there is virtually no bacterial cell division 15-25 μm behind the growing apex. This implies that only 25-50 bacterial cells are actively involved in the growth and propagation of an infection thread. Our studies of root nodule extension indicate an important role for H_2O_2 in the regulation of infection thread growth (Wisniewski et al., 2000). This model is based on the concept that a progressive fluid-to-solid transition of RNE in the infection thread lumen may control the pattern of infection thread growth. The fluid matrix sustains bacterial cell division and apical expansion of the infection thread, the solid phase occludes the lumen and impairs further bacterial division (Brewin, 2004). According to the model (Figure 1), apical growth of the infection thread is sustained by three forces acting simultaneously. First, there is growth and division of bacterial cells within a confined space. Second, there is targeted secretion of RNE and other plant glycoproteins into the lumen of the infection thread. Third, there is membrane growth and cellulose synthesis that is controlled and oriented by the underlying cytoskeleton. For steady state growth, the rate of IT extension should be balanced by the rate of matrix solidification. Over-production of H_2O_2 would lead to abortion of the infection thread.

Future research should focus on: (i) how is RNE secretion targeted to the apex of the infection thread? (ii) what signals (e.g., Nod-factor) control the auto-propagation of the apical growth point? (iii) what signals (perhaps bacterial extracellular polysaccharide) control the rate of peroxide formation in the infection-thread lumen? (iv) what other macromolecules does RNE interact with in the infection-thread lumen? Many of the control points governing infection-thread growth may soon be identified by mutational analysis, which should make it easier to explain the biophysics and biochemistry of infection-thread growth.

Brewin NJ (2004) Crit. Rev. Plant Sci. 23, 293-316.
Gage DJ (2002) J. Bacteriol. 184, 7042-7046.
Hérouart et al (2002) Plant Physiol. Biochem. 40, 619-624.
Rathbun EA et al. (2002) MPMI 15, 350-359.
Wisniewski J-P et al. (2000) MPMI 13, 413-420.

CASTOR AND *POLLUX*, THE TWIN GENES THAT ARE RESPONSIBLE FOR ENDOSYMBIOSES IN *LOTUS JAPONICUS*.

Haruko Imaizumi-Anraku[1], Naoya Takeda[2], Martin Parniske[3], Makoto Hayashi[2], Shinji Kawasaki[1]
[1]NIAS, Kannon-dai 2-1-2, Tsukuba, Ibaraki 305-8602 Japan.
[2]Dept. Biotech., U. Osaka, 2-1 Yamadaoka, Suita, Osaka 565-0871 Japan.
[3]John Innes Centre, Colney Lane, Norwich NR4 7UH, UK.

Among legume mutants blocked at the early stages of *rhizobia* symbiotic interactions, some are also defective in arbusuclar mycorrhizal colonization. The mutated genes of such Nod⁻ Myc⁻ mutants are designated as 'common sym genes', that control both bacterial and fungal endosymbioses.

Lotus japonicus mutants affected in the *CASTOR* or *POLLUX* genes exhibited Nod⁻ Myc⁻ phenotype and no calcium spiking response. These phenotypes indicate that these two genes act very early in a signal transduction pathway of microbial signal perception. *CASTOR* and *POLLUX* were mapped close to the bottom of chromosome 1 and 6, respectively. By positional cloning approaches, we have identified the two genes. Southern hybridization with a *CASTOR* probe indicated that *CASTOR-POLLUX* homologs exist one or more copies in *Medicago truncatula*, Pea, Soybean, *Arabidopsis thaliana* and Rice. Thus, *CASTOR* and *POLLUX* homologous are widely distributed without distinction of leguminous, non-leguminous, and non-mycorrhizal plants, such as *Arabidopsis*. This fact suggests that the twin genes have been derived from an ancient plant gene and have evolved into symbiotic specific, common sym genes.

CASTOR and *POLLUX* encode proteins of 853 and 917 amino acid residues with predicted molecular masses of 95 kDa and 102 kDa, respectively. The twin proteins show high homology with each other and encode a novel class of ion channel, classified into calcium-gated potassium channels such as MTKH of *Methanobacterium thermoautotrophicum*. Protein structure comparison with MTKH revealed a notable difference of amino acid sequences in the filter region, suggesting different ion selectivity.

CASTOR and POLLUX are predicted to possess chloroplast transit peptides at the N-terminus. We confirmed this predicted plastid localisation using fusions of full length CASTOR or POLLUX to green fluorescent protein (GFP). As a positive control, the transit peptide of the *Arabidopsis* plastid protein recA was fused with DsRed2 (AtrecA::DsRed2). When this fusion was transiently expressed in onion epidermis or pea root cells together with CASTOR::GFP or POLLUX::GFP, they co-localised, providing strong evidence for targeting of CASTOR and POLLUX to plastids.

Considering the functional predictions, it would appear that CASTOR and POLLUX control ion fluxes between plastids and cytosol, followed by reception of signal molecules derived from symbionts. Although the twins show high similarity, they cannot complement each other. Accordingly, there is a possibility that a multimetric ion channel consists of hetero-dimer of CASTOR and POLLUX in *Lotus japonicus*. The phenotypes of *castor* and *pollux* mutants have led to the hypothesis that the change of ion fluxes mediated by CASTOR and POLLUX is the key step for the activation of nodulation signalling pathway, including calcium spiking (Figure 1). Plastids have been given the status of the site for multiple biosynthetic functions. However, little is known about its role in endosymbioses. Thus, our results show a new aspect that plastid is prerequisite for the process of endosymbiont infection. Plastids represent the endosymbiotic remnant of a free-living cyanobacterial progenitor. It is fascinating that the 'ancient' symbiont helps 'newcomers' to infect.

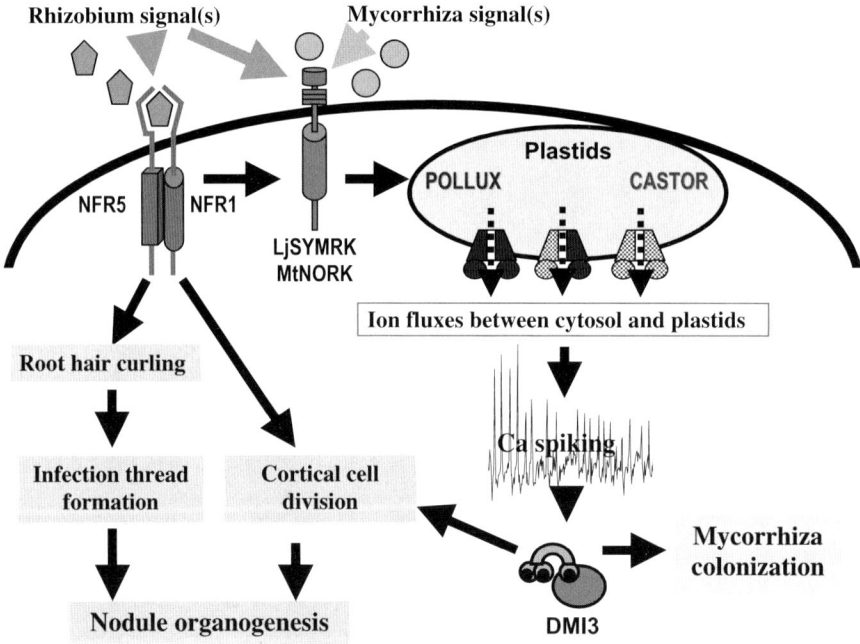

Figure 1. Working model for the functional role of CASTOR and POLLUX in perception of endosymbionts signals.

NFR1/NFR5 may be directly involved in perception of Nod factor. LjSYMRK and its ortholog MtNORK are required for perception of signal molecules, derived from rhizobia and mycorrhiza. DMI3 may act as an effector of Calcium spiking. Considering the functional predictions and mutant phenotypes, it would appear that CASTOR and POLLUX mediate ion-fluxes between plastids and cytosol which are a prerequisite for calcium-spiking.

References:
Imaizumi-Anraku, H. et al. Nature (*in press*)

THE EXPRESSION OF THE *RHIZOBIUM TROPICI GUAB* GENE IS REQUIRED IN THE EARLY STAGES OF BEAN NODULATION

Mónica Collavino[1], Daniel H. Grasso[2], Pablo M. Riccillo[1] and O. Mario Aguilar[1].
[1]IBBM-Facultad de Ciencias Exactas, UNLP. Calle 47 y 115, La Plata, Argentina; [2]Instituto de Suelos, INTA-Castelar, Argentina.

The *R. tropici guaB* mutant strain 899-10.T is unable to grow at high temperature and also to induce effective nodulation on common beans. Nodule cells are devoid of bacteria, blocked in the release from the infection thread. Complementation with the wild type *guaB* sequence restores both thermal tolerance and successful nodulation (Riccillo et al. 2002). In this work we investigated the role of *guaB* in the interaction of *R. tropici* and other host plants and in addition in other rhizobial species.

In order to investigate the temporal pattern of the *guaB* requirement during the *R. tropici*-common beans interaction, we constructed gene fusions between the promoterless *guaB* gene and the symbiotic promoters of *nodA*, *bacA* and *nifH*. The promoter sequences were PCR amplified and cloned upstream of *guaB*. At the 3´ end of the *guaB* sequence, we cloned the reporter gene *gusA*, respectively. Strains carrying these plasmids which were derived from the GuaB⁻ mutant strain *R. tropici* CIAT899 10.T, were used to individually inoculate common bean seedlings. It was found that *guaB* expression driven by the nod promoter allowed the mutant to form effective nodules in which nodule cells were found to be occupied similarly to what found in case of the wild type strain *R. tropici* CIAT899. Contrarily, the symbiotic phenotype of the mutant carrying *guaB* under the control of the *nifH* promoter rendered ineffective nodules which were found to be identical to the mutant GuaB⁻ strain. However, in case of the *bacA* promoter, only a few zones of the nodule were found to show cells with invading rhizobia.

Further analysis of the significance of *guaB* was approached by generating a *S. meliloti guaB* mutant by fragment specific mutagenesis. This *guaB* mutant unlike *R. tropici* was auxotrophic for guanine (Figure 1), although the key enzyme of the alternative via of guanine biosynthesis, the enzyme xanthine dehydrogenase, was demonstrated to be active in *S. meliloti* (Figure 2). The *guaB S. meliloti* induced effective on alfalfa plants.

We concluded that expression is required from early stages of the *R. tropici*-common beans interaction and this requirement is not a general phenomenon.

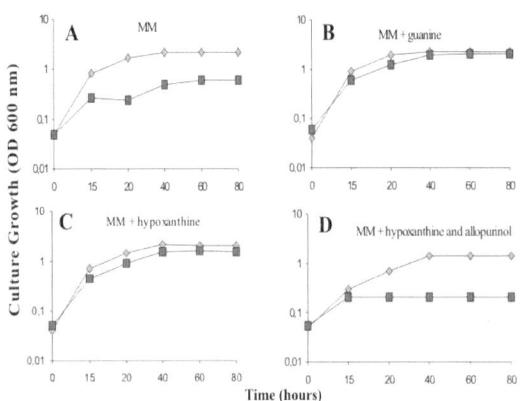

Figure 1. Growth of the *S. meliloti guaB* mutant. The mutant is defective in growth in minimal medium MM (A); growth is restored by guanine supplementation (B) and hypoxanthine, the precursor of xanthine dehydrogenase (C); inhibition by allopurinol (D)

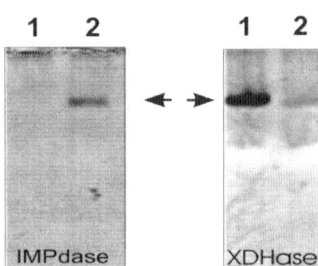

Figure2: Gel staining activity. Activity of inosin monophosphate dehydrogenase (IMPdase) in *S. meliloti* (left panel), and xanthine monophosphate dehydrogenase (XDHase) (right panel), in wild type (lane 2) and *guaB* mutant (lane 1).

Reference
Riccillo P *et al*. (2000) Mol. Plant Microb. Interact. 11, 1228-1236.

Acknowledgement
This work was financed by ANPC y T, Argentina (PICT No.7072/99).

DOES THE TYPE III SECRETION SYSTEM OF RHIZOBIUM SPECIES NGR234 AFFECT THE DEFENCE RESPONSE IN PLANT LEAVES?

O. Schumpp, W.J. Deakin, C. Staehelin and W.J. Broughton
LBMPS, Université de Genève, Sciences III, 30 Quai Ernest-Ansermet, Ch-1211 Switzerland.

A successful interaction between rhizobial cells and their corresponding host plant results in newly formed organs hosting actively nitrogen fixing bacteria. The bacteria signal their presence via Nod factors to initiate specialized plant developmental programs. Infection threads are formed and at the same time there is activation of cell division at the site of the future nodule. The whole process is highly regulated and more is now known to understand why plants initiate root hair deformation only in the presence of their compatible symbiont, more puzzling are the mechanisms contributing to bacteria persistence in plant tissues. Pathogenic bacteria with a biotrophic lifestyle develop strategies to subvert obvious and well characterized plant defences, while they multiply in the host. In rhizobial symbiosis, several reports mention plant responses that are normally associated with molecular defence mechanisms (Mithöfer 2002 for review). However, these showed that a plant defence response may occur in legumes undergoing rhizobial symbiosis. Whether it genuinely limits nodulation or rhizobia progression towards active nodule formation remains unknown.

Rhizobium species NGR234 has a large host-spectrum and is able to nodulate more than 100 plant genera. The symbiotic plasmid has been shown to encode an active type III secretion system (TTSS) which normally reveals pathogenic activity. Such a secretion system is thought to inject effector proteins – Nops, for Nodulation Outer Proteins in rhizobia - from the inside of the bacteria directly into the cytosol of the host cells. These proteins may then interfere with the host cell metabolism to down-regulate the defence response, acting as rhizobial "virulence" factors (Galan and Collmer, 1999). In some cases however, the Nops might also act as flags that enable the host to detect the invading bacteria, turning then into "avirulence" factors.

Accordingly, mutations in a central component of the TTSS of NGR234 that block Nops secretion, produce opposite responses depending on the host plant. One group of plants react negatively to the presence of the TTSS and produce more nodules when inoculated with the mutant strains, perhaps as if they could not detect the invading bacteria. A second group of plants react positively to the TTSS and produces fewer nodules when inoculated by the mutant, maybe because the Nops prevent the activation of plant defences. Some plants do not react to the mutation however, with the same number of nodules produced by the wild type and the TTSS mutants (Marie et al., 2001).

By analogy to virulence factor of phytopathogenic bacteria, Nops are very likely to act redundantly in the plant cell (Badel et al., 2003). However, in *Flemingia congesta*, NGRΩ*nopL* results in an intermediate phenotype in which less nodules are formed than with the wild type NGR234 bacteria but substantially more than when the plant is inoculated with the complete TTSS knock-out (Marie et al., 2003). NopL has also been shown to be specifically phosphorylated in plant cells via serine/threonine kinases. We propose that it might interfere with the MAP kinase-dependant signaling pathways to block defences. Plant defence reactions are far better characterized in the shoot. Therefore we investigated Nop activity in leaves. Tobacco plants overexpressing NGR234 NopL protein under the control of the 35S promoter showed a strong reduction in PR expression after virus challenge, together with a higher sensitivity to the virus infection. In *Lotus japonicus* too, NopL overexpressing plants resulted in a decrease in the basal PR level (Bartsev et al., 2004).

Surprisingly, rhizobia survival in *T. vogelii* leaves shows the opposite effect to nodulation phenotype. After syringe infiltration, the TTSS mutant survives better than the wild type NGR234 strain and forms more colony forming units on plate, after extraction from *T. vogelii* leaf tissues. Whereas in the root, *T. vogelii* reacts positively to the presence of the TTSS, forming more nodules in the wild type bacteria. Apparently, the NGR234 TTSS cannot thwart the plant response in the leaf tissue as efficiently as it does in the root and to the contrary, an active TTSS even contributes to decrease the bacterial population, probably because of a stronger plant reaction. It is likely that the Nops facilitate the recognition of the bacteria as an "alien" component that should be eliminated.

Badel JL et al. (2003) Mol Microbiol 49, 1239-1251.
Bartsev AV et al. (2004) Plant Physiol. 134, 871-879.
Galan JE and Collmer A (1999) Science 284, 1322-1328.
Marie C et al. (2001) Current Opinion in Plant Biology 4, 336-342.
Marie C et al. (2003) Mol Plant Microbe Interact 16, 743-51.
Mithofer A (2002) Trends in Plant Science 7, 440-444.

MATHEMATICAL MODELING FOR NODULE SIZE DISTRIBUTION

Jun-ichi Ikeda, Ayako Fukunaga, Yuko Suga, and Kaneaki Hori.
Ueno 200, Ueno-cho, Ayabe, Kyoto, 623-0035, Japan

Various sizes of nodules are formed on a soybean root. However, the distribution of the nodule size shows a well-organized pattern, a bell-shape curve. Exploring a mechanism underlying the nodule size distribution may enable to explain the pattern. In this study, to describe the nodule size distribution, mathematical equations were derived from a mechanistic model in which photosynthate allocation determined the nodule size after competition for photosynthates among nodules.

1. Development of the model

Some important assumptions, which the model is based on, are following. (1) Photosynthate are supplied from the shoot to the root system, and taken up by each root section according to their local concentration. (2) The photosynthate taken up is supplied to nodules through "a capillary", which is a minimum unit for the supply of photosynthate. (3) The nodules compete with each other for capillaries. The more capillaries the nodule is connected to, the more strongly the new capillary is attracted. (4) The rate of photosynthate supply determines the nodule size. The supply rate over a threshold is required for a nodule to develop into a visible size.

This model contains four input parameters, which represent the rate of photosynthate supply to the root system, distribution density of nodulation sites, a partition coefficient of photosynthate in the root system, and a threshold of photosynthate supply rate to grow up to a visible nodule.

2. Accuracy evaluation

The equation of the nodule size distribution was applied to the nodulation of soybeans (Glycine max L.Merr. cvs. Kitamusume and Toyosuzu) grown under different conditions and for various periods of time. The values of the parameters were determined by simultaneously solving the equation at 2 points on the nodule size distribution. Consequently, complete theoretical nodule size distribution was calculated and reproduced from the obtained values of the parameters. The theoretical distributions were compared with the actual ones to evaluate the accuracy of the equations. They showed a good consistency with the actual ones as did the nodule number and diameter.

3. Simulation

Simulation was carried out using these equations. In the result, the nodule number decreased with the increasing nodule distribution density over the optimum value. The nodule number and weight increased with the increase in the rate of photosynthate supply. The increase in the threshold value decreased both the nodule number and weight, but increased the average nodule weight.

4. Conclusion

The mathematical equations for describing the nodule size distribution were derived from a photosynthate allocation model. The model contributes to data compaction, because only 3 parameters are required to describe the nodule size distribution. The model also provided new physiological concepts, the capillary, which is the minimum unit of photosynthate supply, and their positive feedback competition. Moreover, the model gave the explanation to other phenomenon, over-inoculation, which reduces the nodule number.

References
Ikeda J (2003) Soil Sci. Plant Nutr. 49,805-815

EARLY EVENTS IN NODULATION OF *CASUARINA GLAUCA* BY *FRANKIA*

Laplaze L[1], Svistoonoff S[1], Obertello M[1], Peret B[1], Auguy F[1], Sy MO[1], Hocher V[1], Autran D[1], Nicole M[2], Franche C[1], Bogusz D[1]
[1]UMR 1098, and [2]UMR 1097, Institut de Recherche pour le Développement, 911 Avenue Agropolis, 34394 Montpellier Cedex 5, France.
http://www.mpl.ird.fr/rhizo

Two groups of plants can enter nitrogen-fixing root nodule symbioses with soil bacteria: legumes associate with rhizobia, whereas the so-called actinorhizal plants belonging to eight angiosperm families interact with the actinomycete *Frankia*. Recent molecular phylogeny studies (Soltis et al., 1995) indicate that those plants belong to a same clade. This suggests that a predisposition to form nitrogen-fixing root nodule symbioses originated once in the history of flowering plants. Our group is working on *Casuarina glauca/Frankia* interaction as a model system to study the molecular mechanisms of plant infection by *Frankia*.

1. *cg12* is specifically expressed during plant cell infection by *Frankia*

The *cg12* gene isolated from *C. glauca* encodes a susbtilisin-like protease, or subtilase, specifically expressed in plant cells infected by *Frankia* (Laplaze et al., 2000). Subtilases are a superfamily of proteases widely distributed in diverse organisms including bacteria, fungi and higher eucaryotes (Siezen and Leunissen, 1997). The precise function of plant subtilases is only known for two Arabidopsis genes, *Ale1* and *Sdd1*, which are involved in epidermal surface formation and stomatal distribution respectively (Tanaka et al., 2001; Berger and Altman, 2000). Different functions have been proposed for other plant subtilases including involvement in programmed cell death, cell wall loosening during lateral root development, fruit ripening and response to pathogen attacks. *cg12* is one of the earliest actinorhizal gene induced after *Frankia* infection. Using transgenic *Casuarinaceae* containing the *cg12* promoter fused to a reporter gene, we showed that (a) *cg12* expression is specifically linked to the infection of root hairs and cortical cells by *Frankia* and (b) is not induced in endo- or ectomycorrhizae or by *Frankia* root hair deforming factors (Svistoonoff et al., 2003). This suggests that CG12 plays an important role during plant cell infection by *Frankia*.

Infection-related activation of the *cg12* promoter is conserved between actinorhizal and legume-rhizobia root nodule symbiosis

Transcriptional fusions between the *cg12* promoter and *gus* or *gfp* reporter genes were introduced in the model legume *Medicago truncatula*. We showed that the *cg12* promoter directs reporter gene expression specifically in plant cells infected or about to be infected by *Sinorhizobium meliloti*, but not in mycorrhized roots (Svistoonoff et al., 2004). Using purified Nod factors and bacterial mutants, we also demonstrated that Nod factors are necessary but not sufficient for *cg12* promoter activation (Svistoonoff et al., 2004). This indicates that at least part of the transcriptional environment in plant cells infected by endosymbiotic nitrogen-fixing bacteria is conserved between legume and actinorhizal plants.

2. CG12 is localized at the interface between the plant cell and *Frankia*

In order to cytolocalize CG12 during root invasion by *Frankia*, antibodies were raised against two synthetic peptides corresponding to putative antigenic sites in the predicted protein. In western-blot experiments, a 95-kDa protein was recognized in nodules but not in uninfected roots or aerial part extracts. CG12 immunolabeling was found in infected cells in *Frankia* filaments, in the matrix surrounding *Frankia* infection-threads and vegetative hyphae. A strong labeling was also found in the cell wall of infected cells (Svistoonoff et al., unpublished).

Taken together, our results suggest that CG12 is a subtilase involved either in (a) cell wall remodeling to allow *Frankia* penetration or (b) signal exchanges between the plant cell and the bacteria.

3. Future prospects

cg12 promoter activation in response to bacterial infection is conserved in the model legume *M. truncatula*. We will use genetics in this model plant in order to identify genes involved in the signaling cascade leading to *cg12* expression. Moreover, we are currently setting up a hairy-root based RNAi system for *C. glauca*. This system will enable us to silence *cg12* expression and analyze the effects of the absence of this gene on the infection phenotype.

4. References

Laplaze L et al. (2000) Mol. Plant Microbe Interact. 13, 113-117.
Soltis DE et al. (1995) Proc. Natl. Acad. Sci. USA 92, 2647-2651.
Svistoonoff S et al. (2003) Mol. Plant Microbe Interact. 16, 600-607.
Svistoonoff S et al. (2004) Plant Physiol. 136, 3191-3197.
Siezen RJ and Leunissen JAM (1997) Protein Sci. 6, 501-523.
Tanaka H et al. (2001) Development 128, 4681-4689.
Berger D and Altmann T (2000) Genes Dev. 14, 1119-1131.

GENE EXPRESION IN THE ROOT NODULES OF *ELAEAGNUS UMBELLATA*

Chung-Sun An, Ho-Bang Kim, Sang-Ho Lee, Jung Jang-Hyun, Chang-Jae Oh, Hyoungseok Lee
School of Biological Sciences, Seoul National University, Korea

Over the last twenty years, molecular genetic and biochemical studies on symbiotic N_2 fixation have been focused on the legume-Rhizobium symbiosis and resulted in great advances in this area. However, the studies on the actinorhizal plants-*Frankia* symbiosis are limited, albeit they play crucial roles in the forest ecosystems. These two symbiotic systems show differences in terms of nodule origins and structures, and initial signaling pathways (Pawlowski and Bisseling, 1996).

Our research has been focused on the molecular genetic program of both host plants, *E. umbellata*, and its symbiont, *Frankia* EuIK1 strain, during the symbiotic interaction. To reveal the functional roles of host's genes, we have first isolated and characterized nodule-specific or –enhanced cDNA clones from the root nodule cDNA library of *E. umbellata* by using the hybridization-competition method (Table 1).

Table 1. The root nodule-specifc/enhanced cDNA clones isolated by the hybridization-competition method.

Gene Products	cDNA clones	Nodule vs. Root	Expression zones[a]
Chitinase	*Eu*NOD-CHT1	N > R	MZ
	*Eu*NOD-CHT2	N = R	IC (in FZ)
Polyubiquitin	*Eu*NOD-PUB1	N > R	MZ, IC (in FZ)
Asparagine synthetase	*Eu*NOD-AS1	N	IC (in FZ)
Glutamine synthetase	*Eu*NOD-GS1	N = R	N.D.
S-adenosyl-L-methionine synthetase	*Eu*NOD-SAMS1	N = R	MZ, IC (in FZ)
	*Eu*NOD-SAMS2	N > R	PZ
Chalcone isomerase	*Eu*NOD-CHI1	N > R	N.D.
Auxin-repressed mRNA	*Eu*NOD-AxRm1	N > R	N.D.
Metallothionein	*Eu*NOD-MT1	N > R	N.D.

[a]MZ, meristem zone; FZ, fixation zone; PZ, prefixation zone; IC, infected cells; N.D. not determined

Although two different cDNA clones encoding chitinase, *EuNOD-CHT1* and *EuNOD-CHT2*, showed similar structure of translation products, in situ hybridization showed that *EuNOD-CHT1* and *EuNOD-CHT2* transcripts were detected respectively in the meristem zone and the infected cells of the fixation zone. Plant chitinases have been known as pathogenesis-related (PR) proteins, but recent studies suggest that they might play functional roles during normal plant growth and development (Zhong et al., 2002). To establish the function of two chitinase clones, we constitutively expressed the chitinase genes in *Arabidopsis thaliana*. The transgenic plants expressing either *EuNOD-CHT1* or *EuNOD-CHT2* did not show noticeable morphological changes compared to wild type. However, the transgenic plants showed remarkable resistance against an infecting fungal pathogen, *Botrytis cinerea*, but not against a bacterial pathogen, *Pseudomonas syringae* pv. tomato DC3000. Expression level of *EuNOD-CHT1* was also increased by wounding and jasmonic acid and that of *EuNOD-CH2* was increased only by jasmonic acid. These results suggest that the induced chitinase expression in the root nodules might be attributable to the host plant's defense mechanism against infecting *Frankia*.

To identify genes involved in the symbiotic N_2-fixation in *Frankia* EuIK1 strain, we generated a genomic library into a cosmid vector PWE15. By using heterologous probes and systematic chromosome walking, we determined gene organization of the genomic region containing *nif* gene cluster *nifH, D, K, E. N, X, orf3, orf1, W, Z, B. U, orf2, S, 2-oxoacid ferredoxin oxidoreductase a* and *b*, and *ferredoxin I*. We also isolated *nifV* encoding homocitrate synthase from the strain. In other N_2-fixing bacteria including *Frankia* sp. FaCl and *Frankia alni*, *nifV* gene is clustered with other *nif* genes. However, *nifV* of EuIK1 strain was not found within the *nif* cluster, *nifV* from EuIK1 strain was highly expressed in the root nodules and functionally complemented *nifV* mutant of *Klebsiella pneumoniae* (Oh et al., 2003).

References
Kim HB and An CS (2002) Mol. Plant-Microbe Interact. 16, 209-215.
Oh et al. (2003) Mol. Cells 15, 27-33.
Pawlowski and Bisseling (1996) Plant Cell 8, 1899-1913.
Zhong RQ et al. (2002) Plant Cell 14, 165-179.

Acknowledgments: This work was supported in part by the Korean Ministry of Science and Technology (21C Frontier Microbial Genomics and Applications Program; Project No. MG02-0201-001-2-3-0) to C.S. An.

DIFFUSIBLE SIGNAL FACTORS IN NODULATION OF *DISCARIA TRINERVIS* BY *FRANKIA*

Luciano A. Gabbarini and Luis G. Wall
Biological Interactions Research Program, University of Quilmes, Bernal, Argentina, R. Sáenz Peña 180, B1876BXD Bernal, Argentina, lgwall@unq.edu.ar

The actinorhizal symbiosis between *Discaria trinervis* and *Frankia* is an example of a root intercellular invasion pathway leading to infection and nodulation (Valverde and Wall 1999a). The plant controls nodulation by an autoregulatory mechanism (Valverde and Wall 1999b) that is modulated by N (Valverde et al 2000) and P (Valverde et al 2002). Actinorhizal symbiosis shows specificity between plants and *Frankia* although the bases of specificity are not well understood. A major interest is to discover signals involved in actinorhizal symbiosis since already known signals described in *Rhizobium*-legumes symbioses seems not to operate in it. Some root hair deformation factors (van Ghelue *et al.,* 1997; Cérémonie *et al.,* 1999) and flavonoids (Benoit and Berry, 1997) have been described although the picture with them is not clear enough. We played with two-actinorhizal symbiosis from different cross inoculation groups and with different infection pathways as *Frankia* ArI3 - *Alnus acuminata* (intracellular infection) and *Frankia* BCU110501 - *Discaria trinervis* (intercellular infection). We assayed different experimental designs in order to look for the possibility of a physiological complementation that could override recognition limitation between partners. In those experiments we did not succeed in getting nodulation between symbionts from different cross inoculation groups, although there may be some by pass of the feedback regulation on root hair deformation regardless nodulation of the tap root (Gabbarini and Wall 2001). Nevertheless, those experiments gave us the clues for the existence of diffusible signal factors (DSF) in early interaction between *Frankia* and *Discaria trinervis*. When *Discaria trinervis* nodulation by *Frankia* BCU110501 is analysed in a dose-response experiment, both the nodulation rate and nodulation profile in the tap root appear to be modified by the concentration of bacteria in the inoculum.

In order to look for putative *Frankia* DSFs, *Discaria trinervis* plants grown in pouches were co-inoculated with BCU110501 and other bacteria. Otherwise, either the *Frankia* inoculum or the root was pre-treated with a bacteria dialyzates as a possible source of DSF. After the treatment, the nodulation rate and nodulation profile of *Discaria trinervis* taproots inoculated with *Frankia* isolate BCU110501 were analysed. Nodulation rate was positively modified by co-inoculation with non-infective *Frankia* as ArI3, CpI1, Cj82, or *Streptomyces colecoide*, but was not affected by co-inoculation with other rhizospheric bacteria, as *Azospirillum*, *Pseudomonas* or *Bacillus*. Those bacteria alone produced no effect on development of *Discaria trinervis*. This enhancement of nodulation rate could also be by get pre-treating a diluted BCU110501 inoculum with high concentration of *Frankia* or *Streptomyces* within dialysis bags, but not with γ-ray killed *Frankia*. The cell-to-cell activation for nodulation of *Frankia* BCU110501 increased from 0h to 96h of pre-treatment of the diluted inoculum with the dialyzates of a concentrated suspension. These results suggest that a *Frankia* DSF is common to actinomycetes which can modify the infectivity of *Frankia* for nodulation.

The nodulation profile on the tap root shifted basipetaly depending on the concentrations of *Frankia* BCU110501 dialyzates, or the concentration of *Frankia* BCU110501 culture supernatants, or by pretreatment of the *Discaria trinervis* root with a dialyzate of *Frankia* BCU110501 . The shift of the nodulation profile was not induced by co-inoculation with *Frankia* ArI3. The results suggest that a *Frankia* BCU110501 DSF could specifically modify the transient susceptibility for nodulation of *D. trinervis*.

Alnus acuminata and *Discaria trinervis* were grown in the same pouches so that roots were superimposed and root exudates were shared between plants. Rates of nodulation, nodule distribution and plant growth of each symbiosis we measured after the inoculation with low concentration of either *Frankia* BCU110501, ArI3, or both bacteria. The presence of only the heterologous root or the heterologous bacteria in equal concentration modified neither nodulation kinetics nor nodule distribution in each actinorhizal pair. Nevertheless, the co-existence of both symbiotic pairs in the same pouch significantly enhanced nodulation rate and modify nodule distribution in both plants. These results support the existence of some common signaling at early interactions between different, non-cross infective, actinorhizal symbioses.

Supported by UNQ, CONICET and ANPCyT (Argentina)

Benoit, L. F., Berry, A. M. (1997). *Physiol. Plant.*, *99*: 588-593.
Cérémonie, H. *et al* (1999). *Can. J. Bot.*, *77*, 1293-1301.
Valverde, C., and Wall, L. G. (1999a). *New Phytol.*, *141*, 345-354.
Valverde, C., and Wall, L. G. (1999b). *Can. J. Bot.*, *77*, 1302-1310.
Valverde, C.,*et al* (2002). *New Phytol.*, *153*, 43-52
Valverde, C. *et al* (2000). *Symbiosis*, *28*, 49-62.
van Ghelue, M. *et al* (1997). *Physiol. Plant.*, *99*, 594-600.

CLONING AND EXPRESSION STUDIES OF *FUR*F AND *CAP* F IN *FRANKIA* STRAIN R43.

João Vieira[1], Pedro Moradas-Ferreira[1,2] and Fernando Tavares[1,3]
[1] Instituto de Biologia Molecular e Celular, Microbiologia Celular e Aplicada, Universidade do Porto, Rua do Campo Alegre, 823, 4150-180 Porto, Portugal; [2] Instituto de Ciências Biomédicas Abel Salazar, Universidade do Porto, Portugal; [3] Faculdade de Ciências, Departamento de Botânica, Universidade do Porto, Portugal. E-mail: ftavares@ibmc.up.pt

Despite the fact that the main object of nitrogen fixation studies has been the rhizobia-legume symbioses, increasing attention during the eighties was devoted to the symbiotic nitrogen-fixing root nodules (actinorhizas) induced by the actinobacteria *Frankia* in a wide range of dicotyledonous species (actinorhizal plants).

The process leading to the nitrogen-fixing symbioses has often been compared at the cellular level to phytopathogenic infections. At present, H_2O_2 in phytopathogenic infections is believed to play a crucial, as a cytotoxic molecule against the invader and also as a key regulatory molecule of the plant hypersensitive response. Furthermore, it is believe that a similar mechanism may occur in nitrogen-fixing symbiotic interactions. The elicitation of a hypersensitive reaction involved in the autoregulation of nodulation in *Sinorhizobium meliloti*-alfalfa symbiosis has been supported by several studies. More recently, the role of hydroperoxidases in rhizobia-legume symbioses has been probed by studying the phenotype of *S. meliloti* double mutants affecting two monofunctional catalases (*katA* and *katC*) and a bifunctional catalase-peroxidase (*katB*) (Jamet et al. 2003, Sigaud et al. 1999). These mutants showed a reduced nodulation capacity and a significant decrease in the nitrogen-fixing capacity, indicating the involvement of H_2O_2 scavenging enzymes in the nodulation and nitrogen fixation processes. Concomitantly with the conceivable burst of oxygen radicals by the host plant during the actinorhizal infection, the high-energy demand and the strong reducing conditions required for the nitrogen-fixation process, point towards the existence of an efficient enzymatic removal of oxygen free radicals by symbiotic *Frankia*. Recently two hydroperoxidases were characterized for *Frankia* strain R43: a monofunctional catalase and a bifunctional catalase-peroxidase (Tavares et al. 2003). These enzymes were shown to be cytoplasmatic, growth regulated and expressed mainly during the stationary growth phase. Studies revealed that both enzymes were induced by H_2O_2 and paraquat.

A proteomic survey using 2-D electrophoresis was carried out to identify differentially expressed proteins in *Frankia* cells stressed with H_2O_2 or paraquat. Three proteins were shown to be upregulated: one by treatment with H_2O_2, another by paraquat, and a third by both H_2O_2 and paraquat. The two peptides whose expression increased expressed in presence of paraquat were excised, digested and analyzed by matrix-assisted laser desorption ionization time-of-flight mass spectrometry (MALDI-TOF). However the microsequencing results from each spot did not reveal a homology to any specific protein in databases. The catalase-peroxidase is upregulated under oxidative stress conditions. Then, to evaluate its role in *Frankia*, we proceeded with the cloning of the gene plus the flanking regions. Two genes were identified: the gene coding for *Frankia* R43 catalase-peroxidase designated as *cap*F and upstream a gene coding for a ferric uptake regulator, designated by *fur*F. The gene organization found is identical to what has been described for other actinobacteria. The deduced amino acid sequence of CapF, consisting of 746 amino acids showed 71% of identity to *Streptomyces reticuli* and to *S. coelicolor* A3(2) and 66% identity to *Mycobacterium tuberculosis* and *M. bovis*. FurF, consisting of 145 amino acids, showed the highest sequence identity of 75% to *Mycobacterium tuberculosis* and 66% identity to *Streptomyces reticuli*. Interestingly, Fur is an increasingly recognized transcription factor that generally autoregulates its own expression, function as activator or repressor mediating the expression of a wide range of genes, binds to DNA in the Fur box, and more importantly it has been shown to mediate the regulation of catalase-peroxidase in several actinobacteria. By blast analyses and using a multiple alignment strategy was possible to identify two putative Fur boxes slightly overlapping, one with high similarity to *Streptomyces*, the other more similar to the mycobacteria Fur box. Due to the putative role of FurF as transcriptional regulator of *cap*F in *Frankia* R43, heterologous expression of *fur*F was carried out in *E. coli*. The purified FurF is now being used in mobility bind shift assays to identify the promoter region where FurF may bind and act as regulator of *cap*F.

Expression studies of *cap*F and *fur*F under different oxidative stress conditions by RT-PCR suggested that these genes are transcribed as monocystronic. *cap*F was shown to be constitutively expressed and upregulated under oxidative stress conditions, corroborating the previous studies of activity, on the contrary *fur*F was observed to be expressed only when the cells were challenged with paraquat.

In the near future we expect to clarify the importance of CapF and FurF proteins during the actinorhizal infection, namely how they are regulated, both in free-living *Frankia* cultured under different oxidative stress conditions and in symbiotic *Frankia*, i.e. in the infected root-nodule plant cells by *in situ* studies.

Jamet A et al. (2003) Mol. Plant-Microbe Interact. 16, 217-225.
Sigaud S et al. (1999) J. Bacteriol. 181, 2634-2639.
Tavares F et al. (2003) Plant Soil 254, 75-81.
This work was supported by FCT, Portugal (POCTI/35283/BCI/2000).

CHARACTERIZATION OF MUTANTS FROM *RHIZOBIUM* SP. NGR234 WITH DEFECTIVE EXOPOLYSACCHARIDE SYNTHESIS

C. Staehelin[1], W. D'Haeze[2], M.Y. Gao[2], R.W. Carlson[2], B.J. Pellock[3], G.C. Walker[3], W.R. Streit[4] and W.J. Broughton[1]
[1]LBMPS, Univ. de Genève, Sciences III, CH-1211 Genève 4, Switzerland.
[2]Complex Carbohydrate Research Center, Univ. Georgia, Athens, GA 30602-4712, USA [3]Biology Dept., Massachusetts Inst. Technology, Cambridge, MA 02139, USA [4]Molekulare Enzymtechnologie, Univ. Duisburg-Essen, D-47057 Duisburg, Germany

Rhizobium infection and nodule development are the result of a complex communication between the two partners. Specific surface polysaccharides of rhizobia are crucial for nodule formation on certain host plants. Rhizobial surface polysaccharides include exopolysaccharides (EPS), lipopolysaccharides (LPS), K-antigens and cyclic β-glucans. *Rhizobium* sp. NGR234 has a broad host range and is able to establish symbiosis with more than 112 genera of legumes. An *exo* gene cluster of NGR234 has been identified (Chen et al. 1988) and sequenced recently (Streit et al. 2004). This cluster is localized on the pNGR234*b* megaplasmid and is similar to the *exo* gene cluster of *Sinorhizobium meliloti* 1021. The most striking differences between these two clusters were seen in the region between *exoK* and *exoU*. The *exo* cluster of pNGR234*b* lacks ORFs corresponding to *exoH* and *exoTWV*, suggesting that two deletions occurred. It has been reported that the repeating subunit of the NGR234 exopolysaccharide is a nonasaccharide that consist of a main chain [Gal-(Glu)$_5$] and a side chain [(GlcA)$_2$-Gal] decorated by acetyl groups and a pyruvyl group at the nonreducing terminus (Djordjevic et al. 1986). Based on the known function of *exo* genes in *S. meliloti* (Reuber and Walker, 1993), we suggest that ExoY, ExoF, and ExoL of NGR234 are enzymes required for synthesis of the lipid-linked precursor [Gal-(Glu)$_2$]. ExoP and ExoQ are probably involved in export and polymerization of the repeating subunit, ExoZ is a putative acetyl transferase and ExoK a putative glycanase.

Mutation of various genes on the *exo* cluster (*exoP, L, Y, F* and *Q*) resulted in a set of NGR234 mutants that formed "dry" colonies on agar plates. All these *exo* mutants exhibited significantly lower amounts of carbohydrates in the supernatant of liquid cultures. Glycosyl linkages and composition analyses indicated that the mutants did not produce any detectable amounts of EPS. On *Leucaena leucocephala* plants, the mutants induced nodule-like structures (pseudonodules), whereas nitrogen-fixing nodules were formed with the parent strain NGR234. Nodulation tests with various other legumes did not show differences between NGR234 and these *exo* mutants.

The *exoZ* mutant exhibited a "mucoid" phenotype on agar plates, similar to the parent strain NGR234. Chemical analyses showed that the *exoZ* mutant produced significant amounts of EPS. NMR analysis indicated that EPS from NGR234 has a repeating unit with multiple (possibly three) O-acetylation sites, while EPS from the *exoZ* mutant contains two O-acetylation sites. These data suggest that ExoZ is an acetyl transferase involved in EPS synthesis and that another acetyl transferase must exist outside the sequenced *exo* gene cluster. On *L. leucocephala,* the *exoZ* mutant induced nitrogen-fixing nodules, indicating that the function of ExoZ is not required for establishment of symbiosis.

In *S. meliloti*, ExoK and ExsH are extracellular glycanases that release symbiotically active low molecular weight (LMW) forms (DP 1-3) from high molecular weight (HMW) EPS (York et al. 1998). NGR234 also produces LMW EPS as reported by Djordjevic et al. (1987). Our studies indicated that NGR234 synthesizes LMW EPS at very low rates (at least 100 fold less than HMW EPS). The LMW EPS was purified by fractional precipitation with ethanol, anion exchange chromatography and gel filtration chromatography. Reduction with borohydride followed by hydrolysis and analysis of monosaccharides and their alditols indicated that LMW EPS of NGR234 consists of the repeating subunit (DP 1). Molecular weight determination by MALDI-TOF analysis confirmed these data.

The *exoK* mutant of NGR234 formed "mucoid" colonies on agar plates and produced high amounts of carbohydrates in culture supernatants. We are currently analyzing whether the *exoK* mutant produces HMW EPS or another polysaccharide at elevated levels. LMW EPS could not be isolated from the *exoK* mutant. Molecular weight determination by MALDI-TOF analysis confirmed the lack of LMW EPS in the corresponding fractions after gel filtration chromatography. The *exoK* mutant induced only nodule-like structures on *L. leucocephala*, suggesting that LMW EPS of NGR234 is symbiotically active on this host plant. One possibility is that the nonasaccharide acts as a symbiotic signal, perhaps by binding to a plant oligosaccharide receptor which triggers symbiosis-related processes.

References

Chen H et al. (1988) Mol. Gen. Genet. 212, 310-316.
Djordjevic SP and Rolfe BG. (1986) Carbohydr. Res. 148, 87-99.
Djordjevic SP et al. (1987) J. Bacteriol. 169, 53– 60.
Reuber TL and Walker GC (1993) Cell 74, 269-280.
Streit WR et al. (2004) J. Bacteriol. 186, 535-542.
York GM and Walker GC (1998) Proc. Natl. Acad. Sci. USA 95, 4912-4917.

ISOLATION OF TWO NEW *SINORHIZOBIUM MELILOTI* TRANSCRIPTIONAL REGULATORS REQUIRED FOR NODULATION

Li Luo[2], Shi-Yi Yao[1], Anke Becker[3,4], Silvia Rüberg[4], Guan-Qiao Yu[2], Jia-Bi Zhu[2], and Hai-Ping Cheng[1]
[1]Biological Sciences Dept., Lehman College, The City Univ. New York, Bronx, New York.10468. USA. [2]Lab of Molecular Genetics, Shanghai Inst. Plant Physiology & Ecology, CAS, Shanghai 200032, P. R. China. [3]Inst. für Genomforschung, Univ. Bielefeld, Postfach 100131, 33501 Bielefeld, Germany. [4]Lehrstuhl fur Genetik, Fakultat fur Biologie, Univ. Bielefeld, Postfach 100131, 33501 Bielefeld, Germany

In the absence of fixed nitrogen, *Sinorhizobium meliloti* establishes an effective nitrogen fixing symbiosis with its legume partner, alfalfa (*Medicago sativa*) through a series of intricate signal exchanges and reciprocal structural changes. The signal exchanges between the two organisms starts with the alfalfa flavonoids and the *S. meliloti* nodulation factors. This exchange of signals results in the formation of tightly curled alfalfa root hairs. *S. meliloti* cells elicit the formation of infection threads inside these curled root hairs. The infection threads elongate toward the base of root hairs, penetrating layers of alfalfa root cells, reaching developing nodule primordium, and releasing live bacterial cells into the plant cells inside the nodule primordium. The successful differentiation of the *S. meliloti* cells may depend on another set of signal exchanges between *S. meliloti* and alfalfa involving proteins like BacA. The complexity of the *S. meliloti*-alfalfa interaction raises the possibility that many more signal exchanges between the two organisms remain unknown.

Some of the key regulatory genes involved in the signal exchanges between *S. meliloti* and alfalfa belong to the family of LysR transcriptional regulators. The production of nodulation factor that is part of the signal exchanges that initiate symbiosis, is positively controlled by three different *nodD* genes: *nodD1*, *nodD2* and *nodD3* and the *syrM* gene. The nod factor production is also negatively regulated by the *nolR* gene (9). The *nolR* gene encodes a repressor that negatively regulates the expression of the *nodD1* and *nodABC* operons by binding to the divergent promoter between them.

The genomic sequence has been determined for *S. meliloti* Rm1021 and 90 putative *lysR* family transcriptional regulators have been identified. Each of 90 previously unknown putative *S. meliloti lysR* genes was interrupted by insertions of suicide plasmids into the middle of the open reading frames.

Two of the mutants, *lsrA1* and *lsrB1* (LysR-type Symbiosis Regulator), consistently failed to establish an efficient nitrogen fixation symbiosis with alfalfa, after all 90 mutants were examined. A series of experiments were performed to confirm that the insertion of the suicide plasmid into both *lsrA* and *lsrB* genes were stably maintained by mutant cells, and that the disruption of either one of the two genes prevented *S. meliloti* cells from establishing efficient nitrogen symbiosis with their host plant alfalfa. All findings together suggest that both *lsrA* and *lsrB* genes are involved in the *S. meliloti*-alfalfa symbiosis.

Both *lsrA* and *lsrB* genes were found to be expressed in the free-living wild type strain, but not in either *lsrA1* or *lsrB1* mutants. The loss of the *lsrA* gene didn't change the growth rate the bacterial cells in minimal media or impair cell motility, while the loss of the *lsrB* gene did slow down cell growth and cell motility. These findings suggest that neither LsrA nor LsrB is required for cell growth in the free-living state.

The symbiotic deficient phenotype of the *lsrA1* mutant is relatively more clear-cut. Alfalfa plants inoculated with this mutant were short and yellow with only white nodules. These plants did not have detectable levels of the nitrogenase activity, which was consistent with the white color of the nodules. Since live bacterial cells can be recovered from these white nodules, the *lsrA1* mutant is most likely capable of invading nodules, but is blocked at some point before the induction of leghemoglobin biosynthesis. These findings suggest that LsrA might regulate the expressions of the genes required for nitrogen fixation. To better understand the role of LsrA in symbiosis, studies are being carried out to determine the precise step where the *lsrA1* mutant is blocked during the establishment of an effective nitrogen fixation symbiosis.

Compared to that of the *lsrA1* mutant, the symbiotic phenotype of the *lsrB1* mutant is a bit more complicated. The height of the plant, the color of the leaves, and the percentage of pink nodules indicated that *lsrB1* mutant did not form an efficient nitrogen fixation symbiosis with the plants. These findings, however, have contradicted the findings of the nitrogenase assay, which suggested the overall nitrogenase activities of the roots from plants inoculated with *lsrB1* mutant were even higher than that of the roots from the plants inoculated with the wild type strain. One of the possible explanations is that the symbiosis is also blocked at some point before the fixed nitrogen could reach plant cells. All together, these findings seem to suggest that the symbiosis between the *lsrB1* mutant and alfalfa might be blocked at multiple points.

ACKNOWLEDGEMENTS
This work was supported by grants from NIH (5S06GM08225) and PSC-CUNY (617320030 & 632140032) to H-P.C.; National grants from Bundesministerium für Forschung und Technologie, Germany (0311752 & 031U213D) and the "Bioinformatik Initiative" by Deutsche Forschungsgemeinschaft to A. B. and S. R.; Grants from The National High Technology (863) International Research Program (2001AA214211) and National Science Foundation of China (30170512) to G-Q. Y. and J-B. Z.

RHIZOBIAL CONTROL OF HOST-SPECIFICITY

N.M. Boukli[1], W.J. Deakin[1], K. Kambara[1], H. Kobayashi[1], C. Marie[1], X. Perret[1], A. Le Quéré[1], B. Reuhs[2], M. Saad[1], O. Schumpp[1], P. Skorpil[1], C. Staehelin[1], W. Streit[3] and W.J. Broughton[1].
[1]LBMPS, Univ. de Genève, 30 quai Ernest-Ansermet, 1211 Genève 4, Switzerland. [2]Whistler Center for Carbohydrate Research, Dept. Food Science, Purdue Univ., 1160 Food Science Building, West Lafayette, IN 47907-1160 USA. [3]Inst. für Mikrobiologie und Genetik der Univ., Göttingen, Germany.

Coordination of nodule development requires the exchange of many signals between rhizobia and legumes. Amongst the first signals exchanged are phenolic compounds, mainly flavonoids that are secreted by roots into the rhizosphere. Rhizobial proteins called NodD, function first as environmental sensors of these phenolics, and later as transcriptional activators of a series of rhizobial genes that encode proteins responsible for the production of rhizobial signals. NodD proteins belong to the LysR family of transcriptional regulators, and they have the ability to bind to specific, highly conserved DNA sequences (*nod*-boxes) that are present in the promoter regions of many nodulation genes/loci. *Rhizobium* sp. NGR234 contains three replicons, the smallest of which (pNGR234*a*) carries most symbiotic genes, along with two copies of *nodD* as well as two copies of *syrM*, another LysR-type transcriptional activator. As mutation of *nodD1* (but not *nodD2*, or *syrM1* and *syrM2*) leads to a Nod⁻ phenotype on all plants tested, it is clearly the primary mediator in flavonoid-*nod*-box controlled loci. The number of *nod*-boxes varies in different rhizobia, but there are 19 on pNGR234*a* that help transcriptional regulations of genes involved in synthesis of a range of different signalling compounds (Kobayashi et al. 2004). The first rhizobial signal molecules to be synthesized and secreted are encoded by the nodulation genes (*nod*, *noe*, and *nol*), which are responsible for the synthesis of Nod-factors. Nod-factors provoke deformation and curling of root-hairs and allow rhizobia to enter roots through infection-threads. Infection-thread development requires other sets of rhizobial signals including surface polysaccharides as well as secreted proteins. Surface polysaccharides (SPS) include extra-cellular polysaccharides (EPS), capsular polysaccharides (CPS), lipo-polysaccharides (LPS) as well as cyclic β-glucans. Most of these polysaccharides function during infection-thread development, where they possibly help suppressing plant defence reactions. SPSs have highly diverse structures and they may contribute to rhizobial host-range (for reviews see Broughton et al. 2000 and Perret et al. 2000). Perhaps surprisingly, the flavonoid-NodD1-*nod*-box regulatory system also modulates EPS synthesis and protein secretion. For example, *nod*-boxes NB5 to NB7 delimit a mosaic of genes including some involved in the synthesis of K- and O-antigens. Mounting evidence suggests that these antigens are essential for continuing infection-thread development and bacteroid release. As another example, in conjunction with *ttsI*, a two-component regulator, the flavonoid-NodD1-*nod*-box regulatory system also modulates activity of a symbiotic type-three protein secretion-system (T3SS)(Marie et al., 2004). Mutation of the T3SS in NGR234 causes three host-dependent effects. As compared to the wild-type rhizobia, these mutants can have dramatic affects on nodule development. On *Tephrosia vogelii*, for example, many non-fixing pseudo-nodules are formed. A dramatic enhancement of nodulation is often seen by inoculating these mutants onto *Crotalaria juncea* and *Pachyrhizus tuberosus*, while a third group of legumes seems to be unaffected by the presence/absence of a functional T3SS (e.g. *Lotus japonicus* and *Vigna unguiculata*). Rhizobial nodulation outer proteins (Nops) can be categorised into three groups - external components of the secretion machine, translocators, and effectors. NopX, like HrpF, has two predicted transmembrane domains at the C-terminus, which may anchor it into cell membranes (such as the plant plasma-membrane), and thus help forming the pore through which the pilus enters. Fibrillar surface structures are produced in a flavonoid inducible and NopA/NopB-dependent manner. Mutants in both *nopA* and *nopB* block all protein secretion and consequently permit nodulation of *Pachyrhizus tuberosus* while blocking the interaction with *Tephrosia vogelii*. Microscopic techniques using isolated flavonoid-inducible pili, and *in situ* immuno-gold labelling showed that they contain both NopA and NopB (W.J. Deakin and M. Saad, unpublished). It thus seems that the T3SS pilus is formed of NopA and NopB. Ectopic expression of *nopL* in *Nicotiana tabacum* thwarts the induction of pathogen-related (PR) defence proteins. Similar experiments involving NopP suggest that it too, is a secreted protein that is involved in modulating signal transduction pathways in legumes (Perret et al, 2004; P. Skorpil and W.J. Deakin, unpublished). Thus, NopX is probably an external component of the secretion machine; NopA and NopB are translocators, while NopL and NopP are putative effectors.

References:
Ausmees N et al. (2004) J. Bacteriol. 186, 4774-4780.
Bartsev AV et al. (2004) Plant Physiol. 134, 871-879.
Broughton WJ et al. (2000) J. Bacteriol. 182, 5641-5652.
Kobayashi H et al. (2004). Mol. Microbiol. 51, 335-347.
Marie C et al. (2004) Mol. Plant- Microbe. Interact. 17, 958-966.
Perret XC et al. (2000) Microbiol. Mol. Biol. Rev. 64,180-201.

THE *SINORHIZOBIUM MELILOTI* NIFA EXERTS ITS SIGNALING EFFECT FOR NODULATION IN THE HOST PLANTS

Huasong Zou, Guanqiao Yu and Jiabi Zhu
National Laboratory of Plant Molecular Genetics, Institute of Plant Physiology & Ecology, Shanghai Institutes of Biological Sciences, Chinese Academy of Sciences, Shanghai, China

Previous investigations (Ditta et al.; Vasse et al.) have shown that *nifA* gene is involved in nodulation and symbiotic nitrogen fixation regulation of Rhizobia. Our present investigation is focused on the role of *nifA* on nodulation of leguminous plants.

The result of split-root experiments showed that the wild-type strain *Sinorhizobium meliloti* Rm1021 inoculated one of the lateral roots of the *Medicago sativa* suppressed the nodulation of the other lateral root. The suppression rate was up to 68.3%. In contrast, when the *nifA* mutant strain *S. meliloti nifA*⁻ was used to inoculate one of the lateral roots, the suppression rate was only 6.5%. It indicates that nodulation is feedback controlled by *nifA* and that *nifA* may elicit a systemic signal, which control nodulation on the other lateral root (table 1).

Table 1. Nodulation suppression rate of *nif*A gene

Strain inoculated	Nodules induced on second lateral root	Suppression rate (%)
None inoculation	6.4±0.8	0.0
Sm1021	2.4±0.4	64.0
nifA⁻	5.8±1.5	12.5
nifH⁻	2.8±0.9	56.3

Microscope examination of plant roots infected with *S. meliloti* reveals the occurrence of necrotic cells on root cortical close to the end of an infection thread. Quantitative experiments were carried out using plants infected by Rm1021 and *nifA*⁻ mutant respectively. The results showed that the number of necrotic cells was related to *nifA* copy number in inoculated rhizobia.

For assay of the auto-fluorescent substances accumulated in necrotic cells we

successfully extracted the phenolic compounds from the roots infected by Rm1021 and
nifA⁻ mutant respectively and observed that these phenolic compounds showed the
characteristic of phytoalexin. Using HPLC analysis we found that the amount of
phytoalexin-like substances present in the roots cells infected by Rm1021 was higher
than that of in the roots cells infected by *nifA*⁻ (Fig. 1).

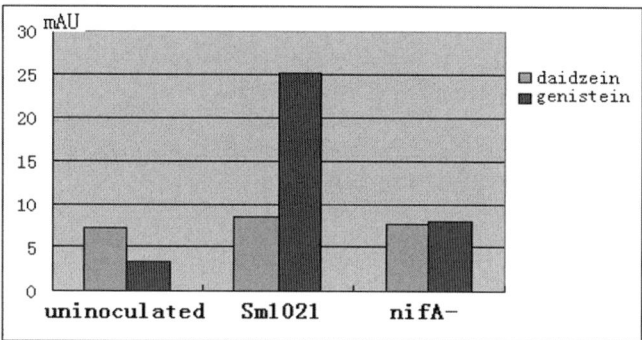

Fig. 1 Daidzein and genistein production in the root inoculated with Rm1021 and *nifA*⁻

In order to know the relationship of necrotic cells formation and NifA activity, we
adapted the semi-quantitative root hair deformation assay to study the activity of Nod
factors secreted by Rm1021 and *nifA*⁻ respectively. It was found that the activities of the
nodule factors were strengthened with the increase of the copies of *nif*A in the rhizobia.

All these results indicate that NifA may elicit a systemic signal which regulates the
nodule formation. The *nif*A gene causes HR in the host plant that leads to necrosis of
infected root-hair cells, thus, terminating the nodule formation. Presumably the signal
elicited by NifA is transmitted through the nodule factors.

References
Ditta G et al. (1987) J. Bacteriol. 169, 3217-3223.
Vasse J et al. (1993) Plant J. 4, 555-566.

EVOLUTIONARY STABILITY OF *RHIZOBIUM* MUTUALISM DEPENDS ON LEGUME HOST SANCTIONS

R. Ford Denison and E. Toby Kiers
Department of Agronomy and Range Science, University of California
1 Shields Ave., Davis, CA 95616, USA

How does fixing N_2 in symbiosis benefit rhizobia? Prior to 2000, there were few explicit attempts to explore this question (Jimenez and Casadesus, 1989; Udvardi and Kahn, 1993; Sprent et al., 1993). Investing carbon (C) substrates in fixing nitrogen (N) for their legume hosts, rather than using the C for their own reproduction, must somehow provide rhizobia with a Darwinian fitness benefit. Otherwise, non-fixing mutants would displace N_2-fixing strains. In other words, genes for infecting legumes and for fixing N_2 in symbiosis persist over the course of evolution only because they enhance survival and reproduction of symbiotic rhizobia (or close relatives) carrying those genes, relative to rhizobia with alternative genes. But how?

A nodule founded by a single rhizobium cell can release millions of descendants into the soil when it senesces. One soybean nodule can contain up to 10^{10} viable rhizobia, all descended from a single foundress (West et al., 2002b), so the evolutionary persistence of genes for nodulation is not surprising. However, the limited field data on increases in soil populations after nodule senescence suggest that only a fraction of these rhizobia escape alive into the soil. Furthermore, competing to found a nodule is like entering a lottery; each participant has a very small chance of a winning a very large fitness prize (Denison and Kiers, 2004). When soil populations of competing symbiotic rhizobia are high, the nonsymbiotic option (not infecting legumes) may lead to greater fitness. This would explain the abundance of nonsymbiotic rhizobia in some soils (Segovia et al., 1991; Laguerre et al., 1993; Sullivan et al., 1996). Despite the long odds, the rewards from symbiosis are large enough, on average, that many rhizobia retain genes for infecting legume roots and founding nodules.

Once inside a nodule, why should rhizobia fix nitrogen? From the perspective of the rhizobia, any C respired to support N_2 fixation could have been allocated to their own reproduction instead. The availability of space inside infected cells may seem to limit opportunities for rhizobium reproduction inside a nodule. But rhizobia might still increase their future representation in soil populations in at least two ways. First, they might be able to trick the host plant into increasing the size of their nodule, perhaps by producing cytokinins (Phillips and Torrey, 1972) or other plant hormones. Second, each rhizobium cell could accumulate resources inside a nodule that would enhance its survival after escaping into the soil. The energy-rich molecule, polyhydroxybutyrate (PHB), can enhance bacterial survival when other C sources are scarce (James et al., 1999). PHB can constitute up to 50% of the weight of rhizobia in soybean nodules "despite the paradox that nitrogen fixation and PHB synthesis both compete for the available reducing equivalents" (Anderson and Dawes, 1990). Consistent with this tradeoff, a non-fixing mutant accumulated more PHB (Hahn and Studer, 1986), whereas a PHB$^-$ mutant fixed more N_2 (Cevallos et al., 1996). In nodules, PHB accumulates mainly in whichever form of rhizobia will have direct descendants, namely, bacteroids in most determinate nodules, but only the undifferentiated, reproductively viable rhizobia in indeterminate nodules (Vance et al., 1982; Denison, 2000). Experiments to confirm or refute the importance of PHB to rhizobium survival in soil are planned.

By supplying their plant host with N, rhizobia typically enhance plant growth and photosynthesis (Bethlenfalvay et al., 1978). Some of the additional photosynthate C is presumably delivered to rhizobia in the nodules. This C could benefit the rhizobia in a given nodule if they fix N_2, but does it also benefit rhizobia in other nodules on the same plant? These other rhizobia (or their descendants) are the most likely future competitors of those in the first nodule. If, by fixing more N_2, rhizobia in a given nodule indirectly increase the reproduction of rhizobia in another nodule, then they will face more competition for the next host plant (Denison, 2000). This is true whether or not the other rhizobia are also fixing N_2.

A gene that indirectly helps others with the same gene (usually relatives) can persist through the evolutionary mechanism of kin selection. However, we doubt that benefits to relatives in other nodules or nearby in the rhizosphere (Olivieri and Frank, 1994; Provorov, 1998; Simms and Bever, 1998) can explain the evolutionary persistence of genes for symbiotic N_2 fixation by rhizobia inside nodules (Denison, 2000; West et al., 2002b). A single plant may be infected by ten or more strains of rhizobia (Hagen and Hamrick, 1996), creating a Tragedy of the Commons (Hardin, 1968). In less-disturbed soils, more of the rhizobia infecting a given plant are likely to share the same genes. However, unmixed soils also increase the chance that two rhizobia from the same plant will compete against each other for their next host. The effects of soil mixing on relatedness and effects on the spatial scale of competition often balance each other (Griffin and West, 2002), greatly reducing kin selection for cooperation. Furthermore, the rhizobia most likely to share a given symbiotic gene are clonemates in the same nodule, not related or unrelated strains nearby. We suggest that, by fixing N_2, rhizobia inside a nodule must usually increase their own Darwinian fitness, relative to rhizobia in other nodules on the same plant. Under this hypothesis, either 1) nodules that fix more N_2 eventually release more rhizobia into the soil, or 2) each rhizobium released from an N_2-fixing nodule has better survival and reproduction in the soil, or both. Because rhizobia typically fix N_2 at rates that vastly exceed their own needs (Kiers et al., 2003), the benefits from fixing N_2 must be indirect.

"Legume sanctions" could reduce the fitness of rhizobia that fix less N_2. Any host-plant response that reduces the reproductive success of less-effective rhizobia would prevent these "cheaters" from displacing more cooperative members of their species over the course of evolution (Denison, 2000; West et al., 2002b). Various mechanisms, from enzymatic digestion of rogue rhizobia to cutting off their access to C or O_2 might work (Udvardi and Kahn, 1993). Genes for imposing such sanctions would only persist in legume populations, however, if imposing sanctions provided an individual benefit to the plant; natural selection always puts the interest of the individual ahead of those of the species. Preventing non-fixing rhizobia from infecting in the first place would be ideal. Some legumes can exclude some bad rhizobia (Devine and Kuykendall, 1996), but legumes cannot consistently recognize and exclude non-fixing rhizobia, perhaps especially those related to their usual symbiotic partners (Amarger, 1981; Hahn and Studer, 1986). Fortunately, models suggest that the savings to an individual plant from cutting off resources to non-fixing nodules can benefit the plant directly, while also providing a future benefit to others of its species, by improving the ratio of good-to-bad rhizobia in the soil (West et al., 2002a). Note, however, that a strain that is less effective on one cultivar might be more effective on another (Lin et al., 1988).

Recent results are consistent with the sanctions hypothesis. We prevented N_2 fixation by one nodule on each of six replicate plants, using an N_2-free atmosphere, whereas a control nodule in air fixed N_2 normally (Kiers et al., 2003). Both nodules contained the same N_2-fixing strain to eliminate confounding differences in recognition signals with differences in actual symbiotic performance. Using non-invasive spectrophotometry of leghemoglobin (Denison and Layzell, 1991), after 48 h in N_2-free air, nodule interior O_2 levels were half those in control nodules (Kiers et al., 2003). Physiological control of nodule gas permeability has been recognized for 20 years (Denison et al., 1983; Sheehy et al., 1983), although there is still some uncertainty about mechanisms (Denison and Kinraide, 1995; James et al., 2000). O_2-saturated respiration, which should decrease if rhizobia were attacked directly or starved for C, did not differ between treatments. Ten days later, nodules that were fixing N_2 contained twice as many rhizobia as those prevented from fixing N_2. Although we thought rhizobia that didn't fix N_2 might accumulate more PHB, rhizobia removed from non-fixing nodules (in a related split-root experiment) actually had lower survival in sand. Maybe there was some reduction in C supply to non-fixing nodules, but not enough to make C more limiting than O_2.

If our results for soybean are typical, it is hard to understand the prevalence of "less-effective" rhizobia in some soils (Erdman, 1950; Moawad et al., 1998; Denton et al., 2000). Maybe less-effective rhizobia share nodules with strains that fix enough N_2 to avoid nodule-level sanctions or produce chemicals that interfere with the host-plant signaling pathways that lead to sanctions. This would be analogous to rhizobium manipulation of nodule number by producing rhizobotoxine (Duodu et al., 1999).

Superior rhizobia introduced via inoculation rarely displace less-beneficial indigenous strains. Genetic manipulation of the sanctions response in legumes may be feasible within the foreseeable future, however. If so, then we could develop crop and forage legumes that would enrich soils with the most beneficial rhizobia selected from among indigenous populations already adapted to local soil conditions.

Amarger N (1981) Soil Biol. Biochem. 13, 475-480.
Anderson AJ and Dawes EA (1990) Microbiol. Rev. 54, 450-472.
Bethlenfalvay GJ et al. (1978) Plant Physiol. 62, 131-133.
Cevallos MA et al. (1996) J. Bacteriol. 178, 1646-1654.
Denison RF (2000) Am. Nat. 156, 567-576.
Denison RF and Kiers ET (2004) FEMS Microbiol. Lett. 237, 187-193.
Denison RF and Kinraide TB (1995) Plant Physiol. 108, 235-240.
Denison RF and Layzell DB (1991) Plant Physiol. 96, 137-143.
Denison RF et al. (1983) Plant Physiol. 73, 648-651.
Denton MD et al. (2000) Aust. J. Exp. Agric. 40, 25-35.
Devine TE and Kuykendall LD (1996) Plant Soil 186, 173-187.
Duodu S et al. (1999) Mol. Plant-Microbe Interact. 12, 1082-1089.
Erdman LW (1950) USDA Farmer's Bull. 2003, 1-20.
Griffin AS and West SA (2002) Trends Ecol. Evol. 17, 15-21.
Hagen MJ and Hamrick JL (1996) Mol Ecol 5, 707-714.
Hahn M and Studer D (1986) FEMS Microbiol. Lett. 33, 143-148.
Hardin G (1968) Science 162, 1243-1248.
James BW et al. (1999) Appl. Environ. Microbiol. 65, 822-827.
James EK et al. (2000) Plant Cell Environ. 23, 377-386.
Jimenez J and Casadesus J (1989) J. Hered. 80, 335-337.
Kiers ET et al. (2003) Nature 425, 78-81.
Laguerre G et al. (1993) Can. J. Microbiol. 39, 1142-1149.
Lin J et al. (1988) Can. J. Bot. 66, 526-534.
Moawad H et al. (1998) Plant Soil 204, 95-106.
Olivieri I and Frank SA (1994) J. Hered. 85, 46-47.
Phillips DA and Torrey JG (1972) Plant Physiol. 49, 11-15.
Provorov NA (1998) Symbiosis 24, 337-368.
Segovia L et al. (1991) Appl. Environ. Microbiol. 57, 426-433.
Sheehy JE et al. (1983) Ann. Bot. 52, 565-571.
Simms EL and Bever JD (1998) Proc. R. Soc. Lond. B 265, 1713-1719.
Sprent JI et al. (1993) In New Horizons in Nitrogen Fixation. Palacios R et al. eds. pp. 65-76, Kluwer Academic Publishers, Dordrecht.
Sullivan JT et al. (1996) Appl. Environ. Microbiol. 62, 2818-2825.
Udvardi MK and Kahn ML (1993) Symbiosis 14, 87-101.
Vance CP et al. (1982) Can. J. Bot. 60, 505-518.
West SA et al. (2002a) J. Evol. Biol. 15, 830-837
West SA et al. (2002b) Proc. R. Soc. Lond. B 269, 685-694.

TYROSINE CROSS-LINKING IN ROOT NODULE EXTENSINS

Sébastien Gucciardo, Jean-Pierre Wisniewski, Lora Mak, Marcus Durrant, Elizabeth Rathbun and Nick Brewin,
John Innes Centre, Norwich, NR4 7UH UK

Root nodule extensins (RNEs) are highly glycosylated plant glycoproteins localized in the extracellular matrix of legumes and in the lumen of infection threads (IT). The amino acid composition of PsRNE-1 includes many tyrosine residues. These offer scope for several different kinds of peroxide-based protein cross-linking in the extracellular matrix (Wisniewski et al., 2000). First, computational modelling indicates that the introduction of iso-dityrosine bridges between near-adjacent Tyr residues is mechanically feasible. Second, end-to-end linkage of RNE molecules probably occurs (at least, in transformed tobacco cells). Furthermore, this concatenation is blocked by introduced epitope tags at either the C-terminus or N-terminus of the coding sequence. Third, a recurrent AGP glycomotif in RNE is always flanked by Tyr residues and associated with His residues that may bind Ni^{2+} ions. So, we synthesised the corresponding peptide VHTYOHOHOVYHSO (O = hydroxyproline) to investigate its properties. We demonstrated a time-dependent synthesis of dityrosine (detectable by its fluorescence at 400nm) in the presence of Ni^{2+} (50μm) and H_2O_2 (0.5 mM). Thus, tyrosine bridges can be formed within and between RNE molecules and these could modify the biophysical properties that support rhizobial cells in the IT lumen. Sources of H_2O_2 in the IT matrix could include a copper-containing diamine oxidase (abundant in the extracellular matrix). Its presence in the IT lumen could correlate with the observed distribution of H_2O_2. Polyamines, the normal substrates for diamine oxidase, could be transported across the IT membrane and regulate peroxide in the luminal matrix. A polyamine transporter has been cloned from pea nodules (Butelli and Brewin, unpublished). Other peroxide sources are a NADPH oxidase or a cell-wall peroxidase.

Gucciardo S et al. (2005) MPMI, in press.
Brewin NJ (2004) Crit. Rev. Plant Sci. 23, 293-316.
Rathbun EA et al. (2002) MPMI 15, 350-359.
Wisniewski J-P et al. (2000) MPMI 13, 413-420.

ENODDR1 REGULATES NODULE NUMBER THROUGH THE CONTROL OF ABSCISIC ACID CONCENTRATION

Akihiro Suzuki[1], Mitsumi Akune[1], Yoshihiro Imagama[1], Ken-ichi Osuki[1], Toshio Aoki[2], Toshiki Uchiumi[1] and Mikiko Abe[1]
[1]Kagoshima University, Kagoshima, Japan. [2]Nihon University, Fujisawa, Japan

EnodDR1 gene was isolated from cDNA library of white clover. The expression of this gene was suppressed by the inoculation of *Rhizobium leguminosarum* bv. *trifolii* strain 4S. In order to know the function of *enodDR1* gene, *Lotus japonicus* transformed by the chimeric gene of CaMV35S promoter and *enodDR1* was prepared using *Agrobacterium*. The number of root nodule of *enodDR1* transformant was drastically reduced compared with control transformant. Moreover, abscisic acid (ABA) concentration of *enodDR1* transformants was increased. We have already reported that exogenous ABA decrease nodule number in white clover and *Lotus japonicus*. So, we concluded that the reduction of nodule number in the *enodDR1* transformant was due to ABA concentration.

Abamine is known as an inhibitor of NCED in the indirect pathway of ABA synthesis. Then, this reagent was adapted to decrease the *in vivo* level of ABA. Twenty eight days after inoculation of *Mesorhizobium loti*, nodule number of 10 μM abamine treated *enodDR1* transformant was almost recovered to the level of control transformant.

These results suggest that down-regulation of *enodDR1* gene positively affect on the nodulation due to the maintenance of low ABA concentration by the inoculation of *Rhizobium*.

References
Suzuki A et al. (2001) Gene 266, 77-84.
Suzuki A et al. (2004) Plant Cell Physiol. 45, 914-922.

ISOLATION, SCREENING AND IDENTIFICATION OF NODULIN GENES IN *ASTRAGALUS SINICUS*

Minxia Chou and Junchu Zhou
State Key Laboratory of Agricultural Microbiology, Huazhong Agricultural University Wuhan, 430070, P. R. China

To understand the molecular events governing legume nodule formation processes, a multiple technique is being used to identify the genes differentially expressed in the root nodules formed by symbiotic interactions of *Astragalus sinicus* with *Mesorhizobium huakuii*. This technique is a combination of suppression subtractive hybridization (SSH), SMART cDNA synthesis and RACE-PCR. SSH is a new technique that enriches the amount of cDNA fragments from transcripts present in one population of mRNAs (tester) but absent from another (driver). The technique simultaneously reduces fragments of equally expressed genes. Subtraction was performed in both forward (tester: root with nodules mRNA) and reverse (tester: root without nodules mRNA) orientations.

We constructed the subtractive cDNA library specific for nodule-forming root after the target fragments had been identified by SSH. Subsequently, the differential cDNA fragments were screened by dot blotting and sequenced. The results of their BLAST program indicate that 17 fragments have high degree of homology with leghemoglobin genes, 3 fragments being identical with *A. sinicus* AsAS mRNA, and 5 fragments similar to genes of ENOD2, nodulin-26, *Sesbania rostrata* phosphoenolpyruvate carboxylase, *Capsicum annuum* putative cystein proteinase and *Lupinus luteus* diphosphonucleotide phosphatase 1, respectively. In addition, 32 fragments are novel. The candidate clones are being verified further by RT-PCR and virtual Northern blotting analysis. Then, the fully length cDNAs will be isolated by RACE-PCR.

NODULE-SPECIFIC EXPRESSION OF A NOVEL NODULIN GENE GMN479 IN THE TRANSGENIC SOYBEAN *GLYCINE MAX*

Xian-Guo Cheng[1], Shigeyuki Tajima[2], and Hiroshi Kouchi[3]
[1]Insititute of Soils and Fertilizers, Chinese Academy of Agricultural Sciences, 100081 Beijing, China. [2]School of Agricultural Sciences, Kagawa University, Kitagun, Kagawa, Japan. [3]Institute of Agrobiological Resources, Kannodai 305, Japan.

Nodulin genes are highly regulated both spatially and temporally during nodule development. Regulation of organ-specific gene expression in the plant is controlled by the interaction of *cis*-regulatory sequences and *trans*-acting factors. Promotor analysis for nodule-specific expression has been extensively (1). Although more than 30 genes encoding the C_2H_2 zinc-finger protein have been characterized from higher plants, little data is available on nodulin genes with conserved a zinc-finger C_2H_2 domain (2). A GmN479 cDNA clone was isolated from mature soybean nodules (3) and a chimeric-gene harboring a 2090-bp 5'-upstream promoter fragment of the GmN479 gene and a GUS reporter gene was constructed and transferred into *Glycine max* (4). GmN479 contained a typical C_2H_2-type zinc-finger domain with a conserved QALGGH motif and a LDLELRLGL motif close to C-terminus (5). GmN479 may be a homolog of *SUPERMAN* in *Arabidopsis* (6). GUS-activity staining suggested that GUS was expressed only in the infected cells of the nodule, indicating that GmN479 may function as a transcriptional regulator for nodule morphogenesis. GmN479 is a symbiotically induced host gene, which may help to identify *cis*-acting elements involved in nodule-specific gene expression and general transcriptional control.

1. De Bruijn FJ et al. (1990) Dev. Genet. 11, 182-196.
2. Chrispels HE et al. (2000) Plant Mol. Biol. 42, 279-290.
3. Kouchi H and Hata S (1995) Molecular Plant-Microbe Interactions 8, 172-176.
4. Petit A et al. (1987). Mol. Gen. Genet. 207, 245-250.
5. Takatsuji H et al. (1994) The Plant Cell, 6, 947-958.
6. Sakai H et al. (1995) Nature 378, 199-203.

HOMOLOGUES OF NODULIN GENES IN *ORYZA SATIVA*

Yanzhang Wang, Guanqiao Yu and Jiabi Zhu
National Laboratory of Plant Molecular Genetics, Institute of Plant Physiology & Ecology, Shanghai Institutes of Biological Sciences, Chinese Academy of Sciences, Shanghai, China

The release of rice genome sequences offers valuable resources to assess the nitrogen-fixation potential of rice. With the continual discovery of nodulin genes, which are specifically expressed and play important functions in the development of leguminous root nodules, there is a great need to know whether orthologs of these nodulin genes exist in rice genome and then explore their potential functions related to nodule organogenesis. Seventy-five nodulin genes, found in nucleotide databases, were used as probes to scan the rice genome by the application of bioinformatics methods. Thirty-one homologues of nodulin genes have been detected in the rice genome, all exhibit more than 35 % identities with those of the leguminous nodulin genes in amino acid sequences. It indicates that leguminous nodulin homologous genes exist extensively in rice genome. In comparison with the leguminous nodulin gene *enod40*, sucrose synthase gene and *Rab* gene, their corresponding homologues found in rice reveal that they all belong to a class of orthologous genes, presumably originated from the common ancestral genes. However, the leguminous nodulin genes mutated during in the early evolution stage so as to meet the requirements of nodule organogenesis. The other forty-four leguminous nodulin genes without their homologs in rice genome, may play an essential roles for establishing symbiosis between the *Rhizobia* and the host legumes. We thus assume that lack of these leguminous nodulin genes in rice results in its inability of nodulation and nitrogen fixation. Anyway our present investigation on the occurrence of nodulin homologs in rice may provide clue to the nitrogen-fixing potential of rice.

References
1. Reddy PM et al. (1999) Biochem. Bioph. Res. Co. 258, 148-154.
2. Verma DPS et al. (1986) Plant Mol. Biol.7, 51-61.

PROMOTER OF THE SOYBEAN EARLY NODULIN GENE *ENOD2B* IS INDUCED BY RHIZOBIAL NOD FACTORS IN TRANSGENIC RICE

Yanzhang Wang, Guanqiao Yu and Jiabi Zhu
National Laboratory of Plant Molecular Genetics, Institute of Plant Physiology & Ecology, Shanghai Institutes of Biological Sciences, Chinese Academy of Sciences, Shanghai, China

The signaling molecule Nod factors secreted by Rhizobia are the principal determinants of host specificity in *Rhizobium*–legume symbiosis. Nod factors elicit a variety of responses on a legume host, such as activation of the expression of certain nodulin genes etc. Our present investigation has focused on the question whether the rice plant would respond to the signaling molecules issued by the rhizobia. The promoter of the soybean (*Glycine max*) early nodulin gene *Gmenod2B* fused to the β-glucuronidase (GUS) reporter gene was used as a molecular marker to explore whether Nod factors can be recognized by rice cells as signaling molecules. Transgenic rice plants harboring the chimeric gene *Gmenod2BP-GUS* were obtained via an *Agrobacterium tumefaciens*-mediated system. NodNGR factors produced by a broad-host-range *Rhizobium* strain NGR234(pA28) were used as probes to investigate the activity of the *Gmenod2B* promoter in rice. Our results showed that the early nodulin gene *Gmenod2B* promoter was induced by NodNGR factors in transgenic rice. Moreover, the early nodulin gene was specifically expressed in rice plant roots and was regulated by nitrogen status. These findings thus indicate that rice plant perceives the signaling effect of Nod factor, and that the way for signal transduction system seems present in cells of rice plant.

References
1. Ardourel M. et al. (1994) Plant Cell, 6, 1357-1374.
2. Denarie J et al. (1996) Annu. Rew. Biochem. 65, 503-535.
3. Franssen H J et al. (1987) Proc. Natl. Acad. Sci. USA, 84, 4495-4499.
4. Reddy P M et al. (1998) The Plant J. 14, 693-702.

INVOLVEMENT OF POLYAMINE IN BRASSINOLIDE-DIRECTED NODULE REGULATION IN SOYBEAN PLANT

JunkoTerakado[1,2], Tadakatsu Yoneyama[3], and Shinsuke Fujihara[1]
[1]National Agricultural Research Center, Kannondai 3-1-1, Tsukuba, Ibaraki 305-8666, Japan. [2]Japan Society for the Promotion of Science, Tokyo 102-0082, Japan. [3]Department of Applied Biological Chemistry, University of Tokyo, Tokyo 113-8657, Japan.

In legumes, the number of root nodules is controlled by the mechanism called auto-regulation. Several reports implicate the presence of nodulation-inhibiting signals in the shoot. However, the principal compound in nodule regulation has not been identified.

Previously, we found that the tissue contents of spermidine (spd) and spermine (spm) in super-nodulating soybean mutant (En6500) (1) were always lower than those in wild type soybean (Enrei) while putrescine (put) levels were considerably higher, indicating the repression of spd and spm biosynthesis from their precursor put in En6500. In order to ascertain the possibility whether polyamines (PAs) mediate nodule regulation in soybean plant, we investigated the effect of foliar treatment with PAs and/or PA-inhibitor on root nodule formation using soybean Enrei and En6500. Foliar application of PAs, especially spd and spm, led to a significant reduction of nodule formation in En6500, while neither spd nor spm caused such effects on Enrei. By contrast, foliar treatment with MDL74038 (2), a specific inhibitor which represses spd biosynthesis, apparently increased root nodule production in Enrei.

We also demonstrate that foliar treatment of plant hormone brassinolide not only increased the levels of spd in the leaf but also reduced the root nodule formation of En6500. These data suggested that brassinolide-induced spd synthesis in the leaf might closely associate with the suppression of root nodule formation in soybean plant.

References
1. Akao S and Kouchi H (1992) *Soil. Sci. Plant Nutr.* 38, 183-187.
2. Wright PS et al. (1991) *Biochem. Pharmacol.* 41,1713-1718.

USING CDNA-AFLP DISPLAY TO IDENTIFY NOVEL INDUCED GENES DURING ALFALFA NODULE DEVELOPMENT

Zhishui He, Rong Xie, Jiabi Zhu and Guanqiao Yu
National laboratory of Plant Molecular Genetics, Institute of Plant Physiology & Ecology, Shanghai Institutes for Biological Sciences, Chinese Academy of Sciences, Shanghai, China

To identify the induced genes during alfalfa nodule development, patterns of cDNA-amplified fragment length polymorphism (AFLP) nodules were used to compare with those from uninfected roots. More than 3,000 cDNA fragments were examined in both nodules and uninfected roots. 70 fragments were found only unique to nodules. Analysis of these cDNA sequences showed that most of the transcript-derived fragments (TDFs) shared little homology with those previously characterized as nodulins or nodule-induced genes. These TDFs appear as novel nodule-induced genes, which may play important roles in signal transduction, gene expression or regulation. Results of northern gel blot analysis are consistent with that of cDNA-AFLP differential pattern, indicating the availability of cDNA-AFLP to identify novel nodule-induced genes. Further studies of three TDFs showed that the expression of *MsHSP70-1* (RX89) and *MsEHs-1* (RX32) required the action of *bacA* gene of rhizobia. Some TDFs maybe induced at the stage of alfalfa nodule development when bacteroids begin to differentiate.

References
Bachem CWB et al. (1996) Plant J. 9, 745-753.
Fedorova M et al. (2002) Plant Physiol. 130, 519-537.
Gyorgyey J et al. (2000) Mol. Plant-microbe Interact. 13, 62-71
Nollen EAA and Morimoto R I (2002) J. Cell Sci. 115, 2809-2816.

CHARACTERIZATION OF A NOVEL *LOTUS JAPONICUS* SYMBIOTIC MUTANT, *LOT1*, THAT SHOWS REDUCED NODULE NUMBER AND DISTORTED TRICHOMES

Y. Ooki[1], M. Banba[1], K. Yano[2], J. Maruya[3], S. Sato[4], S. Tabata[4], K. Saeki[3], M. Hayashi[2], M. Kawaguchi[5], K. Izui[1], and S. Hata[1]
[1]Graduate School of Biostudies, Kyoto University, Kyoto 606-8502, Japan.
[2]Graduate School of Engineering, Osaka University, Osaka 565-0871, Japan.
[3]Graduate School of Science, Osaka University, Osaka 560-0043, Japan. [4]Kazusa DNA Research Institute, Chiba 292-0812, Japan. [5]Graduate School of Science, University of Tokyo, Tokyo 113-0033, Japan.

Through EMS mutagenesis and visual screening for a symbiotic phenotype, we isolated a recessive symbiotic mutant of *L. japonicus* that defines a new genetic locus, *LOT1* (for low nodulation and trichome distortion). The nodule number of the mutant was only about one-fifth of that of the wild type. The *lot1* mutant showed a moderate dwarf phenotype in both the presence and absence of exogenous nitrate. It also showed distorted trichomes, but its root hairs showed no apparent differences to those of the wild type. Root hair deformation and infection thread formation after inoculation of *Mesorhizobium loti* occurred in the mutant as usual but the entire process of nodule primordia formation seemed to be repressed, compared to that in the wild type. The nodule primordia of the *lot1* mutant did not result in any aborted nodule-like structure, all nodules becoming mature and exhibiting high nitrogen-fixation activity. The mutant was colonized by AM fungi as effectively as the wild type. The *lot1* mutant showed higher sensitivity to nitrate than wild type. However, the mechanism by which the high nitrate sensitivity of *lot1* results in low nodulation remains to be clarified. The sensitivity of *lot1* to ethylene was similar to, or lower than, that of the wild type. Thus, the low nodulation of *lot1* is not caused by higher ethylene sensitivity. Grafting experiments with *lot1* and wild-type seedlings indicated that the root genotype determines the low nodulation phenotype of the mutant, whereas the trichome distortion is regulated by the shoot genotype.

We next examined whether or not the *LOT1* gene product is involved in the well-known *HAR1* autoregulation pathway as related to nodule number. The level of the *HAR1* transcript in the *lot1* mutant was nearly the same as that in the wild type. Grafting of *har1-4* shoots to *lot1* roots resulted in a medium nodule number, i.e., more than that of *lot1* and less than that of *har1-4*. These results indicate that *LOT1* is involved in a distinct signal-transduction pathway independent of *HAR1*.

A LOW EXPRESSION OF *PURL* IN FREE-LIVING CONDITIONS WAS REQUIRED FOR ESTABLISHMENT OF EFFICIENT SYMBIOSIS BETWEEN *SINORHIZOBIUM. FREDII* AND *GLYCINE MAX*

Bo Xie and Junchu Zhou
State Key Laboratory of Agricultural Microbiology, Huazhong Agricultural University Wuhan, 430070, P. R. China

Most purine auxotrophs of rhizobia species are unable to form effective nodules on their host plants. Previous reports showed that *purL* mutant of *Sinorhizobium fredii* HH103 induced pseudonodules on *Glycine max* and altered in its lipopolysaccharide (Buendia-Claveria 2003). The plant host might not supply enough purine or purine precursors to the mutant during the nodulation process. However, although the addition of AICA-riboside or adenine to the plant nutritive solution enabled the *purL* mutant to form infected nodules, the symbiotic properties were still not restored. The precise role of *purL* gene in rhizobium-plant interactions is not known. We present here evidence on the relationship between the establishment of efficient symbiosis and *purL* gene expression by using different deletions of *purL* coding sequence. A 1.98kb deletion mutation in *purL* open reading frame of *S. fredii* was constructed. Plasmid pBBR-PG was used as a positive control of *purL* expression. Another plasmid pHN811 had a low expression of *purL-gusA*. Several promoters of genes expressed during symbiosis, fixk (S. meliloti), fixR (B. japonicum), nifH (S. meliloti), fixN (S. meliloti) and nifQ (S. fredii), were used as to monitor expression of *purL* during symbiosis. It was observed that reduced expression or suitable symbiotic specific expression of *purL* (such as under the control of *nifH*, or *nifQ* promoters) could meet the efficient symbiosis establishment of *S.fredii* on *Glycine max* without the exogenous supplementation of any adenine or of a purine precursor. However, the competition ability of the recombinant strains was significantly reduced. A low expression of *purL* in free-living conditions was required.

Buendia Claveria A M et al. (2003) Microbiology, 149, 1807–1818

FUNCTIONAL ANALYSIS OF SNARE GENE IN *LOTUS JAPONICUS*

Mika Nomura, Mai Ha Thu, Shigeyuki Tajima
Faculty of Agriculture. Kagawa Univ., Miki-cho, Kita-gun, 761-0795 Japan.

Vesicle traffic underpins cell homeostasis, growth and development in plants. Traffic is facilitated by a superfamily of proteins known as SNAREs (soluble N-ethylmaleimide sensitive fusion protein attachment protein receptors) that interact to draw vesicle and target membrane surfaces together for fusion of the bilayers.

In plants, as in mammalian tissue, exocytotic and endocytotic events have been identified with stepwise changes in capacitance that accompany the increase or decrease of membrane surface area during vesicle membrane fusion and removal, respectively. To find the SNARE protein for the nodule formation in *Lotus japonicus*, we have screened 12 SNARE-like clones that have high homologies to yeast SNARE genes. One (Sn6gene) of the 12 genes expressed higher in nodules than another organs. The Sn6 gene has high homology to the yeast Sed5 gene required for protein transport from ER to Golgi. To analyze the function of Sn6 gene, we constructed the antisense and sense Sn6 gene fused to 35S promoter and introduced them into the *Lotus japonicus* by *Agrobacterium* mediated transformation. As a result, the growth rate of the antisense Sn6 plants was much slower than non-transformant. Some of the nodules in antisense plants jointed each other. *In situ* hybridization data indicated that the signals of Sn6 gene appeared at the top of nodule primodia and in the cortex of surrounding the nodules. These data suggest that the Sn6 protein might be important SNARE for the formation in the direction of the division of the cell.

STUDY ON MECHANISM FOR A TRANSIENT DOWN-REGULATION OF TRANSCRIPTIONAL LEVEL IN SOYBEAN MICRO-CALLUS CELLS INDUCED BY NOD FACTOR OF *B. JAPONICUM* USDA 110.

Tadashi Yokoyama [1], Tsuneo Hakoyama [2], Hiroshi Kouchi [3], and Yasuhiro Arima [1]
1. Faculty of Agriculture, Tokyo University of Agriculture and Technology; 2. Aichi University of Education; 3. Plant Physiology Department, National Institute of Agrobiologic Sciences.

To know what kinds of gene expression are induced in leguminous plant cells after Nod factor perception, changes in transcriptional level in response to Nod factor in soybean suspension-cultured cells were surveyed with time in terms of differential display. Seven fragments were successfully cloned. The transcriptional levels of the seven fragments were clearly repressed within 2hr after initiation of the treatment. In further experiment to confirm this observation, of 13 known-genes, the 12 genes excluding GTP-binding protein showed clear suppression at transcriptional levels after Nod factor addition. Regarding correlation between chemical structure and gene expression, Nod factor only induces down-regulation of the mRNAs in soybean cultured cells, while 5mer of N-acetyl-glucosamine and elicitor purified by *Phytophthora infestans* failed to induce that. BAPTA treatment with Nod factor inhibited down-regulation of mRNAs in soybean cells, while addition of K252a with Nod factor induced a transient down-regulation in soybean cells. To evaluate number of mRNA suppressed in cultured cells with Nod factor, Total RNAs were isolated from the cultured cells with and without Nod factor at 0, 10, and 90 after initiation of the treatment. Target cDNA labelled with ^{33}P were hybridized with the cDNA macro array membrane constructed from cDNA libraries of *Lotus japonicus*. At 10 min after Nod factor addition, many clones in the scatter plots showed down-regulation. Furthermore, at 30 and 90 min after Nod factor addition, large number of clones showed down-regulation. To elucidate the mechanism of a transient down-regulation of transcriptional level in soybean micro-callus cells induced by Nod factor, we tried to gel retardation assay using transcriptional regulatory DNA regions of *CDPK*, *ß-tubulin*, *TFIIB* and nucleic proteins isolated from the Nod factor treated and untreated soybean micro-callus cells. The nucleic proteins of soybean micro-callus at 30 min after the Nod factor addition had clear retardation activities to the DNA fragments tested. After retardation assay, the nucleic proteins binding to the DNA fragments were collected from the gels using dialysis method. The protein spot located in pH10.5 and at 31KDa in two-dimensional gel electrophoresis.

TRANSCRIPTIONAL CHANGES OF *LOTUS JAPONICUS* DURING ARBUSCULAR MYCORRHIZAL FUNGI DEVELOPMENT

Y. Deguchi[1], Y. Shimoda[2], S.A. Checchetka[1], M. Bamba[1], Y. Ooki[1], A. Suzuki[2], T. Uchiumi[2], S. Higashi[2], M. Abe[2], H. Kouchi[3], K. Izui[1], and S. Hata[1]
[1] Graduate School of Biostudies, Kyoto University, Japan. [2] Department of Chem. BioSci. Kagoshima University, Japan. [3] National Institute of Agrobiol. Sciences, Japan.

It is well known that arbuscular mycorrhizal (AM) fungi perform mutually beneficial symbiosis with about 80% of vascular flowering plants. The mycorrhizal association is believed to have originated more than 400 million years ago on fossil evidence. However, the molecular mechanism of AM symbiosis is poorly understood especially in terms of plant gene expression. We analyzed gene expression profiles in *Lotus japonicus* roots during the development of AM symbiosis with *Glomus mosseae* by means of a cDNA macroarray of about 19,000 non-redundant ESTs isolated from various organs of L. japonicus. Because commercial inoculum containing spores of *G. mosseae* (Idemitsu Kosan, Tokyo) are contaminated with other microorganisms, we also used *Gigaspora margarita* (Central Glass, Tokyo). The results indicated that several defensive genes, such a PAL (phenylalanine ammonia-lyase), were down-regulated and that several genes for cysteine proteinase and alpha-mannosidase, which appear to be involved in decomposition of arbucules, were up-regulated. Apparently, the symbiotic system maintains an exquisite balance between keeping alive or killing microorganisms.

CHARACTERIZATION OF A *LOTUS JAPONICUS* PHOSPHATE TRANSPORTER (*LJPT4*) THAT IS SPECIFIC FOR ARBUSCULAR MYCORRHIZAL SYMBIOSIS

D. Maeda, K. Ashida, K. Iguchi, Y. Deguchi, K. Izui, S. Hata
Graduate School of Biostudies, Kyoto University, Kyoto 606-8502, Japan

Arbuscular mycorrhizal (AM) fungi are ancient symbionts on plant roots. AM fungi uptake phosphate through extraradical mycelium that extend beyond Rhizosphere and supply phosphate to plants. In return, plants supply photosynthates to AM fungi. Recently, AM root-specific phosphate transporter genes of potato (*StPT3*) (Rausch, C., et al, 2001), rice (*OsPT11*) (Paszkowski, U., et al., 2002) and *Medicago truncatula* (*MtPT4*) (Harison, M.J. et al., 2002) were identified. Analysis of genes that are important for AM symbiosis in plants will give new insights in molecular evolution of plant-microbe symbiosis. By PCR cloning, we isolated three phosphate transporter genes from AM roots of *Lotus japonicus* (*LjPT4*, *LjPT8*, and *LjPT16*). Each encoded a protein that consisted of 12 conserved transmembrane domains, two phosphorylation sites, and an N-linked glycosylation site. Expression analysis of the three genes by real time RT-PCR revealed that only *LjPT4* was specifically expressed in AM roots. *LjPT4* coding region was introduced into a yeast mutant lacking phosphate transporters (PAM2, a gift from Dr. Bengt Persson), and we confirmed the phosphate transport activity. In addition, *LjPT4* knockdown transformants by RNAi showed reduced number of arbuscules, indicating the importance of *LjPT4* function for the symbiotic relationship.

ANALYSIS OF GENE LOCI ASSOCIATED WITH NODULE MATURATION IN *AZORHIZOBIUM CAULINODANS* ORS571.

Maru Yukihiro, Miwa hiroki, Aono Toshihiro, and Oyaizu Hiroshi
Biotechnology research center, The University of Tokyo:maru@leaveanest.com

In the symbiotic interaction between the tropical legume *Sesbania rostrata* and the bacterium *Azorhizobium caulinodans* ORS571, root and stem nodules are formed. Recently, intensive studies have been reported on early stage (entry mechanism) of nodulation. However, little is known about the late stage or maturation stage of nodulation. In this study, we intended to study the late stage of nodulation by establishing non-maturating mutants of *A.caulinodans* and find new genes for symbiosis. *A.caulinodasn* was mutagenized with the Tn5 transposon derivative mTn5*gusA*-*pgfp21* carrying combinations of a constitutively expressed gfp gene and a promoter less gusA gene. We screened about 1,000 Tn5 insertion mutants, which non-maturating nodules developed, and of transposon insertion sites showed that for three mutants, ORS571-C1, ORS571-C12 and ORS571-C24. Total DNA from the selected mutants were digested with *Xho*I and ligated into the *Xho*I site of pBluescript. Inserts containing part of the mTn5*gusA*-*pgfp21* transposon were selected. From the cloned fragment, the partial sequence flanking the mTn5*gusA*-*pgfp21* insertion was determined, with using the *gusA* primer (5'-GGTGGGGTTTCTACAGGACGTAACAT -3'). To isolate the corresponding wild-type *A.caulinodans* gene, the sequenced fragment was amplified by PCR .The resulting about 400-bp PCR fragment was used as a probe to hybridize a charomid gene library. Sequencing of a 4181bp fragment gene locus hit by the Tn5 in ORS571-C24, revealed two open reading frame (ORFs), designated *omclA* and *omclR*. The *omclA* gene, which codes for a protein of 511 amino acids and located in the region upstream from *omclR gene*, is homologous to the Outer membrane efflux protein belonging to The OEP family form trimeric channels protein that allow export of a variety of substrates in Gram negative bacteria. The *omclR* gene contains 252 amino acids homologous to transcriptional regulators belonging to the TetR family of bacterial regulatory proteins (45% identical with *Bradyrhizobium japonicum*). Sequence conservation is particularly strong in the helix-turn-helix structure localized in the N-terminal region of the protein and responsible for DNA binding in the TetR protein.

ADENYLATE-COUPLED ION MOVEMENT: A MECHANISM FOR THE CONTROL OF NODULE PERMEABILITY TO O_2 DIFFUSION

Hui Wei and David B. Layzell*
Dept. Biology, Queen's University, Kingston, ON K7L 3N6, Canada

Introduction. Legume nodules have physiological control over their permeability to O_2 diffusion, a mechanism to protect nitrogenase from O_2 inhibition. Various treatments are known to decrease or increase nodule permeability, alter the infected cell O_2 concentration and change the adenylate pools in cytosolic and bacteroid fractions[1,2,3]. However, the mechanism for diffusion barrier control is unknown. A hypothesis (Fig. 1) was proposed for the adenylate-coupled movement of ions and water into and out of infected cells as a possible mechanism for diffusion barrier control in legume nodules.

Methods. To test this hypothesis, nodulated soybean plants were exposed to a range of treatments known to alter nodule O_2 permeability. Then the nodules were rapidly frozen (lqd N_2), freeze-dried, dissected into cortex and central zone (CZ) fractions and assayed for K, Mg and Ca ion concentrations.

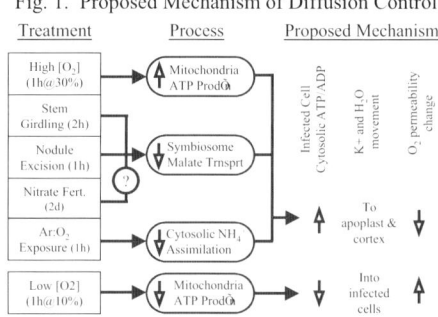

Fig. 1. Proposed Mechanism of Diffusion Control

Results. Treatments like high pO_2 and $Ar:O_2$ exposure known to decrease nodule permeability and increase cytosolic ATP/ADP ratio[1,3], were associated with a significant increase in $[K^+]_{cortex}:[K^+]_{CZ}$. However, low pO_2, known to increase nodule permeability and decrease cytosolic ATP/ADP ratio[1,3] was correlated with a decrease in the $[K^+]_{cortex}:[K^+]_{CZ}$. No changes were observed in [Mg] or [Ca].

Conclusions. The observed treatment-induced changes in $[K^+]_{cortex}:[K^+]_{CZ}$ were consistent with the proposed mechanism for O_2 permeability control in nodules.

References. 1. Kuzma MM et al. (1999) Plant Physiol. 119, 399-407.
2. Wei H et al. (2004a) Plant Physiol. 134, 801-812.
3. Wei H et al. (2004b) Plant Physiol. 134, 1775-1783.

CLONING, REGULATION AND FUNCTIONAL ANALYSIS OF *noeAB* FROM SINORHIZOBIUM *MELILOTI 042BM*

Bing Hai Du, Lei Wang, Su Wei Qi, Ju Quan Jiang, Su Sheng Yang
Department of Microbiology, College of Biological Sciences, China Agricultural University, Beijing 10094, China

The Tn5-1063 was used to mutagenize *S. meliloti* 042BM. One mutant, 042BMR5, involved in nodulation was obtained. DNA sequences flanking Tn5-1063 of 042BMR5 were amplified using inverse PCR, and it was found that Tn5-1063 was inserted into *noeB* gene. Hydrophobicity analysis of the NoeB showed that it is a transmembrane protein and includes four transmembrane regions at N-terminal.

Further, 042BM *noeA* was obtained by PCR. It was found that the similarity of 303-362 region of the NoeA to the 160-220 region of L11 methyltransferases of *E.coli* (PrmA) is 41%. Compared to 042BM, the *noeA* deletion mutant 042BMA-Km showed different degree of increase in number of nodule, fresh weight of nodule and plant top dry weight on alfalfa cultivars of Putong Zihua, Baoding, Ningxia, Baifa and Aohan, but decrease on Milu. However, this mutant has no significant change in ability to nodulate cultivars of Huanghou and Zahua. Hence, *noeA* is involved in alfalfa cultivar-specific nodulation.

The inducer trigonelline could not elevate the level of *noeAB* expression, which indicated that these genes are not regulated by *nodD2*. And also, they are not controlled by *nodD3-syrM* system either. However, induction of luteolin resulted in increase of *noeAB* expression, which indicated that *noeAB* are regulated by *nodD1*. Most interestingly, β-galactosidase activity assay showed that more than 30 times increase in its expression was observed on TY medium without any flavonoid. Thus, it was suggested that *noeAB* probably controlled by another unknown mechanism.

Proteomic analysis of 042BMA-Km was carried out using two-dimensional gel electrophoresis. Deletion of *noeA* had a pleiotropic effect on protein synthesis levels. A minimum of 44 protein differences were observed for the *noeA* mutant. Among them, at least 15 proteins were up-regulated, over 28 proteins were newly synthesized, and one protein was down-regulated.

THE ROLES OF arg-tRNA AND *mincDE* ON MEGAPLASMID pSymB IN *SINORHIZOBIUM MELILOTI* 1021

J. Cheng, C. D. Sibley, T. Landry, P. S. G. Chain, S. Lehman, G. B. Golding and T. M. Finan
Department of Biology, McMaster University, Hamilton, Ontario L8S 4K1, Canada

Sinorhizobium meliloti is a free-living bacterium that induces N_2-fixing root nodules on legume alfalfa (*Medicago sativa*). The organism possesses three replicons: a 3.65-Mb chromosome and two megaplasmids pSymA (1.35-Mb) and pSymB (1.68-Mb). We have made defined deletions of pSymB by either homologous recombination between IS50 elements of Tn5-derivative transposon insertions or an FRT/Flp recombination system, and identified several loci involved in exopolysaccharide synthesis, C4-dicarboxylate transport, lactose metabolism and replication.

In this work, the FRT-Flp system was employed to detect a region carrying the only copy of the arg-tRNA (CGG) and *mincDE* between $\Omega 5015$ and $\Omega 5177$ on pSymB, and the function of these genes was studied. We defined the transposon insertion site by inverse PCR and DNA sequencing. A Flp recombinase-mediated deletion of the 194-kb FRT-flanked region was obtained only when the arg-tRNA was provided *in trans*, suggesting that the $^{Arg}tRNA_{CCG}$ is essential but the *mincDE* is not.

Three defined mutants in *mincDE* were created in wild type Rm5000 background to assess the function of the *S. meliloti* Min system. The *mincDE* is transcribed as a single unit and *S. meliloti* Min proteins function in *E. coli*. Deletion of *mincDE* (SmK690) or *minDE* (SmK691) did not have obvious morphological consequences. However, removal of *minE* (SmK692) resulted in large, swollen and branched cells with single or multiple asymmetric septa. Deletion of *minE* also delayed symbiotic N_2 fixation. These results indicated that the *mincDE* is required for normal cell division. The complex cell division phenotype resulting from lack of *minE* expression is consistent with other cell division phenotypes in *S. meliloti* such as: overexpression of DNA methyltransferase CcrM and FtsZ, and treatment with DNA-damaging agents.

Acknowledgement: We would like to acknowledge Genome Canada for funding this project.

MOLECULAR CHARACTERIZATION OF GENES INVOLVED IN BIOTIN UPTAKE IN *RHIZOBIUM ETLI* CE3

Karina Guillén-Navarro, Gisela Araíza and Michael F. Dunn
Programa de Ingeniería Metabólica, Centro de Investigación sobre Fijación de Nitrógeno. UNAM. Cuernavaca, Morelos, México.

Rhizobia contain several important metabolic enzymes dependent on biotin as a prosthetic group. *Rhizobium etli* is able to synthesize biotin *de novo* only under certain growth conditions and so is normally dependent on a source of exogenous biotin. We found that orthodox biotin biosynthetic genes were not present in *R. etli*. We thus undertook a molecular genetic analysis of biotin uptake in this organism. By complementation of a *Sinorhizobium meliloti bioM* mutant we isolated an *R. etli* chromosomal region encoding homolog of the *Sinorhizobium meliloti bioMNB* genes, whose products have been implicated in intracellular biotin retention in that organism. Disruption of the *R. etli bioM* resulted in a strain defective in biotin transport during growth. The mutant took up biotin at a lower rate and accumulated significantly less biotin than the wild type in biotin uptake assays with non-growing cells. As in *S. melioti*, the *R. etli bioMN* resemble the ATPase and permease components, respectively, of an ABC-type transporter. The *bioB* gene is in fact similar to members of the BioY family, which has been postulated to function in biotin transport. We report here, the first experimental evidence that the *R. etli* BioY functions as a permease involved in biotin transport. The *bioMNY* operon is transcriptionally repressed by biotin as would be expected for a gene encoding a component of a biotin transporter. An analysis of the competitivity of the wild-type strain versus the *bioM* mutant showed that the mutant had a diminished capacity to form nodules on bean plants.

Acknoledgements: Supported by grant 31711-N from CONACyT. K. G.-N. thanks CONACyT (grant 138526), DGEP-UNAM and DGAPA-UNAM (projects 202327 and 202363) for financial support.

Encarnación et al. (1995) J. Bacteriol. 177:3058-3066.
Entcheva et al (2002) Appl. Environ. Microbiol. 68:2843-2848.
Rodinov et al (2002) Genome Res 12, 1507-1516.

RHIZOBIAL SYMBIOTIC PLASMID TRANSFER INTO AGROBACTERIUM CHROMOSOME

Hiroki Nakatsukasa, Toshiki Uchiumi, Akihiro Suzuki, Shiro Higashi, Mikiko Abe
Lab. Plamt & Microbe, Dept. Chem. & BioSci., Kagoshima U., Japan.

The symbiotic mega-plasmid (pRt4Sa::Tn5-mob, 315 kb) of Rhizobium leguminosarum bv. trifolii 4S5 was transferred into pTi-cured Agrobacterium tumefaciens A136 using triparental mating technique.

Many transconjugants maintained nodulation ability to white clover root and transferred pRt4Sa was determined as plasmid form. However, one transconjugant strain could not be detected plasmid, even it nodulated on the clover roots and could be confirmed the existence of nod genes of strain 4S on Afcs1 chromosome fragments by Southern hybridization. In the strain Afcs1, the pRt4Sa is expected to integrating into Agrobacterium chromosome.

To identify the integration site of pRt4Sa into Agrobacterium chromosome, the genomic DNA of strain Afcs1 was analysed by Pulsed-Field Gel Electrophoresis (PFGE). The integration of pRt4Sa::Tn5-mob in the strain Afcs1 could be clarified by Southern hybridization against PFGE profiles. From the reported total DNA sequences of A. tumefaciens C58, the integration site of pRt4Sa was estimated as putative P4-family integrase gene.

Inverse PCR technique was applied to analyze the insertion locus of pRt4Sa in the linear chromosome of strain Afcs1. Up to now, It the right junction which is exhibited 406 bp region from 5' terminal nucleotide sequences of Agrobacterium integrase gene and reverse repeated sequences of Tn5 (IS50R) including in pRt4Sa::Tn5-mob.

APPARENT INCOMPATIBILITY OF PLASMID PSFRYC4B OF *SINORHIZOBIUM FREDII* WITH TWO DIFFERENT PLASMIDS FROM ANOTHER STRAIN

Lihong Miao [1,2], Kui Zhou [1], Junchu Zhou [1], Fuli Xie [1]
[1]State Key Laboratory of Agricultural Microbiology, Huazhong Agricultural Univerisity, Wuhan, 430070; [2]Biological and Chemical Engineering Department, Wuhan Polytechnic University, Wuhan 430023, P.R China

Fast-growing rhizobia often contain one or more large indigenous plasmids, some of them could be transferred among *Rhizobium* species or between *Rhizobium* and *Agrobacterium*, or into some soil bacteria. Nevertheless, the molecular basis for rhizobia plasmid incompatibility is poorly understood. Gene similar to the plasmid replication gene *repC* were detected and characterized in *Rhizobium leguminosarum* and *Sinorhizobium* strains by conserved PCR amplification. A large genetic diversity was found among *repC* gene sequences. It has been reported that more than one *repABC* replicons may coexist in a given strain. However, no report indicates that two stable indigenous rhizobial plasmids in one strain could be cured simultaneously by the introduction of another rhizobial plasmid.

Most of rhizobial plasmids, show high stability that imply the presence of an accurate maintenance mechanism. *Sinorhizobium fredii* strain HN01 harbors three large plasmids, pSfrHN01c, pSfrHN01b, which is the symbiotic plasmid, and pSfrHN01a. Our results showed that these plasmids were stably inherited through generations without apparent loss. This report describes a simultaneous curing of two stable indigenous plasmids pSfrHN01a and pSfrHN01b of *S. fredii* strain HN01 by the introduction of pSfrYC4b, the second largest plasmid of *S. fredii* strain YC4B. Recent studies demonstrate that a site-specific recombinase (RinQ) is essential for the incompatibility with the symbiotic plasmid of *Rhizobium etli* (Quintero et al. 2002). This work may provide evidence on recombination and incompatibility for indigenous plasmids in rhizobia.

THE INDUCED EXPRESSION OF *SINORHIZOBIUM MELILOTI* C_4-DICARBOXYLATE TRANSPORT SYSTEM (DCT) IS REGULATED BY OXYGEN CONCENTRATION

Jin Wen[1], Bei-Yan Nan[1], Fergal O'Gara[2] and Yi-Ping Wang[1]
[1]National laboratory of Protein Engineering and Plant Genetic Engineering, College of Life Sciences, Peking University, Beijing 100871, China, [2]Microbiology Department, University College Cork, National University Ireland, Cork, Ireland.

The *Sinorhizobium meliloti* C_4-dicarboxylate transport (Dct) system is essential for symbiotic nitrogen fixation. The *dctA* gene (coding for the C_4-dicarboxylate transport protein) has a promoter dependent on activation by the two-component sensor-regulator pair DctB and DctD, upon induction by C_4-dicarboxylate. During symbiosis, the *dctA* promoter expression was higher under conditions compatible with nitrogen fixation, when the oxygen tension was low, and when all the *nif* genes were highly expressed (Boesten et al. 1998). In order to investigate the effect of oxygen concentration on the induction of the *dctA* promoter, *E. coli* DH5α strain carrying the *dct* genes was inoculated in semi-solid medium. It was found that the specific induction of the *dctA* promoter occurred at a certain depth under the surface of M63-0.6% agar media. This indicated that Dct system could also respond to oxygen as an alternative signal, upon succinate-induced expression. When induction levels of *dctA* were determined under conditions of different oxygen concentrations, a two fold increase was observed at 2% oxygen concentration when compared with those observed at higher oxygen concentrations. Inactivation of *dctB* or *dctD* abolished the expression from the *dctA* promoter under all physiological conditions tested. Thus, we predict that the DctBD system, in addition to detecting C_4-dicarboxylates, may have a novel function of oxygen-sensing.

Reference: Boesten B et al. (1998) Mol. Plant-Microbe Interact. 11, 878-886.

Acknowledgements:
Work supported by 973 (No. 2001CB108903) and 863 (No. 2001AA214021) programs from the MOST, and the NNSF (No. 39925017) of China. CI-2003-07 from the MOST of China and SFI of Ireland.

SINORHIZOBIUM MELIOTI DCTB MAY CONTAIN A PAS DOMAIN SENSING BOTH C_4-DICARBOXYLIC ACIDS AND OXYGEN SIGNALS

Bei-Yan Nan, Jin Wen, Wen Yue, Bi-Qing Wen and Yi-Ping Wang*
National laboratory of Protein Engineering and Plant Genetic Engineering, College of Life Sciences, Peking University, Beijing 100871, China.

Sinorhizobium melioti DctBD is a two-component system (TCS). DctB senses C_4-dicarboxylic acids (dCA) and phosphorylates DctD, which in turn, activates transcription of *dctA*. Previous studies indicate that the DctBD system is also regulated by oxygen (Wen J et al., in press). In this study, sequence analysis and structural prediction suggest that the arrangement of secondary structure modules in DctB is similar to that of FixL. Further more, the key residues homologous to H194, I209 and I210 in FixL (Miyatake H et al., 2000) are also found in corresponding modules of DctB, as H167, L178 and I180 respectively. The predicted sensing domain of DctB has a similar "glove" conformation which appears in FixL PAS domain: the β-barrel and α-helices form a space which hold the ligands such as dCA, while the highly flexible F/G loop is on the opposite site of the four β-sheets, forming a hydrophobic dome of the ligand binding domain which provides a proper location for Leu178, Ile180, L191 and V193 to monitor the ligand binding state and to trigger the conformation changes. Upon mutagenesis of H167 to alanine, the dCA induced activity of DctB is marginally affected, but its oxygen sensing ability is drastically destroyed; In contrast, when both L178 and I180 of DctB are replaced by alanine, it become totally non-inducible by both signals. These results suggest that DctB might detect oxygen tension through the spin state shift of iron-containing ligand as FixL. There might be a ligand (such as heme) which could incidentally bind to imidazole of H167, while L178 and I180 are the key residues in F-G loop involved in conformation change required for sensing of both signals. Taken together; DctB may contain a PAS domain where its dCA and oxygen sensing functions are achieved. This could be a novel example that a sensor histidine kinase of TCS detects two different kinds of signals by a single domain.

References: Wen J et al., Chinese Sci. Bul. (in press).
Miyatake H et al., (2000) J. Mol. Biol. 301, 415-431.

Acknowledgements: Work supported by 973 (No. 2001CB108903) and 863 (No. 2001AA214021) programs from the MOST, and the NNSF (No. 39925017) in China.

THE *SINORHIZOBIUM MELILOTI* EXOR AND EXOS PROTEINS REGULATE BOTH SUCCINOGLYCAN BIOSYNTHESIS AND FLAGELLA PRODUCTION

Shi-Yi Yao[1], Li Luo[2], Katherine J. Har[1], Anke Becker[3], Guan-Qiao Yu[2], Jia-Bi Zhu[2], and Hai-Ping Cheng[1]
[1] Biological Sci. Dept., Lehman College, City Univ. New York, Bronx, New York 10468, USA, [2] Lab. Mol. Genet., Shanghai Inst. Plant Physiology and Ecology, CAS, Shanghai 200032, P. R. China, [3] Inst. für Genomforschung, Univ. Bielefeld, Postfach 100131, 33501 Bielefeld, Germany

The production of *Sinorhizobium meliloti* exopolysaccharide, succinoglycan, is required for the formation of infection threads inside root hairs, a critical step during the nodulation of alfalfa (*Medicago sativa*) by *S. meliloti* (Leigh et al; Reed et al). Two bacterial mutations, *exoR95::Tn5* and *exoS96::Tn5*, resulted in the overproduction of succinoglycan and a reduction in symbiosis. Systematic analysis of the symbiotic phenotypes of the two mutants demonstrated reduced efficiency of root hair colonization. In addition, the *exoR95* mutation dramatically reduced long-term nodulation efficiency, while the *exoS96* mutation did not. Both the *exoR95* and *exoS96* mutations suppressed the biosynthesis of flagella and abolished the ability of the cells to swarm and swim. Succinoglycan overproduction did not appear to be the cause of the suppression of flagella biosynthesis. Further analysis of suggested that both *exoR95* and *exoS96* mutations suppressed the expression of the flagella biosynthesis gene but not chemotaxis genes. These findings suggest that both ExoR protein and the ExoS/ChvI two-component regulatory system are involved in the regulation of succinoglycan and flagella biosynthesis. These findings provide new avenues to understand the physiological changes the *S. meliloti* cells go through during the early stages of symbiosis and the signal transduction pathways that mediate such changes.

References
Leigh JA et al. (1985) Proc. Natl. Acad. Sci. USA 82, 6231-6235.
Reed JW et al. (1991) J. Bacteriol. 173, 3789-3794.

PHOSPHATE ASSIMILATION AND REGULATION IN *SINORHIZOBIUM MELILOTI*

Ze-Chun Yuan, Richard A. Morton, Rahat Zaheer, Adrian Rybak, Turlough M. Finan.
Department of Biology, McMaster University, 1280 Main Street West, Hamilton, Ontario, L8S 4K1, Canada.

Phosphorus is an essential nutrient and, in bacteria, its transport and assimilation are regulated by the PhoR-PhoB two-component system. PhoR is a transmembrane sensor histidine kinase and PhoB is a transcriptional activator. Under phosphate-limiting conditions, PhoR autophosphorylates itself and phosphorylates PhoB. Phosphorylated PhoB binds at the pho box and thus regulates transcription. The classical pho box is 18 nucleotides with two 7-bp repeats of 5'-CT{T/G}TCAT-3'. *Sinorhizobium meliloti* has two phosphate transporters: PhoCDET and the Pit system.

GusA gene fusions indicate that, in strain Rm1021, *pstS* expresses at the same level in both limiting (5µM) and high (2mM) phosphate, whereas in strain RCR2011, it is induced by limiting phosphate but repressed by high phosphate. Two putative pho boxes were identified in the *pstS* promoter region. By comparison with the *Agrobacterium tumefaciens* C58 *pstSCAB* sequence, a single nucleotide deletion in *pstC* occurs in Rm1021; however, in RCR2011, Rm5000, and 9 other rhizobia species, the *pstC* gene does not contain the deletion. The *phoCDET* mutation in RCR2011 can fix nitrogen, but the *phoCDET* mutant in Rm1021 cannot. *pstSCAB* over-expression complements the phenotype of the *phoCDET* and *pit* double mutant, allowing N_2 fixation. We conclude that the frame shift in *pstC* in Rm1021 still allows limited phosphate transport. Using previously identified pho boxes from *S. meliloti* and *E. coli*, we derived a 18-bp weight matrix for the PhoB recognition motif to demonstrate a consensus sequence of two tandem 5'-CT{T/G}TCAT-3' motifs separated by a 4-bp linker. We then used this to identify several new *pho* regulons. A putative pho box was found in the promoter region of both the *S. meliloti* and *P. aeruginosa katA* genes, which are highly induced under low phosphate conditions and require a functional PhoB. Similarly, a putative pho box was found in the *afuA* promoter region in *S. meliloti* and *A. tumefaciens*, which was also strongly induced by low phosphate condition and required functional PhoB. Maybe phosphate limitation causes H_2O_2 over production and bacteria express catalase to keep a reasonable H_2O_2 level to prevent cell damage.

EXPRESSION ANALYSIS OF HEMOGLOBIN GENES IN ACTINORHIZAL PLANT *ALNUS FIRMA*

Fuyuko Sasakura, Katsumi Takenouchi, Toshiki Uchiumi, Akihiro Suzuki, Shiro Higashi and Mikiko Abe
Graduate School of Science and Technology, Kagoshima University, Kagoshima, 890-0065 Japan

The actinomycetes *Frankia* can form nitrogen-fixing root nodules with some non-leguminous woody plants, so-called actinorhizal plants (e.g. *Alnus, Elaegnus, Casuarina*). In the legumes-*Rhizobium* symbiosis, a nodule specific protein, so-called leghemoglobin, is synthesized abundantly and cooperates in nitrogenase activity. Several hemoglobin genes have been identified in not only legumes but also non-legumes and they are classified into two groups, symbiotic type and non-symbiotic type.

In this study, two hemoglobin genes (*AfNhb1, AfNhb2*) were isolated from the cDNA library of *A. firma* nodules by plaque hybridization using *Lotus japonicus* hemoglobin genes (*LjLb3, LjNSG1*) as a probe. Deduced amino acid sequences of AfNhb1 and AfNhb2 show 100% identity and these genes were classified into the non-symbiotic hemoglobin group. Dominant expression of *AfNhb1* and *AfNhb2* was detected in the nodules and leaves respectively from real-time RT-PCR analysis. In addition, the expression of these genes were analyzed under various stress treatments (hypoxia, 4°C and high osmolality) and nitrogen compounds (NO_3^-, NO_2^- and NO). Both genes were expressed at similar levels against nitrogen compounds. When whole plants were treated simultaneously with nitrogen compounds and NO scavenger, both genes were suppressed.

These results suggest that *AfNhb1, AfNhb2* may be involved in nitrogen metabolism or/and role as NO scavenger. *AfNhb1* expressed in nodules may be involved especially in interaction with *Frankia*.

Furthermore analysis of promoter sequences and transgenic roots that up- or down-regulated AfNhb will be required to understand the function of hemoglobin on actinorhizal-*Frankia* symbiosis.

SECTION V.

TOWARDS SUSTAINABLE AGRICULTURE AND PROTECTION OF THE ENVIRONMENT

ECOLOGICAL SIGNIFICANCE OF LUMICHROME AND RIBOFLAVIN AS SIGNALS IN THE RHIZOSPHERE OF PLANTS

Felix D Dakora[1], Sheku Kanu[2] and, Viviene N Matiru[3]
[1]Research Development and Technology Promotion, and [2] Faculty of Applied Science, Cape Peninsula University of Technology, District Six, Keizersgracht, PO Box 652, Cape Town 8000, South Africa. [3] Botany Department, University of Cape Town, Rondebosch 7701, South Africa.

About 50 years ago, West and Wilson (1938) reported on the ability of root-nodule bacteria to synthesize the vitamins riboflavin and thiamin for cell growth. Since then, various studies have shown that several bacterial species including *Sinorhizobium meliloti*, *Pseudomonas fluorescens*, *Rhizobium leguminosarum bv viceae*, *Azospirillum brasilense* and *Azotobacter vinelandii* (Rodelas et al. 1993; Sierra et al. 1999; Yang et al. 2002) produce vitamins (e.g. thiamin, riboflavin, niacin, biotin, and panthothenic acid) and amino acids (e.g. glutamate, lysine, arginine, tryptophan, methionine, etc) for their own use. Subsequently, the presence of riboflavin was detected in soil and its uptake and translocation to shoots of plants demonstrated by Capenter (1943). Thirty years later, Rao (1973) reported on riboflavin stimulation of plant growth.

Recently, a number of biologically-active novel molecules have been purified and identified from rhizobial exudates that stimulate cell growth and nodule organogenesis (De Jong *et al*. 1993; Daychok *et al*. 2000). Further studies have revealed the ability of these compounds to stimulate plant growth. Phillips *et al*. (1999) have, for example, shown that lumichrome isolated from *Sinorhizobium meliloti* culture filtrate increased root respiration by 11 – 30%, and promoted plant growth by 8 – 18% when applied to seed or seedling roots. The enhanced growth was apparently due to increased net C assimilation, possibly via PEP carboxylase activity. Lipo-chito-oligosaccharide (LCO) molecules isolated from rhizobia have also been shown to stimulate seed germination (Zhang and Smith 2001) and seedling development in both monocots and dicots (Smith et al. 2002). Taken together, these findings show that bacteria are capable of producing various organic molecules that alter plant development. However, because bacteria also produce the commonly known phytohormones such as auxius, gibberellins, cytokinins and abscicic acid (Phillips and Torrey 1970; Dart 1974; Dakora 2003), many workers have in the past tended to attribute bacterial ("PGPR") influences on plants to unexplained phytohormone effects. From the recent discoveries of active bacterial metabolites, it has become clear that, besides the classical phytohormones, additional signalling molecules exist that orchestrate plant development.

1. Ecological significance of rhizobial exudation of lumichrome and riboflavin in the rhizosphere

Thin layer chromatographic (TLC) analyses of rhizobial strains isolated from indigenous *Psoralea* species have revealed considerable variation in their sythesis and release of riboflavin. As shown in Table 1, the rhizobial isolate from *P. pinnata* did not only produce the highest concentration of lumichrome, it also produced more riboflavin in culture filtrate than any of the other 7 strains. Interestingly, the bacterial isolate from *P. aphylla* was the most limited in its production and release of both lumichrome and riboflavin (Table 1). Although the ecological consequence of low or high production of lumichrome and riboflavin have not yet been assessed, *Sinorhizobium meliloti* strains carrying extra copies of the riboflavin biosynthesis gene *rib* BA were found to release 15% more riboflavin than wild-type and were 55% more efficient in alfalfa root colonization (Yang *et al*. 2002). The increased root colonization by mutant *S. meliloti* strain was, no doubt, a first step towards better nodulation. Although riboflavin is less efficient in inducing CO_2 production by roots (Phillips *et al*. 1999), its accumulation and degradation into lumichrome has the advantage of the latter inducing an increase in concentration of rhizosphere CO_2, which is needed for growth of rhizobial populations (Lowe and Evans 1962).

Table 1. Production of lumichrome and riboflavin by root-nodule bacteria isolated from eight *Psoralea* species. Values (Mean $\pm$ SE n = 4) followed by dissimilar letters in a column are significantly different at $P \leq 0.05$.

Psoralea species	Metabolite concentration (ng.mL)	
	Lumichrome	Riboflavin
P. laxa	5.04 $\pm$ 0.99ab	17.92 $\pm$ 2.10bc
P. aphylla	3.10 $\pm$ 0.09c	14.48 $\pm$ 2.95c
P. asarina	3.77 $\pm$ 0.39bc	15.01 $\pm$ 1.48c
P. repens	4.50 $\pm$ 0.47ab	25.05 $\pm$ 1.50b
P. pinnata	5.82 $\pm$ 0.63a	51.81 $\pm$ 4.56a
P. monophylla	3.47 $\pm$ 0.29bc	16.99 $\pm$ 2.25bc
P. aculeata	4.28 $\pm$ 0.45abc	15.1 7 $\pm$ 1.13c
P. restioides	3.89 $\pm$ 0.72bc	22.82 $\pm$ 3.67b

Furthermore, the increase in rhizosphere CO_2 concentration from lumichrome and riboflavin can stimulate growth of vesicular-arbuscular fungi (Becard and Piche 1989; Becard *et al*. 1992) and increase the incidence of mycorrhizal symbiosis. These indirect benefits of lumichrome and riboflavin to legume symbioses via their effects on the plant are essential for enhanced N and P nutrition. Also, the mere fact that most root-colonizing bacteria commonly produce and release lumichrome and riboflavin (Phillips *et al.* 1999), and some bacteria also produce 8 times more extracellular riboflavin in exudates relative to internal cellular concentration suggests that these molecules have evolved directly or indirectly as rhizosphere signals influencing outcomes of plant-bacterial interactions.

2. Species-wide effects of lumichrome on plant growth and organ development
An earlier study by Phillips *et al.* (1999) showed an increase in growth of alfalfa plants following lumichrome supply and this was attributed to increased net assimilation of CO_2 release from lumichrome-induced root respiration. Whether this growth response of alfalfa to lumichrome is unique to this species, remains unknown. So we tested four legume and four cereal species for their growth response to 0, 5 and 50 nM lumichrome. The legume species included cowpea (*Vigna unguiculata* L. Walp.), soybean (*Glycine max* L.), Bambara groundnut (*Vigna subterranea* L Verdc.) and common bean (*Phaseolus vulgaris* L.), while the cereals were finger millet (*Eleucine coracana* L.), sorghum (*Sorghum bicolor* L. Moench), maize (*Zea mays* L.) and Sudan grass (*Sorghum bicolor* subsp. *Sudanensis* L.). Of the four legumes, only cowpea and soybean responded to the application of 5 nM lumichrome, but not common bean or Bambara groundnut. Culturing cowpea with 5 nM lumichrome depressed root growth but increased shoot dry matter and total biomass as a result of early initiation of trifoliate leaf development. With soybean, the application of 5 nM lumichrome also caused early initiation of trifoliate leaf development, an expansion in unifoliate and trifoliate leaves, an increase in stem elongation, which together lead to increased shoot dry matter and plant total biomass. Of the four cereal species tested, only sorghum, millet and maize responded to lumichrome, but not Sudan grass. Supply 5 nM lumichrome to roots increased sorghum shoot and root dry matter relative to control. With millet, 5 nM lumichrome increased root mass but not shoot dry matter, while it decreased root and total dry matter at 50 nM concentration. Applying 5 nM lumichrome to roots of maize increased total leaf area measured on per-plant basis by 26%, resulting in significantly increased shoot, but not root, biomass. Unlike millet however, maize plants did not show any decrease in shoot or root growth at the higher 50 nM lumichrome concentration. Taken together, our data show that plants differ in their response to the signal lumichrome possibly as a consequence of species-specific diversity and/or physiological variation.

3. Effects of lumichrome on plant function
Supplying 5 nM lumichrome to roots of inoculated and uninoculated soybean and cowpea plants increased the concentration of this molecule in xylem and its accumulation in leaves relative to control, indicating that the metabolite gets taken up by plant roots and transported to shoots where it alters leaf and stem development. Treating cowpea, Bambara groundnut, soybean, sorghum and maize plants with 10 nM lumichrome, infective rhizobial cells (10 mL of 0.2 OD600) and 10 nM abscicic acid

altered stomatal conductance and leaf transpiration. While in cowpea the application of these treatments increased stomatal conductance leading to increased rates of leaf transpiration; it decreased stomatal conductance in Bambara groundnut, soybean and maize and resulted in reduced leaf transpiration. These differences in plant response to root application of rhizobium or its exudation product lumichrome have implications for nutrient uptake and whole-plant water relations. An increase in stomatal conductance would favour increased nutrient uptake by cowpea in its transpirational stream, while a decrease in stomatal conductance would reduce water loss and improve water relations as a measure against drought.

4. Conclusions
- Bacterial exudation of the rhizosphere signals lumichrome and riboflavin can vary with rhizobial strain, soil temperature and pH.
- There is a species-wide promotional effect of lumichrome (and possibly riboflavin) on plant growth and organ development at low nanomolar (5 nM) concentrations.
- Lumichrome taken up by plant roots and transported to the shoot probably elicits cell division, cell expansion and cell extensibility, leading to an increase leaf expansion, and stem elongation.
- A direct correlation exists between plant response to rhizobial inoculation and root application of lumichrome in respect of stomatal function. Both treatments consistently increased or decreased stomatal conductance and transpiration rates in responsive test species.
- Plant roots seem capable of collecting environmental signals from soil in the form of simple organic molecules released by microbes, and using them to adapt to their environment, e.g. for drought tolerance when these molecules cause decreased stomatal conductance and consequently reduced water loss.

References
Becard G et al. (1992) App. Environm. Microbiol. 58, 821-825.
Capenter CC (1943) Science 98, 109-110.
Becard G and Piche Y (1989) App. Environm. Microbiol. 55, 2320-2325.
Dakora FD (2003) New Phytologist 158, 39-49.
Dart PJ (1974) In: The Biology of Nitrogen Fixation (Quispel A, ed.), Amsterdam, The Netherlands, North-Holland Publishing Co., pp. 381-429.
Daychok JV et al.(2000) Plant Cell Rept. 19, 290-297.
De Jong AJ et al. (1993) Plant Cell 5, 615-620.
Lowe RH and Evans HJ (1962) Soil Sci. 94, 353-356.
Phillips DA and Torrey JG (1970) Physiologia Plantarum 23, 1057-1063.
Phillips DA et al. (1999) Proc. Natl. Acad. Sci. USA 22, 12275-12280.
Smith DL et al. (2002) In: Nitrogen Fixation: Global Pespective (Finan TM et al., eds.), CABI Publishing, Willingford, UK, pp. 327-330.
West PM and Wilson PW (1938) Nature 142, 307-398.
Zhang F and Smith DL (2001) Adv. Agron. 76, 125-161.

RHIZOSPHERIC PLANT – MICROBE INTERACTIONS FOR SUSTAINABLE AGRICULTURE

Kauser A. Malik, F.Y.Hafeez, M.S.Mirza, S. Hameed, G. Rasul and R. Bilal
National Institute for Biotechnology & Genetic Engineering (NIBGE),
P.O.Box 577, Jhang Road, Faisalabad, Pakistan. kamalik@comsats.net.pk

Rhizosphere appears as the most important niche for microbial activity. It is of direct relevance to nutrient cycling affecting plant nutrition. Rhizosphere can be defined as the root surface and the soil adhering the root, that is the zone directly influenced by root exudation, which provides nutriments required for microbial activity. Diazotrophs constitute an important component of the rhizospheric bacteria. They have been extensively studied in relation to Legume – *Rhizobium* symbiosis. However, after the development of the acetylene reduction technique for detecting nitrogenase activity there have been several reports of nitrogen fixation associated with non-legumes. With availability of various immunological and molecular biology techniques, such associations are being better explored.

Such microbial associations are important for sustainable agriculture. Legumes are important components of all cropping system but in overall economics of agriculture, they constitute nearly 10% of all the major crops. Rice and wheat are nearly 50% of all the major crops grown. Therefore any developments to meet their nutritional requirements through biological means will not only greatly impact their productivity but will drastically reduce the input costs by conserving natural resources. This is the essence of sustainable agriculture.

During the past two decades progresses on nitrogen fixation associated with grasses have been achieved in Pakistan. Initially, acetylene reduction assays with excised roots from various grasses and other non-legumes growing on saline lands were performed to estimate their nitrogenase activity. Extensive studies of Kallar grass *(Leptochloa fusca)* rhizosphere established that nitrogenase activity was present in the excised roots. Microbial strains were isolated from root surface and root interior and several strains of diazotrophs were characterized. Quantification of nitrogen fixation was performed using the ^{15}N isotope dilution method. Kallar grass can fix around 60KgN/ha (Bilal et.al 1993; Malik et al 1991, 1997)

Nitrogen-fixing bacteria-plant associations have also been studied with cereals and other grasses, as wheat, rice, maize and sugarcane. More recently studies on cash crops such as cotton was undertaken. Several microbes from different fractions of the rhizosphere have been identified (Ladha et al 1997)

Root inoculation studies of all the isolated diazotrophs and other rhizospheric bacteria have indicated beneficial effects (Gull et al 2004; Hafeez et al 2004 a, b). In order to elucidate such effects, screening of several of the isolated organisms have been carried out for the following properties:
- Nitrogen fixation
- Phytohormone production
- Phosphorous solubilization
- Zn mobilization
- Bacteriocin production

All the screened strains, which showed some of these properties were considered as Plant Growth Promoting Rhizobacteria (PGPR). In order to further elucidate their role in the rhizosphere, studies with electron microscopy and fluorescent *in situ* hybridization (FISH) methodology have been performed. In most cases colonization of the root surface, root tissues, in particular into the intracellular spaces was found. In some cases, bacteria have been seen inside the cells using immunogold labeling. However, there is still no conclusive evidence to term these associative bacteria as being true 'endophytes. Several strains have been shown to produce phytohormone and also solubilize phosphorous. In addition, several studies have focused on the properties of some *Rhizobium* to also live in association with non-legumes such as cotton and rice.

As a result of these experiments, a biofertilizer has been developed in Pakistan and is presently being marketed under the trade name of "BioPower". Biofertilizers are defined as preparations containing live, non-pathogenic microorganisms, which are beneficial to agricultural production in terms of nutrient supply. Developing a beneficial bacterial strain into a biofertilizer needs several steps. In addition, the strains suitable for producing biofertilizers must possess some if not all of the following properties:
- Ability of survival in a wide range of environments
- Competitive ability
- Nitrogen fixing, phosphorous solubilizing and plant growth promotion
- Infectivity and efficiency in the presence of soil N
- Ability to multiply in broth
- Survival on the seed carrier
- Survival under adverse environmental conditions
- Tolerance to pH changes, agrochemicals (pesticides)
- Genetic stability during growth and storage

After selection of the strains, an appropriate consortium has to be developed which should include the requisite properties and should be crop specific. For this purpose appropriate quality control and methods for monitoring of the inoculated organisms is essential. Several methods including serological techniques, tagging of the strains with detectable markers and PCR based 16sRNA probes are being developed.

Some of the steps that are essential for the development of a commercial fertilizer are as follows:
- Selection of the strains
- Large scale production of the selected strains
- Carrier selection and its preparation
- Mixing and curing
- Maintenance of appropriate number of bacterial cells
- Stringent quality control

Following all these steps Biofertilizer produced under trade name of BioPower has been extensively tested on the following crops in the green house, in the fields and on the farmers' fields rice, legumes (chick pea, mung bean, lentils, pea, *Medicago*), wheat, cotton, and maize. The provincial agriculture extension department of Punjab also independently evaluated BioPower for rice.

Based on several field inoculation trials and feedback from the farmers, an approximate cost benefit by using BioPower for different crops has been tabulated as under. It shows substantial savings in fertilizer consumption and increase in the income of the farmers, which is one of the objectives of Sustainable Agriculture.

Table 1: Evaluation of the use of the BioPower fertilizer

	Cost benefit of BioPower / Ha		
	Yield Increase%	Fertilizer Saving %	Benefit in US $
Legumes	60-80	70-90	62- 288
Rice	20	50	82
Maize	20	50	102
Wheat	15	30	53
Cotton	30	50-70	255

References.
Bilal R et al. (1993) World J. Mirobiol. Biotechnol. 9, 61-69.
Gull M et al. (2004) Aust. J. Exp. Agriculture 44, 623-628.

Hafeez FY et al. (2004 a) Aust. J. Exp. Agricult. 44, 617-722.
Hafeez FY et al. (2004 b). Environ. Exp. Botany (In Press).
Ladha JK et al. (1997) Plant Soil 194, 1-10.
Malik KA et al. (1991) Plant Soil, 137, 67-74.
Malik KA et al. (1997) Plant Soil 194, 37-44.

GENETIC PROGRAMS FOR DEVELOPMENT OF NODULES AND ARBUSCULAR MYCORRHIZA IN LEGUMES: SOLID FACTS AND UNSOLVED PROBLEMS

Borisov A.Y., Tsyganov V.E., Ovtsyna A.O., Provorov N.A. and Tikhonovich I.A.
All-Russia Research Institute for Agricultural Microbiology, Podbelsky Sh. 3, St.-Petersburg, 196608, Russia; fax: 7-812-470-43-62

Endosymbiosis formed by legume plants with rhizobia (N_2-fixing nodules) and glomalean fungi (arbuscular mycorrhiza, AM) constitute the unique models for developmental biology and genetics of symbiosis. The plant objects for this research are the legume crops (*Pisum sativum* L.; *Glycine max* (L.) Merr.) in which the symbiosis genetics were initiated, and the model legumes (*Lotus japonicus* (Regel.) K. Larsen, *Medicago truncatula* Gaertn.) which facilitate the progress in molecular research of symbiotic (*Sym*) genes. Up to now, more than 130 *Sym* genes were identified in legumes using classical genetic methods and more than 15 genes were cloned and sequenced.

Among the two symbiotic systems, nodulation is studied in more details since it is facultative for both partners and may be arrested at any stage without a decrease of survival in plants (if combined N is provided) and in bacteria. Two developmental subprograms are involved in nodulation: (1) building the system of symbiotic compartments for transfer of rhizobia from root surface into symbiosomes leading to transformation of bacterial cells into bacteroids; (2) organogenesis of the nodule - the novel organ containing novel cell/tissue structures. Both subprograms are initiated by rhizobial Nod factors encoded by *nod* genes and released in response to plant flavonoids.

The most complete set of identified *Sym* genes is available in pea. They control root hair curling, Hac (*Sym8, Sym9, Sym10, Sym19, Sym30*), infection thread initiation, Iti (*Sym7, Sym14, Sym35*), infection thread growth in root hair, Ith (*Sym2, Sym36, Sym37, Sym38*), infection thread growth in the root, Itr (*Sym5, Sym16, Sym34*), infection thread growth in the nodule, Itn (*Sym33*), infection droplet differentiation, Idd (*Sym40, Sym41*), bacteroid development, Bad (*Sym31, Sym32*) and nodule persistence, Nop (*Sym13, Sym25, Sym26, Sym27*). Using the model legume approach, some of these genes were cloned and sequenced encoding receptor-like kinases (*LjNFR1~MtLYK3~PsSym2, LjSYMRK=MsNORK=MtDMI2=PsSym19, LjNFR5=PsSym10, LjHar1=PsSym29=*

GmNARK), transcriptional activator (*LjNin=PsSym35*), ligand gated cation channel (*MtDMI1~PsSym8*) and Ca^{2+}/calmodulne dependent kinase (*MtDMI3=PsSym9*) [1-13].

However, it looks surprising that for all studied genes, the orthologs were identified readily in the legumes representing different cross-inoculation groups. From the rhizobial side, host specificity is correlated to great differences in the spectra of *nod* genes for different cross-inoculation groups. Therefore, genetic determination of host specificity from the plant side looks to be an important problem for further research.

The *Sym* genes are common for development of symbiotic compartments and nodule organogenesis. For analyzing the order of gene functioning we used the approach based on developmental epistasis in the double recessive mutant lines. It is important that the orders for their functioning are different for the two subprograms. Therefore, further research should resolve an alternative: if the same or different gene products (mRNAs, proteins) are involved in determining the fates of bacterial cells and the plant cells/tissues during the endosymbiosis development in the nodules?

In contrast to root nodule symbiosis, AM is obligatory for endosymbont development. All its stages, including spore germination, root penetration and colonization, development of intracellular structures (arbuscules) and spores are dependent on the plant signals. From the plant side, the deep reorganizations occur in the cells occupied by arbuscules. Dissection of genetic programs for development of nodules and AM in legumes demonstrated a range of common genes involved in signal transduction within the host and the development of symbiotic compartments. It is generally believed that these common genes had been recruited from the ancestral AM symbiosis (originated 400-500 Myr ago together with terrestrial plants) to the younger nodulation symbiosis (originated 70-80 Myr ago in the Rosid I dicots). This concept stimulates us to address several exiting questions.

1. What happened to the *Sym* gene structures and functions due to recruiting from AM to nodule systems? Comparison of AM genes from nodulated legumes to their orthologues in non-legumes and non-nodulated legumes will facilitate greatly analysis of this question.

2. What mechanisms are responsible for differential regulation of *Sym* genes after the double inoculation of a legume host by rhizobium and AM-fungus leading to development of alternative symbiotic structures in the nearby portions of the same root?

3. How the same plant genes can promote quite different reorganizations in the bacterial and fungal symbionts?

These questions have not only basic but also the clear applied aspects. After the double inoculation of legumes by rhizobia and AM fungi, a tripartite symbiosis is formed that is highly effective in the plant nutrition under N and P limitation. We demonstrate that an effective selection of plants is possible for simultaneous improving both symbioses. This data convince us that legumes possess the mechanisms for coordinated regulation of both bacterial and fungal symbionts. What may be the nature of this regulation?

Since both symbionts demand the sufficient investment of the plants' C compounds, much attention should be paid to optimization of photosynthetic activity under conditions of symbiotrophic N+P nutrition. The preliminary data on the increasing the leaf surfaces and chlorophyll contents upon rhizobial or AM fungal inoculation should stimulate us to look for the systemic signals which may be responsible for these effects. Up to now the research of autoregulation of symbiosis was concentrated on the nodule number and its regulation by combined N: the pathway "nodule-leaf-nodule" is in the focus of attention. This pathway may be common for two symbioses since an enhanced mycorrhization was revealed in Nod^{++} mutants (in *LjHar1=PsSym29=GmNARK* genes). One should suppose a symmetrical regulation: does a "leaf-nodule-leaf" circuit exist to coordinate activities of N_2-fixing, AM mediated plant nutrition and CO_2-fixing machineries?

A swift progress in symbiosis genetics in legumes allows us to think that at least some of the mentioned problems may be solved in the nearest future.

Acknowledgements. The research was implemented in collaboration with: E.M. Barmicheva - V.L. Komarov's Institute of Botany (Saint-Petersburg); A.F. Topunov - A.N. Bach's Institute of Biochemistry (Moscow); S.M. Rozov - Institute of Cytology and Genetics (Novosibirsk); T.S. Naumkina - Institute of Grain Legumes and Groats Crops (Orel); J. Olivares, J. Sanjuan - EEZ-CSIC (Granada); U.B. Priefer - RWTH-AACHEN; J. Stougaard - University of Aarhus; N. Brewin - John Innes Centre (Norwich); T. Bisseling - Wageningen Agricultural University; V. Gianinazzi-Pearson - INRA (Dijon).

References
1. Ané J.M. et al. 2004 Science. 303: 1364.
2. Borisov A.Y. et al. 2003. Plant Physiol. 131: 1009.
3. Endre G. et al. 2002. Nature. 417: 962.
4. Krusell L. et al. 2002. Nature. 420: 422.
5. Lévy J. et al. 2004. Science 303: 1361.
6. Limpens E. et al., 2003. Science. 302: 630.
7. Madsen E.B. et al., 2003. Nature. 425: 637.
8. Mitra R.M. et al., 2004. Proc. Natl. Acad. Sci. USA. 101: 4701.
9. Nishimura R. et al. 2002. Nature. 420: 426.
10. Radutoiu S. et al. 2003. Nature. 425: 585.
11. Schauser L. et al. 1999. Nature. 402: 191.
12. Searle I.R., et al. 2003. Science. 299: 109.
13. Stracke S. et al. 2002. Nature. 47: 959.

MICROBIAL INOCULANTS: A CHALLENGE TO TUNE THE MICROBIAL METABOLITES AND SIGNALS FOR PLANT RESPONSIVENESS IN THE FIELD

R. Remans, C. Snoeck, E. Luyten, S. Dobbelaere, E. Somers, A. Croonenborghs, J. Michiels, and J. Vanderleyden
[1] Centre of Microbial and Plant Genetics (CMPG), KULeuven, Belgium
Roseline.Remans@agr.kuleuven.ac.be

Biological nitrogen fixation, solubilization of phosphorus, mineralization, among others are microbial processes with agronomical relevance and subject of mechanistic studies. As a result, signals that are required to initiate, sustain or enhance these processes have been gradually discerned (for a recent review, see Somers et al. 2004). Nod factors, quorum sensing signals, phytohormones, lipopolysaccharides, and exopolysaccharides are among the best-studied microbial signals in microbe-plant interactions. Blocking the synthesis of these signals has a major impact on the outcome of the microbe-plant interaction in which they have been identified, thereby illustrating their importance. Fewer are the studies in which the signaling pathways are modified in such a way that microbe-plant interactions are improved and thereby resulting in better plant productions. Moreover the successful cases are mostly limited to lab scale or at the most, greenhouse conditions. Generally, it is accepted that the plant is in control of microbial interactions. The study of *Rhizobium*-plant symbiosis has led to the concept of co-evolution. This means that, in terms of improving plant-microbe interactions, the genotypes of both the plant and the interacting microbes have to be taken into account Snoeck et al. 2004). In this report, we discuss three reported and ongoing studies in which bacterial inoculants have been modified and used for improving plant production.

A first case shows how nitrogen fixation can be enhanced by modification of Nod factor decoration in *Sinorhizobium* sp. BR816. Key compounds in eliciting nodules on legume roots within this broad host range strain were intensively studied, such as the elucidation of a peculiar mixture of sulfated and methylated Nod factor signaling molecules (Snoeck et al. 2001. A unique mutual exchange of household and symbiotic APS/PAPS pools in BR816 allowed us hypothesize that repartitioning of activated sulfate can be a key factor in maintaining sulfate or methyl substitutions of a wide array of bacterial determinants for symbiosis, hereby influencing symbiotic interactions (Laremans et al. 1997; Snoeck et al. 2003). Evidence came from bean inoculation experiments with

BR816 WT and derived mutants in sulfate activation genes, hereby shifting the balance of activated sulfate toward either sulfate assimilation, Nod factor biosynthesis or other cellular sulfation and methylation processes.

Mutational analysis of the *nodH* encoding sulfotransferase, revealed that knocking out *nodH* enables inoculated bean plants to fix significantly more nitrogen in hydroponic system. Similar results were obtained in greenhouse trials using perlite and in pot experiments using soil from to different areas in Mexico. Bacteroid expression studies confirmed a correlation between the observed higher nitrogen fixation capacity and *nifH* expression. Functional complementation with *nodH* and/or the sulfate activation genes *nodPQ* together with nodule microscopy, demonstrated that the increased nitrogen fixation capacity is likely due to a higher amount of available activated sulfate (APS/PAPS).

In a second study the effect of inoculation of *Rhizobium* in combination with other plant growth promoting rhizobacteria (PGPR) on nitrogen fixation under low phosphorus stress is evaluated. Soil is a complex dynamic habitat of rhizobacteria continually interacting with many different species. Different efficient tri-partite symbioses have already been described. *Azospirillum brasilense* has been shown to increase exudation of flavonoids in the bean rhizosphere, hereby positively influencing the nodulation by rhizobia (Burdman et al.1996). Co-inoculation of *Bacillus* and *Rhizobium tropici* ameliorates the nodulation on bean (Camacho et al. 2001).

Phosphous deficiency specifically is a nutritional limiting factor under nitrogen fixing conditions, as compared to plant growth in the presence of mineral nitrogen. It is of our interest to investigate if these tri-partite symbioses can enhance nitrogen fixation under low P stress and in combination with different bean genotypes. Inoculation experiments with *Rhizobium etli* CNPAF512 on two contrasting bean lines BAT477, selected as a line efficient in symbiotic nitrogen fixation (SNF) and P-use (CIAT), and DOR364, a SNF and P not efficient line (CIAT), revealed that *Azospirillum brasilense* (Sp245) has a negative effect on nodule number and plant dry weight of BAT477 under low P stress. Replacing the wildtype strain Sp245 by the indole-3-pyruvate decarboxylase (*ipdC*) mutant, reduced in auxin production, partly restored this effect. On the nodulation of DOR364, on the contrary, there was no significant effect of *Azospirillum* observed. Co-inoculation of *Rhizobium* with *Bacillus* increased the nodule number of both bean lines under low P stress. However, this effect was much more pronounced for DOR364 (144% of single *Rhizobium* inoculation) than for BAT477 (113% of single *Rhizobium* inoculation). On plant dry weight a similar effect of *Bacillus* under low P stress was observed.

In a third attempt to improve the plant growth promoting effect of rhizobacteria, the expression of the *ipdC* gene, encoding a key enzyme in indole-3-acetic acid (IAA) biosynthesis, in *Azospirillum brasilense* Sp245 was altered by replacing the endogenous *ipdC* promoter with a constitutive or plant inducible promoter. Inoculation experiments *in vitro* and in the greenhouse revealed that introduction of both recombinant *ipdC* genes into the wild type strain *A. brasilense* Sp245 further improved the plant growth promoting effect of this strain.

References
Burdman *et al.* (1996) Appl. Environ. Microbiol. 62: 3030-3033.
Camacho *et al.* (2001) Can. J. Microbiol. 47:1058-62.
Laeremans *et al.* (1997) Microbiology 143, 3933-3942.
Snoeck C *et al.* (2001) Mol. Plant-Microbe Interact. 14, 678-684.
Snoeck C *et al.* (2003) Appl. Environm. Microbiol. 69, 2006-2014.
Snoeck C *et al.* (2004) Plant Breeding Rev. 23: 21-72.
Somers E *et al.* (2004) Critical Rev. Microbiol., 30:1-36

ARE LEGUMES DOING THEIR JOB? THE EFFECT OF HERBICIDES ON N_2 FIXATION IN SOUTHERN AUSTRALIAN AGRICULTURAL SYSTEMS

Elizabeth A Drew, Vadakattu VSR Gupta, David K Roget
CSIRO Land and Water, PMB2, Glen Osmond, 5064, South Australia
Email: elizabeth.drew@csiro.au

Introduction

Grain and pasture legumes are an integral part of cropping systems in southern Australia. They can (i) act as a break crop for cereal root diseases, (ii) offer an opportunity to control grass weeds through the use of herbicides and (iii) increase soil nitrogen (N). The culmination of these benefits generally results in improved cereal yields following a legume crop (Khan *et al.* 2003). However, in the dryland farming regions of southern Australia (the Mallee), legumes have failed to provide the 'expected' N benefits, due primarily to low N_2 fixation (about 25% that of expected) (Gupta, V.V.S.R., CSIRO, personal communication). The aim of this research was to determine if recommended post-emergent herbicides affect the legume-rhizobia symbiosis, and thereby the soil N balance.

Materials and Methods

Field trials were set up in South Australia (4 sites) to test the effect of commonly used post emergence herbicides (Group A, B, C and D) on the growth and N_2 fixation by legume crops (peas, vetch and medic). Soil types ranged from alkaline sands to sandy loams and annual rainfalls from 250–450 mm. In this paper we present data for peas (*Pisum Sativum* cv. Parafield) inoculated with BioCare Group E (SU303) *Rhizobium leguminosarum* bv. *viciae* and grown at Waikerie, South Australia in 2003 (calcareous sandy loam, pH 8.1). The growing season rainfall (April-Oct) for 2003 was 160mm. Peas were either left unsprayed (control) or sprayed with one of six herbicides (Table 1) at the 3-4 node crop stage as per herbicide label directions. Three weeks after herbicide application, chlorophyll content, dry matter production and the number of 'effective' (pink) nodules were determined. At anthesis the %Ndfa (N_2 derived from fixation) and total N_2 fixed by the crop ($\delta^{15}N$ natural abundance technique) were measured. At harvest, yield, grain N, %Ndfa in the grain were measured.

Results

With the exception of Butroxydim + Fluazifop-P, all herbicides tested significantly affected above ground crop parameters, including shoot dry matter (DM) and crop colour (yellowing). This trend was also evident below ground, with significant reductions in the number of effective nodules per plant. However, herbicide application had no significant effects on above ground dry matter production or shoot [N] at anthesis (data not shown). While only the application of Flumetsulam reduced crop yield significantly, three of the herbicides significantly reduced the %Ndfa (Table 1). Further analysis showed that legumes treated with herbicides such as Haloxyfop, Fulazifop-P and Diflufenican contributed significantly (P<0.05) less to the soil N budget than unsprayed plants.

Table 1. Effect of post emergence herbicides on various parameters of the pea-rhizobia symbiosis relative to the unsprayed control treatment. * indicates significance at P<0.05. Values in parenthesis indicate non-significant trends.

Herbicide	Group	Mode of Action	Shoot DM	Yellowing	Yield	Effective Nodules	%Ndfa	Total N Fixed	Total Grain N	Grain Ndfa
Haloxyfop	A	AACase	-22%*	-14%*		-23%*	-60%*	-51%*		-18%*
Fluazifop-P	A	AACase		-14%*		-31%*	-36%*	(-37%)		-24%*
Butroxydim + Fluazifop-P[1]	A	AACase				(-29%)				(-12%)
Flumetsulam	B	ALS	-30%*	-26%*	-28%*	-43%*	(-21%)	(-27%)	-27%*	-20%*
Imazethapyr[1]	B	ALS		-26%*		-34%*	(-27%)	(-27%)		(-12%)
Diflufenican	F	PDS	-32%*	-14%*		-38%*	-34%*	(-37%)		-24%*
1. 2 replications per treatment. The control and all other treatments in the trial had 4 replications.										

Discussion and Conclusions

Our results indicate that some grass and broadleaf herbicides applied at recommended rates to grain legumes (e.g. field peas) can negatively affect the legume-rhizobia symbiosis in southern Australian agricultural systems. While the herbicides generally had little impact on grain yields or anthesis shoot [N] contents, they significantly reduced N_2 fixation. Furthermore, the reduced N_2 fixation due to some herbicides (e.g. Haloxyfop) meant that the legume crop (peas) was not contributing to the soil N pool and in some cases was potentially mining soil N. This corresponds to reports that %Ndfa in field peas ranged from 31-95% in southern Australia with only slightly positive to substantially negative contribution to the N balance (Unkovich *et al.* 1997). Finally, results presented in this paper provide explanation for the lack of rotational benefits observed in Mallee farming regions following grain legume crops. Soil type and climate moderate the effect herbicides have on the legume-rhizobia symbiosis. The current data suggests herbicide effects are more prominent in stressed environments (low soil moisture, low nutrients).

References

Khan DF et al. (2003) Aust J. Agric. Res. 54, 333-340.
Unkovich M et al. (1997) Aust J. Agric. Res. 48, 267-293.

NON-SYMBIOTIC BACTERIAL DIAZOTROPHS IN CROP-FARMING SYSTEMS: CAN THEIR POTENTIAL FOR PLANT GROWTH PROMOTION BE BETTER EXPLOITED?

Ivan R. Kennedy, A. T. M. A. Choudhury, Mihály L. Kecskés, Rodney J. Roughley and Nguyen Thanh Hien
SUNFix Centre for Nitrogen Fixation, Faculty of Agriculture, Food and Natural Resources, The University of Sydney, NSW 2006, AUSTRALIA

Biological N_2 fixation (BNF) by associative diazotrophic bacteria is a spontaneous process where soil N is limited and adequate C sources are available. Yet the ability of these bacteria to contribute to yields in crops is only partly a result of BNF. Hypothetically, diazotrophic plant growth-promoting rhizobacteria (PGPR) may have advantages in their specific interactions with C_3 and C_4 crop plants (rice, wheat, maize, sugarcane, cotton), significantly increasing the crops vegetative growth and grain yield.

We have reviewed the potential of these bacteria to contribute to yield increases in a range of field crops and outline possible strategies to obtain such yield increases more reliably (Kennedy et al., 2004) The mechanisms involved have a significant plant growth-promoting potential (PGPP), retaining more soil organic-N and other nutrients in the plant-soil system, thus reducing the need for fertiliser N and P. Bowen and Rovira (1999) included a discussion on possible mechanisms for the PGPR response including plant growth regulating effects (phytohormones), both positive and negative, induced systemic resistance to microbial pathogens, siderophore production aiding plant nutrition by chelation, P solubilisation and root-associated N_2 fixation. They drew attention to the role of these PGPR microbes as yield-increasing bacteria (YIB), a term favoured by Chinese workers in the area and a particular aim for field crops in this review.

Kennedy and Islam (2001) reviewed the possible contribution by non-symbiotic bacteria to crop growth from BNF with a focus on the historical evidence as well as some justification for the mechanisms involved. Dobbelaere et al. (2003) reviewed the diazotrophic PGPR in details, highlighting their mechanisms of action including BNF, plant growth promotion by production of auxins, cytokinins, gibberellins and ethylene, P-solubilisation, increased nutrient uptake, enhanced stress resistance, vitamin production and biocontrol. We have advanced the thesis that PGPR may promote more sustainable crop yield increases by modifying soil-plant processes so that N and other nutrients are more completely retained in the plant-soil system (Kennedy et al., 2004).

In Vietnam the performance of a multi-strain biofertiliser (BioGro) designed for rice culture has been statistically assessed in field trials over 3 y. This biofertiliser contained three strains of bacteria, originally selected from rice rhizospheres in the Hanoi area of Vietnam (Nguyen et al., 2003). One strain (*Pseudomonas fluorescens/putida,* 1N or 2N) was selected for its ability to reduce C_2H_2 to C_2H_4, as an indication of its potential for N_2 fixation. A second strain (*Klebsiella pneumoniae,* 4P), also a diazotroph, was selected for its ability to solubilise precipitated $Ca_3(PO_4)_2$ in an agar medium. The third strain (*Citrobacter freundii,* 3C), also a diazotroph, produces toxic extra-cellular compounds which inhibited 50% of a test group of 100 rhizosphere organisms but to which the inoculum strains are resistant. A new project on rice in Vietnam is investigating the effect of varying combinations of microbial strains, the need for repeated inoculation with each successive rice crop, the interaction of BioGro with rates of N and P application and the capacity of the biofertiliser to provide needs of high quality rice.

Rice, wheat and maize are the three major staple food crops for the world's population. A rice crop removes around 16-17 kg N to produce 1 t dry weight of rough rice, including straw (Ponnamperuma and Deturck, 1993). A wheat crop requires about 26-28 kg N to produce 1 t of rough grain including straw (Angus, 2001). Maize plants require 9-11 kg N to produce 1 t biomass (Anuar et al., 1995). Legumes nodulated by rhizobia would need N of the order of 60-80 kg N per t of grain. The much lower N demand by cereals compared to legumes of only 20-40% is a significant fact, requiring that to fix significant N_2, PGPR organisms need only occur at about 1-2% the frequency per g fw of crop plants as rhizobia in legume nodules (ca. 10^{11} per g nodule tissue *versus* 10^9 per g fw in cereals), given that PGPRs can occur as endophytes throughout the plant tissues or in the rhizosphere. Other PGP effects such as phytohormone production can presumably occur at much lower frequencies (ca. 10^5-10^6 per g fw).

Economic and environmental benefits include increased income from higher yields, reduced fertiliser costs and reduced emission of the greenhouse gas, N_2O or leaching of NO_3^--N to ground water. Obtaining maximum benefits on farms will require a systematic strategy of quality control designed to fully utilise all these beneficial factors, allowing crop yields to be maintained or even increased while fertiliser applications are reduced.

References
Angus JF (2001) Aust. J. Exp. Agric. 41, 277-288.
Anuar AR Shamsuddin ZH Yaacob O (1995) Soil Biol. Biochem. 27, 595-601.
Bowen GD and Rovira AD (1999) Ad. Agron. 66, 1-102.
Dobbelaere S et al. (2003) P Crit. Rev. Plant Sci. 22, 107-149.
Kennedy IR and Islam N (2001) Aust. J. Exp. Agric. 41, 447-457.
Kennedy et al. (2004) Soil Biol. Biochem. 36, 1229-1244.
Nguyen TH et al. (2003) Symbiosis 35, 231-245.

EFFECT OF LEGUME NODULE HYDROGEN UPTAKE STATUS ON HYDROGEN-OXIDIZING RHIZOBACTERIA AND THE ROTATION OF CROPS

Zhongmin Dong
Department of Biology, Saint Mary's University, Halifax, Nova Scotia,
B3H 3C3 Canada

Legume crops have been often used in crop rotations with cereal crops because of their positive effects on cereal growth and yields. Many believed that fixed-nitrogen left over by legumes is responsible for the beneficial effect of the rotation with legumes. Careful experiments showed, however, that nitrogen fertilizer can compensate for some, but not all, of the rotation benefit. Among many proposed factors to explain the beneficial effect of legumes, hydrogen gas released by legume nodules has been suggested recently as an important "fertilization" in rotation with legume crops (Dong and Layzell, 2002).

Hydrogen gas, a by-product of nitrogen fixation (Hunt and Layzell, 1993), can be oxidized inside the nodule where it is produced when the bacterial symbiont expresses an uptake hydrogenase (HUP) enzyme. In the case where nodules lack HUP activity (HUP^-) (Uratsu, et al., 1982), hydrogen gas diffuses into the soil and is consumed there (Conrad, and Seiler, 1979). This hydrogen oxidation causes increased rates of O_2 consumption and is coupled with chemoautolithotrophic CO_2 fixation (Dong and Layzell, 2001). It is believed that hydrogen oxidation also causes lasting soil changes that benefit the legume symbiosis as well as subsequent cereal crops. Studies have shown that hydrogen treatment of soils improved the growth performance of spring wheat, canola, barley and soybean (non-symbiotic) when compared with untreated soils or soils pretreated with air under growth chamber and field conditions (Dong and Layzell, 2002; Dong et al., 2003). The agents responsible for the hydrogen uptake and the plant-growth promotion of hydrogen-treated soil have been suggested as bacterial (McLearn and Dong, 2002; Irvine, et al., 2004). In these studies, hydrogen gas from electrolysis or commercial tanks was used to treat the soil samples although the experiments were designed to show the effect of hydrogen from the legume nodules.

To test the effect of the HUP status of symbioses on crop rotation, two isogenetic strains of *Bradyrhizobium japonicus*, JH andJH47, were used to inoculate soybean plants in field trials. The two strains only differ in that the JH47 strain has aTn5 insert in the hydrogenase gene to make it HUP⁻ (Hom *et al.*, 1988). Soil adjacent to the HUP⁻ nodules (nodulated by JH47) had a much higher hydrogen uptake rate compared with the soil surrounding HUP⁺ nodules (nodulated by JH). The rotation crop, barley, growing in soil, which had HUP⁻ soybean, had a significant higher tiller number and yield than that in soil, which had HUP⁺ soybean (Dean and Dong, unpublished).

Many hydrogen uptake bacterial strains were isolated from both hydrogen-treated soils and rhizospheric soil surrounding HUP⁻ nodules by an improved isolation method. Most of these bacteria belong to the beta-proteobacteria. Besides oxidizing hydrogen gas, another common characteristic of these strains is that they can reduce the 1-amino-cyclopropane-1-carboxylic (ACC) concentration in plant roots. Some of these strains express ACC deaminase, and others produce rhizobitoxine, which inhibits ACC synthase activity. ACC is the precursor of ethylene and ACC synthase is the rate-limiting enzyme in the biosynthesis of ethylene in higher plants (Yang and Hoffman, 1984). So, reducing ACC concentration is one of the most effective ways to reduce ethylene production in plant roots. Ethylene plays an important role in controlling the rhizobial infection of legumes. Nodule formation can be stimulated by ethylene inhibition (Yuhashi *et al.*, 2000, Ma *et al.*, 2002). Lower ethylene concentrations could lead to increased nodule formation and possibly more hydrogen gas evolution from nodules.

Ethylene has been shown to inhibit the root elongation. Some hydrogen-oxidizing bacteria can reduce ethylene production by decreasing ACC content through deaminase activity or ACC synthase inhibition and promote the root elongation. Our preliminary results showed that inoculation of hydrogen-oxidizing bacteria from soil adjacent to HUP⁻ nodules significantly promoted root elongation of young spring wheat seedlings (Maimaiti and Dong, unpublished). Very likely, any rhizobacteria that can reduce ethylene production in roots should also be plant growth-promoting rhizobacteria (PGPR).

Another interesting activity of these hydrogen-oxidizing bacteria is degrading N-acyl homoserine-lactone (HSL or NAHL), which is a quorum-sensing signal and a key regulator of the community behavior of many proteobacteria. Some rhizosphere bacteria need coordinated activation of diverse bacterial functions. HSL allows the induction of bacterial function(s) in a synchronous way and when the "appropriate" cell density is reached. This quorum-sensing regulates many gram-negative bacteria. Very likely, the NHSL-degrading activity of hydrogen-oxidizing bacteria around roots gives these bacteria potential to compete with gram-negative bacteria in natural ecosystems and to attenuate the plant pathogenicity of other soil bacteria as an "antivirulent" agent to counteract the pathogen.

Legume crops have many long lasting effects on soil. One of them is a change in the rhizobacterial population structure and an influence on plant growth. It is known that the soil microflora can improve plant growth through providing 'plant-available' nutrients, increasing the uptake of mineral nutrients, and protecting plants against pests and diseases. Our study has investigated the effect of hydrogen released from nodules on the hydrogen-oxidizing bacterial population in soil and showed that these organisms act as PGPR, regulate plant growth regulator balance, and promote plant growth.

References
Conrad R and Seiler W (1979) FEMS Microbiol. Let. 6, 143-145.
Dong Z and Layzell DB (2001) Plant and Soil. 229, 1-12.
Dong Z and Layzell DB (2002) Nitrogen Fixation, Global Perspectives. Eds. Finan T et al., CABI Publishing, New York, p.331- 335.
Dong Z et al. (2003) Plant, Cell and Environ. 26, 1875-1879.
Hom et al. (1988) Appl. Environ. Microbiol. 54, 358-363.
Hunt S and Layzell DB (1993) Annu. Rev. Plant Physiol. 44, 483-511.
Irvine P et al. (2004) ISHS Acta Horticulturae 631, 239-242.
Ma W et al. (2002) Can. J. Microbiol. 48, 947-954.
McLearn N and Dong Z (2002) Biol. Fertil. Soil. 35, 465-469.
Uratsu SL et al. (1982) Crop Sci. 22, 600-602.
Yuhashi KI et al. (2000) Appl. Environ. Microbiol. 66, 2658-2663.
Yang SF and Hoffman NE (1984) Annu Rev Plant Physiol 35, 155-189.

PHOSPHORUS USE EFFICIENCY FOR SYMBIOTIC NITROGEN FIXATION IN COMMON BEAN (*PHASEOLUS VULGARIS*) AND ITS CONSEQUENCE ON SOIL P DYNAMIC.

Jean-Jacques Drevon[1], Nora Alkama[1], Mathew Blair[2], Aurelio Garcia[3], German Hernandez[3], Philippe Hinsinger[1], Benoit Jaillard[1], Aline Lopez[4], Paula Rodino[5]

[1]ENSA.M-INRA UMR Rhizosphère & Symbiose, Place Viala, 34060 Montpellier, France; [2]CIAT, AA6713 Cali, Colombie; [3]Estación Experimental La Renée, Quivicán, La Habana, Cuba; [4]CIFN Cuernavaca, Mexico; [5]CSIC, Pontevedra, Espana.

Grain legumes can contribute to cropping systems through their ability to fix nitrogen from air. However, the symbiotic nitrogen fixation process requires additional phosphorus in plant for optimal nodule formation and N_2-dependent crop growth. Thus, P-deficiency limits the production of grain legumes in many soils of tropical and Mediterranean areas. The present work with *Phaseolus vulgaris* as a model grain legume, show genotypic variability in the phosphorus utilization efficiency for symbiotic nitrogen fixation, and the increase in proton efflux by nodulated-roots under P deficiency, a mechanism that may contribute to soil phosphorus availability.

Variability in phosphorus utilization efficiency for symbiotic nitrogen fixation, defined as the ratio of plant biomass per plant P content, was found among recombinant inbred lines (RILs) from the cross of BAT477 and DOR364, like previously among CIAT genetic resources, using an intensely aerated nutrient solution, namely hydroaeroponic culture, under P sufficiency versus sub-deficiency, i.e. 250 versus 75 µmol Pi week^{-1} plant^{-1}, and inoculation with *Rhizobium tropici* CIAT899. In figure 1 are shown the numerous of most contrasting RILs, located at the mean values of growth at flowering stage for both P supplies, RILs 28, 34, 75, 104 and 115 being significantly more efficient than RILs 7, 35, 70, 83, 124 and 147, and the parents.

Nodulated-root proton efflux, for RIL115 in hydroaeroponics, was higher under P limiting supply (figure 2). This resulted in a decrease (not shown) of 1.2 unit pH in a 1 mm layer of a calcareous soil, with initial pH of 7, on which the nodulated roots were grown for 2 weeks after their transfer from hydroaeroponics at vegetative stage. Lines contrasting in the phosphorus utilization efficiency for symbiotic nitrogen fixation are now available for studying the legume yield stabilization in low fertility soils.

Figure 1. Contrasting lines from the cross of BAT477 with DOR364 in g shoot dry weight per plant at 42 days after sowing

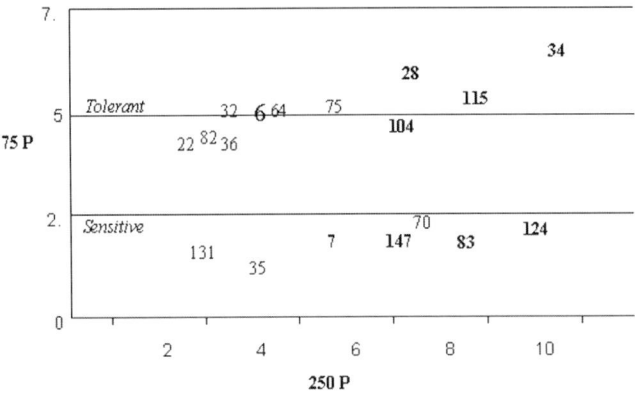

Figure 2. Proton efflux of RIL 115 – *R tropici* CIAT899 in daily efflux and during growth cycle

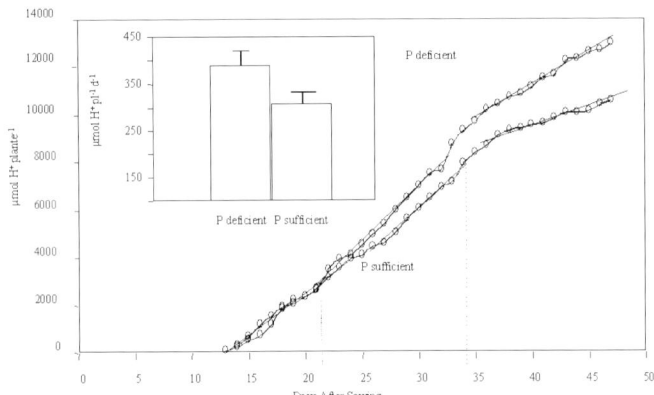

Further work in our laboratory is addressing whether their rhizospheric exchanges may increase P availability from soil or from a cheap source of fertilizer as rock phosphate. Simultaneously, these lines are cooperatively used for searching genes and nodule functions involved in the phosphorus utilization efficiency for symbiotic nitrogen fixation.

References
Vadez et al., 1999, Euphytica, 106, 231-242.

MUTATIONAL ANALYSIS TO STUDY THE ROLE OF GENETIC FACTORS IN PEA ADAPTATION TO STRESSES DURING DEVELOPMENT ITS SYMBIOSES WITH *RHIZOBIUM* AND MYCORRHIZAL FUNGI

Viktor E. Tsyganov[1], Alexander I. Zhernakov[1], Anna V. Khodorenko[1], Pavel Y. Kisutin[1], Andrei A. Belimov[1], Vera I. Safronova[1], Tatyana S. Naumkina[2], Alexey Y. Borisov[1], Peter Lindblad[3], Karl-Josef Dietz[4], Igor A. Tikhonovich[1]
[1] All-Russia Research Institute for Agricultural Microbiology, Podbelsky ch. 3, Pushkin 8, 196608, St.-Petersburg, Russia; [2] Institute of Grain Legumes and Groat Crops, Orel, p/b Streletskoe, Russia; [3] University of Uppsala, Sweden; [4] University of Bielefeld, Germany.

Beneficial endosymbioses formed by legumes with nodule bacteria and arbuscular mycorrhizal fungi are sensitive to different kind of stresses but at the same time they can relieve stress effect on plants. That is why the identification of plant genes involved in control over adaptations of legume symbioses to stresses is important. Two mutants with changed reactions to stresses: SGEcrt (*crt*) with hypersensitivity of roots to mechanical impedance (Tsyganov et al., 2000) and SGECd^t (*cdt*) with increased tolerance to cadmium (Cd) (Tsyganov et al., 2004) have been analysed by their ability to interact with *Rhizobium* and mycorhhizal fungi.

Previously it was shown that due to changed ethylene status the SGEcrt mutant has decreased nodulation ability in comparison with wild-type line SGE (Lutova and Pavlova, 1999; Tsyganov et al., 2002). Here the influence of mutation *crt* on mycorrhiza formation was analysed.

The plants were inoculated with *Glomus intraradices* Schenck and Smith isolate CIAM8 (BEG144) from the Collection of ARRIAM and grown in root/soil mixture with low phosphorus content. At the first time point (35 day after inoculation (DAI)), in mutant SGEcrt the frequency of mycorrhiza in the root system (F%) was 2-fold increased in comparison with wild-type (98.9% in mutant against 41.1% in wild-type). At the second time point (63 DAI), there was no difference in the frequency of mycorrhiza in the root system between mutant and wild type. So, with respect to mycorrhiza the mutant SGEcrt (*crt*) is characterized with accelerated mycorrhizal colonization. Thus, it seems that ethylene controls root, nodule and mycorhhiza development through common signal transduction pathways.

The mutant SGECd^t (*cdt*) was characterized with increased tolerance to Cd stress in comparison with wild type. Here the influnce of mutation *cdt* on nodulation ability in presence of Cd was analysed. The plants were grown in hydroponic culture (from 0.25 to 2 µM CdCl$_2$). Seedlings were inoculated with *R. leguminosarum* bv. *viciae* strain VF39gusA (kindly provided by Prof. Ursula Priefer, RWTH, Germany). Under control conditions without Cd the mutant and wild-type plants did not differ by nodule number, nodule weight, root and shoot biomass. In wild-type the number of nodules was gradually decreased with increase of Cd concentration and at 1.5 µM of CdCl$_2$ the nodulation was completely inhibited (Figure 1). In mutant the decrease in nodule number was less pronounced and mutant was still able to form nodules at concentration 2 µM of CdCl$_2$ in hydroponic culture (Figure 1). Light and electron microscopy have not revealed differences in histological and ultrastructural nodule organization between mutant and wild type under control conditions without Cd. The presence of Cd (0.5 µM of CdCl$_2$) led to decrease of nodule size and violation of histological zoning and bacteroid differentiation in wild type. Under the same Cd concentration nodules in mutant maintain their histological organization and the level of bacteroid differentiation similar to untreated control nodules.

Thus, mutation in the gene *cdt* determines changes in some mechanism(s) making the plant and its nodulation tolerant or insensitive to increased Cd concentration in its tissues. Introduction of this mutation into pea genotype with high level of biomass production will lead to creation of a novel pea variety which can be used for bioremediation of soils with elevated Cd content enhancing soil with useful symbiotic microbes and bioavailable nitrogen and phosphorus.

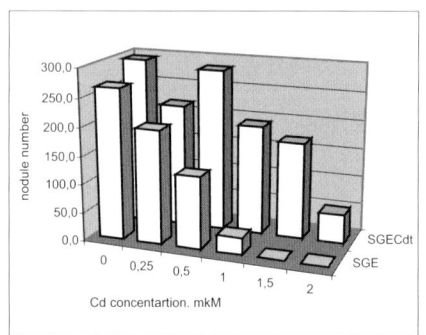
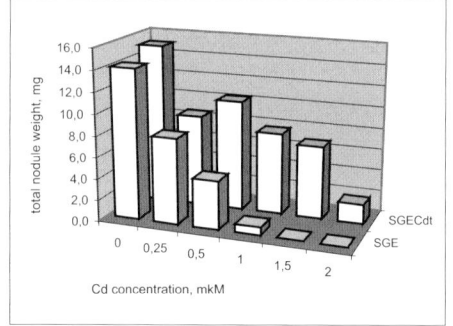

Figure 1. Effect of Cd on nodulation of wild-type SGE and mutant SGECd^t plants inoculated with *Rhizobium leguminosarum* bv. *viciae* strain VF39gusA in hydroponic culture

Acknowledgments
Supported by Russian Academy of Agricultural Sciences, Russian Fund for Basic Research (04-04-48462-a), St. Petersburg Governement (2004), INTAS (01-270) and for VET by St. Petersburg Government fellowship (PD04-1/4-230).

References
Lutova L and Pavlova Z (1999) Pisum Genet. 31, 48-50.
Tsyganov V et al. (2000) Ann. Bot. 86, 975-981.
Tsyganov V et al. (2002) Book of abstracts of 5th European Nitrogen Fixation Conference, Norwich, UK, p. 10.5.
Tsyganov V et al. (2004) Biology of Plant-Microbe Interactions, Vol. 4, pp: 506-509.

EFFECTS OF WATERLOGGING ON NITROGEN FIXATION OF A SUPERNODULATING SOYBEAN GENOTYPE, SAKUKEI 4

T. Matsunami, G. H. Jung, Y. Oki, W. H. Zhang and M. Kokubun
Graduate School of Agricultural Sciences, Tohoku Univ., 1-1 Tsutsumidori-amamiyamachi, Aobaku, Sendai 981-8555, Japan)

The supernodulating soybean cultivar Sakukei 4 was previously characterized by its superior ability to maintain a high leaf nitrogen content and high photosynthetic rate, and thereby is potentially high yielding. In Japan, prolonged rainy period or heavy rainfall in the field with poor soil drainage converted from paddy field often hinders the vegetative growth and thereby causes a reduction of the seed yield in upland crops. The objective of the present study was to examine the effect of waterlogging on nitrogen fixation and the recovery from damage in Sakukei 4.

Materials and Methods
Experiment 1: The supernodulating soybean genotype Sakukei 4 and its parental cultivar Enrei were grown in 1-liter pots filled with 0.8 kg of a fine-textured clayey Terrace Yellow soil. On 32 days after sowing (DAS), the flooding treatments were imposed by submerging pots to the level of soil surface. Dissolved oxygen concentration of the flooding treatment with (FA) and without (FN) aeration was 10.2 % and 8.6%, respectively. Acetylene reduction activity (ARA) of nodules was measured on 0, 3, 6 and 9 days after the initiation of the treatment.
Experiment 2: The same materials as in Exp.1 were grown in 4.8-liter pots filled with 4.0 kg of soil as in Exp.1 and the same treatment was given for 10 days during the different three vegetative growth stages. ARA and specific ARA (ARA per nodule dry mass) were measured after cessation of the treatment (pod-expansion stage; R4).

Results and Discussion
ARA and specific ARA declined rapidly with time after flooding treatment in both cultivars (Fig.1). The decline of specific ARA was more pronounced than that of ARA in Sakukei 4, but it was vice versa in Enrei (Fig.2). After cessation of the treatment, ARA of both cultivars was about 80% of their unstressed plants (Table 1). The recovery of specific ARA after the treatment was up to 90 % in Enrei, whereas it was more pronounced in Sakukei 4.

The damage caused by flooding from 39 DAS was prolonged after cessation of the treatment in both cultivars. Fig. 3 shows the time course of nodule dry mass per root dry mass (N/R ratio) during the whole growth period for three years in unstressed plants. The N/R ratio of both cultivars rapidly increased from 30 to 50 DAS and exhibited the peak around 50 DAS. In conclusion, Sakukei 4 and Enrei responded differently to waterlogging, the recovery of specific ARA was more pronounced in Sakukei 4, and the damage caused by flooding was most critical when it was given from 30 to 50 DAS.

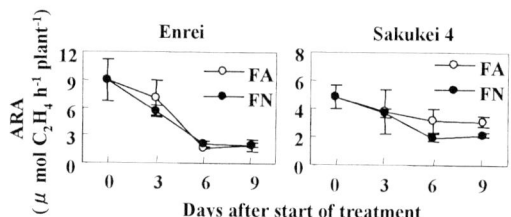

Fig. 1 Effects of waterlogging on acetylene reduction activity (ARA) (Exp. 1). Vertical bars represent standard error (n=3).

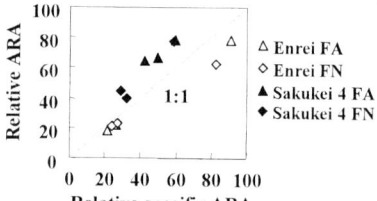

Fig. 2 Relative acetylene reduction activity (ARA) versus relative specific ARA during waterlogging treatment (Exp. 1). Each value is relative to control at each measuring day. A dotted line in panel is the 1:1 relationship between relative ARA and relative specific ARA.

Table 1 Recoveries of acetylene reduction activity (ARA) and specific ARA after cessation of treatment (Exp. 2).

	Treatment period	ARA (μ mol C_2H_4 h^{-1} $plant^{-1}$)		Specific ARA (μ mol C_2H_4 h^{-1} g dry weight^{-1})	
Enrei	Control	128		56	
	14-24DAS	110	(0.86)	53	(0.95)
	29-39DAS	86	(0.67)	46	(0.83)
	42-52DAS	102	(0.80)	51	(0.91)
	Average	99	(0.78)	50	(0.90)
Sakukei 4	Control	62		17	
	14-24DAS	57	(0.93)	23	(1.36)
	29-39DAS	40	(0.64)	24	(1.39)
	42-52DAS	52	(0.84)	16	(0.96)
	Average	50	(0.80)	21	(1.24)

Each values is the mean of five replicates.
Figures in the parentheses are relative values to the control plot.

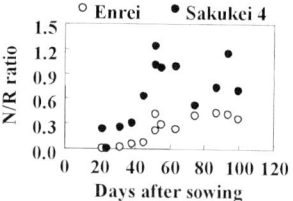

Fig. 3 Time-course of nodule dry mass per root dry mass (N/R ratio) in unstressed plants during the whole growth period in 2002, 2003 and 2004 (Exp. 2). Each value is the mean of five replicates.

EFFECT OF N APPLICATION RATE ON NITROGEN FIXATION AND TRANSFER FROM VETCH TO BARLEY IN MIXIED STAND

Hyowon Lee[1], Wongo Kim[2], Hyungsu Park[3], and Sugon Kim[3]
1 Korea National Open University. 2 National Livestock Reseach Instiute.
3 Seoul National University.

Introduction

Consumers in Korea are interested in so called Well-being agricultural products and they prefer organic farming products than conventional to products. In order to reduce chemical fertilizers or adopt Zero fertilizer, Farmers try to provide the crops with N by introducing legumes such as vetch clover in cropping system. Dairy farmer in Korea begin to think production of organic milk and some farmers are considering use of mixtures of legumes with cereals. The forage can supply higher dry matter roughage, fixation of N_2 and roughage with nutritionally balanced feed between energy and protein. The findings of N_2 fixation and N transfer from vetch to barley in mixed stands is lacking in Korea. The major objective of this research was to measure N_2 fixed in vetch mixed with barley and estimate the amount of N transferred from vetch to barley by 15N dilution method.

Material and Methods

The studies reported here were conducted in the mid-part of South Korea from mid October in 2001 to May in 2002. Four levels of N fertilizer 0, 75, 113 and 150kg N/ha as urea were treated. P and K applied to all treatment at the rate of 200, 200kg/ha. The experimental design was complete randomized with three replications plus only barley sowing plot as reference crops. The plot was 2×3m and $(^{15}NH_4)_2 So_4$ solution at 99.7 atom% 15N excess was applied to $0.1m^2$ area at the center of plot. A total of 900ml of solution was in distributed over the treated area. In early April in 2002 and marked with iron pegs at the four corners. Herbage was harvested from each microplot at ground level and separated into grass and clover fraction. A sub sample of the herbage was taken for analysis of 15N concentration by a continuous flow stable isotope ratio mass-spectrophotometer. Biological nitrogen fixation by vetch was estimated by peoples et al

(1989) method, using barley as the control. The transfer of biologically fixed N was by chalk (1996) with difference method and 15N dilution.

Results and Discussion
The proportion of clover N derived from N_2 fixation was dependent on the level of N nitrogen application. The rate of proportion was from 95.5 to 84.8. The result was higher than that of peoples's, which reported 78.6 to 84.5%. But Laidlaw et al.(1990) who applied 0, 1.5, 3.0, 4.5 or 6.0g N m^2 on the pasture had similar results of 95, 91, 89, 87 and 83%. They reported different results which revealed 89.5% in summer and 55.9 in Autumn (Laidlaw et al.,1996). In another paper, Chen et al.(1998) also found out the fact that N derived from N_2 fixation reached peak levels of 85-95% during the spring. When we applied the N-difference method to calculate the estimation of N-transfer, we had the proportion of N_2 fixation from 58% to 49% depending on N application level and the amount of transfer was 94kg to 68kg/ha. Even though Fujita et al.(1990) reported similar result of N-transfer, some researchers pointed out the possibility of overestimation of transfer. The 15N dilution method showed different proportion of N_2 fixation with various N fertilization, ranging from 38 kg/ha to minus fixation. This results was similar to Papastylianou's result, showing high level N fertilization suppressed N_2 fixation and transfer. Papastylianou and Danso(1991) reported that vetch derived on the average over 70% of its N from fixation while N transfer from vetch to oats was not significant. But they revealed also different results representing 9kg/ha N transfer from vetch (1992).

4. Conclusion
Estimates of proportion of N transfer to barley were investigated. The proportion by N-difference method with increment of N fertilizer were from 0.58 to 0.49 by difference method while 0.39 to 0.23 by ^{15}N-dilution method. The amount of nitrogen fixation per ha was from 150kg/ha to 219 kg/ha by different method instead of 49 to 105kg N/ha by ^{15}N-dilution.

Reference
Peopes M.B et al (1989) ACIAR. Canberra. Australia.
Laidlaw ASP et al (1990) Grass and Forage Science 45. 295-301
Laidlaw ASP et al (1996) Plant and soil 179. 234-255
Chen DJ et al (1998) Biological Nitrogen Fixation for 21st century 599-604
Fujita K et al (1990) Soil Science and Plant Nutrition.
Papastylianou I and Danso SKA (1991) Soil Biology & Biochemistry 23:447-452
Chalk PM (1996). Nitrogen transfer from Legumes to Cereals in intercropping Ito et al dynamics of roots and nitrogen in cropping systems of the semi-Arid Tropics. Japan International Research Center for Agricultural Science

ANALYSIS OF PROMOTIVE EFFECTS OF DEEP PLACEMENT OF SLOW RELEASE FERTILIZERS ON GROWTH AND SEED YIELD OF SOYBEAN BY ^{15}N DILUTION METHOD

Tewari Kaushal[1], Masaru Onda[2], Sayuri Ito[1], Akihiko Yamazaki[1], Hiroyuki Fujikake[1], Norikuni Ohtake[2], Kuni Sueyoshi[2], Yoshihiko Takahashi[2] and Takuji Ohyama[2]
[1]Graduate School of Science and Technology, Niigata University, [2]Faculty of Agric. Niigata University, 2-8050 Ikarashi, 950-2181, Niigata, Japan

Soybean (*Glycine max* [L] Merr.) plants can utilize N_2 fixed in the root nodules. Also, they can use the N absorbed from roots, from soil mineralized N or from fertilizer N when applied either in the form of nitrate or ammonium (Ohyama *et al*. 1989). However, the nodulation and nitrogen fixation of soybean is depressed by the application of combined N such as chemical N fertilizers. Takahashi *et al.* (1991) initially reported the promoted effects of deep placement of slow release N fertilizer, 100-day type coated urea on soybean seed yield. Deep placement of slow release N fertilizers, such as coated urea and lime nitrogen promotes soybean growth and seed yield without depressing N_2 fixation by root nodules (Tewari *et al*. 2002, 2003, 2004). In the present study, the effect of deep placement of various fertilizers was compared by the analysis of N derived from atmospheric N_2 *(Ndfa)*, N from soil *(Ndfs)* and N from fertilizer *(Ndff)* in the rotated paddy field of Niigata Agricultural Research Institute using ^{15}N dilution method.

Soybean (cv. Enrei) and its non-nodulating isoline En1282 were planted with planting density 8.9 plants m^{-2} in rotated paddy field in Niigata. The ^{15}N-labeled N fertilizers (100 kgN ha^{-1}), ammonium sulfate (AS), urea (U), coated urea (CU), and lime nitrogen (LN) was applied separately at the depth of 20 cm for Enrei and En1282 just below seedling line. Whole plants were sampled at R7 stage, and dry weight, ^{15}N abundance, and N concentration in each plant part were determined. The evaluation of the *Ndfa, Ndfs,* and *Ndff* was conducted by ^{15}N dilution method using En1282 as a reference plant. Both in nodulating and non-nodulating soybean, plant growth was highest with the application of LN and CU compared to U and AS (Figure 1). Non-nodulated line accumulated only 36-40% of total DW and 16-30% of total N in comparison to the nodulated line in all the fertilizer treatments, due to the lack of nitrogen fixation resulting in N deficiency. Both, in the nodulated and non-nodulated soybean, plant growth and total accumulated N were highest with the application of LN followed by CU, U and AS respectively.

The *Ndfa* was higher in LN (32 gN m^{-2}) and CU (26 gN m^{-2}) compared to U (21 gN m^{-2}) and AS (18 gN m^{-2}). In addition, the amount of N derived from fertilizer determined by ^{15}N abundance was also highest in LN and CU (Figure 2). The recovery rate of N fertilizers in Enrei was also highest in LN (42.5%) followed by CU (36.3%), U (20.7%) and AS (8.5%). The root growth and *Ndfs* was similar between nodulated Enrei and non-nodulated En1282 by deep placement of LN and CU, suggesting that continuous supply of N from slow release N fertilizers might contribute to the development of the root system. These results indicated that the use of LN and CU is effective to improve soybean growth, whereas AS and U was less effective. The combination of "deep placement" and "slow release N fertilizers" efficiently supply N without depressing the nitrogen fixation activity (%*Ndfa*) during the maturation stage, consequently resulting in the increase of seed yield. Between LN and CU, LN tended to give higher *Ndfa*, *Ndfs* as well as *Ndff*. This further suggests that LN, which is cheaper than CU, has a good future scope for increasing the soybean yield.

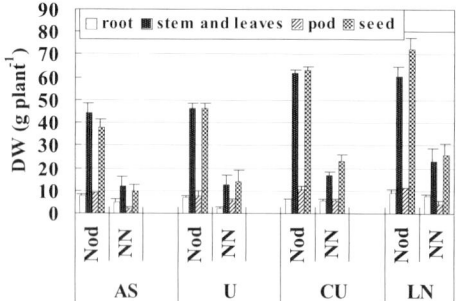

Fig 1: Dry matter accumulations in roots, leaves and stems, pods, and seeds of soybean plants (Nod- Enrei and NN- non nodulating En1282) cultivated with various ^{15}N fertilizer treatments, AS (Ammonium sulphate), U (Urea), CU (Coated urea CU-100), and LN (Lime nitrogen). Column and bars on the top denote the mean and standard deviation of four plants respectively (n=4).

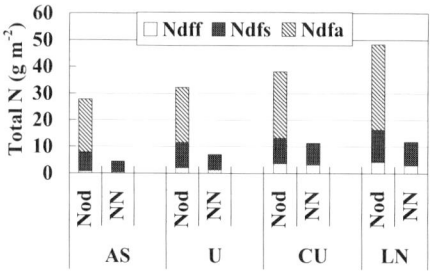

Fig 2: Estimation of N derived from various sources, N derived from fertilizer (*Ndff*), from soil (*Ndfs*) and from atmospheric N$_2$ (*Ndfa*) by soybean plants cultivated at Nagaoka using ^{15}N dilution method. Column denotes the mean of four plants (n=4).

Ohyama T et al. (1989) Soil Sci. Plant Nutr., 35, 9-20
Takahashi Y et al. (1991) Soil Sci. Plant Nutr., 37, 223-231.
Tewari K et al. (2002) Soil Sci. Plant Nutr., 48, 855-863
Tewari K et al. (2003) Jpn. J. Soil Sci. Plant Nutr., 74, 183-189
Tewari K et al. (2004) J. of Agronomy and Crop Science, 190, 46-58

ROLE OF RHIZOBIAL ENDOPHYTES AS NITROGEN FIXER IN PROMOTING PLANT GROWTH AND PRODUCTIVITY OF INDIAN CULTIVATED UPLAND RICE (*ORYZA SATIVA* L.) PLANTS

R. K. Singh, R. P. N. Mishra and H. K. Jaiswal
Microbial Genetics Laboratory, Department of Genetics and Plant Breeding, Institute of Agricultural Sciences, Banaras Hindu University, Varanasi-221005, India. rksbhu@yahoo.com

Rhizobia are bonafide member of symbiotic N_2-fixing microorganisms. They can exclusively induce root or stem nodules on leguminous hosts. Recently it has been explored that rhizobia can make an intimate association with non-legumes such as rice, wheat, maize, etc as endophyte, without forming any nodule like structure or causing any disease symptoms (Yanni *et a*l 1997; Chaintreuil *et al*., 2000; Gutierrez-Zamora and Martinez-Romero 2001; Tan *et al* 2001). The infection process of rhizobia to rice is *nod* gene independent, non-specific and does not involve the formation of infection threads which is a characteristic feature of legume-rhizobia symbiosis. Although rhizobial colonization extend from rhizosphere to inner into the epidermis, endodermis and cortex, but the main site of colonization is intercellular spaces of rice root (Reddy *et al* 1997). Rhizobial interaction with rice may be manifested a full spectrum of growth responses. Rhizobia may promote, inhibit or have no effect on rice plant growth (Yanni *et al*, 2001, Prayitno *et al.*, 1999; Perrine et al., 2001 Biswas *et al* 2000). In this study, we have reported the isolation of rhizobial endophytes from rice roots of cultivated upland Indian rice plant and their ability to fix nitrogen in association with the rice as host.

Materials and Methods
Rice root samples were collected at heading stage. Roots were thoroughly washed and surface sterilized as described by Yanni *et al*., (1997). Samples were macerated aseptically and one ml of suspension was inoculated to a number of leguminous plants like, *Trifolium, Phaseolus, Cicer, Vigna, Sesbania, Pisum,* etc. Data on nodule induction was recorded after 45 days of inoculation. In order to study endophytic colonization of rice root the best performing rhizobial isolate (BHUE-3) was conjugated with *E. coli* strain S17-1lambda pir (m*Tn*5Ss*gus*A21) harboring *gus*A reporter gene. Gus tagged rhizobia was inoculated to axenically growing rice plants. Gus assay was carried out after 30 days of inoculation according to Verma *et al*, (2001). Plant growth contributing

factors like nitrogen fixation, phosphate solubilization, phytohormone production and secretion of cell wall degrading enzymes were determined for rhizobial isolates. Nitrogenase activity was measured according to Hardy *et al.* (1973). Phosphate solubilization, cellulase and pectinase activities were detected in YEM plates supplemented with $CaHPO_4$, corboxymethyl cellulose (CMC) and pectin, respectively (Verma *et al*, 2001; Yanni *et al*., 2001). Green house experiment was performed in pots containing oven dried field soil. Plants were inoculated with 1 ml (~10^8 CFU) rhizobial culture and observation was recorded related to plant growth at maturity. Number of replications was six for each treatment and statistical analyses were done in Microsoft EXCEL computer software package.

Results and Discussion
Using the legume trap technique, we have isolated nine endophytic rhizobial strains and designated them as BHUE-1 to BHUE-9. These isolates followed the Koch postulates and expressed the ability to nodulate commonbean (*Phaseolus vulgaris* L.) plants under axenic conditions. Out of the nine isolates, four (BHUE-2, 3, 4 and BHUE-5) were selected for plant growth response studies under laboratory and green house conditions. All four isolates gave a positive response in enhancing the plant growth measured in terms of plant height and shoot dry weight. BHUE-3 and BHUE-5 showed the best response (Table 1). In addition to other plant growth promotion mechanisms such as phosphate solubilization and phytohormone production, these isolates expressed nitrogenase activity both *ex planta* and in association with rice plants. Such activity was not detected when plants inoculated with USDA 2695 or ANU843 were used for assay.

Table 1. Effect of rhizobial inoculation on growth and productivity of rice (cv Sarjoo-52) in green house conditions.

Strains	Characters			
	Plant height (cm)	Shoot dry weight (g hill^{-1})	Grain yield (g hill^{-1})	Nitrogenase activity §
Control	89.8	20.76	12.98	Nil
BHUE2	94.6^{ns}	24.72**	14.4*	0.81 ± 0.34*
BHUE3	96.0*	25.65**	15.53**	0.96 ± 0.42
BHUE4	93.1^{ns}	22.60*	14.25*	0.84 ± 0.40
BHUE5	95.3*	26.80**	15.60**	0.92 ± 0.38
USDA2695	92.2^{ns}	21.75^{ns}	13.32^{ns}	Nil
ANU843	91.6^{ns}	21.68^{ns}	13.00^{ns}	Nil

§ (micro mol C_2H_4 g^{-1} dry weight h^{-1})
Mean values followed by * or ** are significantly higher than control at the 95% and 99% levels, respectively and 'ns' indicates non-significant.
Values followed by ± indicates standard error mean

Although ANU843 is known to promote rice plant growth, its inability to express nitrogenase activity with rice cultivar Sarjoo 52 indicate that the conditions required for nitrogenase activity is not available whereas in the similar condition, the expression of nitrogen fixation using endophytic rhizobial isolates as source of inoculation indicate that these isolates create a better atmosphere for expression of *nif* genes.

The expression *nif* gene in "rice born" rhizobial inside the rice root and in free-living conditions also indicates that these isolates might be having some additional gene(s) that provide a protection for their nitrogenase enzymes from oxygen damage. Mutational and transcriptional analysis of nitrogen fixing machinery of these isolates will provide more authentic evidences for endophytic colonization, nitrogen fixation and plant growth promotion.

Acknowledgments

Authors are highly thankful Dr. Peter van Berkum, USDA, Beltsville, USA and Prof. B.G.Rolfe, Australia for providing USDA2695 and ANU843 cultures respectively. Department of Science and Technology and "TSBF/UNEP/GEF project on conservation and sustainable management of Belowground Biodiversity" are sincerely acknowledged for financial assistance.

References

Biswas et al., (2000)   Agronomy Journal 92: 880-886.
Chaintreuil et al., (2000) Appl.Environ. Microbial. 66(12):5437-5447.
Hardy et al, (1973). Soil. Biol. Biochem. 43, 47-87.
Perrine et al., (2001) Aust J. plant. Physiol 28: 923-937.
Prayitno et al., (1999) Aust J.Plant Physiol 26: 521-535.
Reddy et al., (1997) Plant Soil 194:81- 98
Tan et al.,(2000). Appl.Environ. Microbiol.67(8):3655-3664
Verma   et al (2001)   J.Bio/Technol. 91, 127-141.
Yanni et al.,(1997). Plant Soil 194: 99-114.
Yanni et al.,(2001) Aust J. Plant Physiol. 28: 845-870.

GROWTH, PHOTOSYNTHSIS, NODULE NITROGEN AND CARBON FIXATION IN DESI AND KABULI GENOTYPES OF CHICKPEA (*CICER ARIETINUM L.*) UNDER SALT STRESS

Garg Neera and Ranju Singla
Department of Botany, Penjab University, Chandigarh-160014, India

Soil salinity is a major agricultural problem in the world which significantly reduces the productivity of a broad range of crops. Legumes have long been recognized to be highly salt sensitive. Chickpea is a premier pulse crop amongst the one dozen pulse crops under cultivation and accounts for 28 percent of the total production of pulses grown in India. Two commercial types of chickpea, kabuli and desi, are known depending upon the differences in the seed size and geographic origin. The kabuli are large seeded (macrosperma) and of Mediterranean origin and the desi are small seeded (microsperma) and of Indian origin. Chickpea is known to be highly susceptible towards soil salinity causing one of the major agronomic problems of the Indian subcontinent.

Exploitation of genetic variability in cultivated species like chickpea offers the possibility of developing some salt tolerant genotypes. In order to have a better understanding of the influences of salinity stress, investigations were carried out to study morphophysiological and biochemical response of different genotypes of chickpea, associated with salinity tolerance. The main objectives were to i) evaluate the growth response of different genotypes of chickpea, ii) examine the effects of salinity on the symbiont by assessing infection, nodule function, nitrogen fixation and carbon metabolism, and iii) to ascertain whether the negative effects of salinity on nitrogen fixation were due to limitation on the nodule number and nodule mass which sustains nitrogenase activity.

Four genotypes, two each of desi {DCP 92-3, CSG 8962} and kabuli {BG 267, CSG 9651} differing in their salt sensitivities were selected. The seeds were surface sterilized and were inoculated with a salt tolerant strain of *Mesorhizobium ciceri*. Plants were grown in the greenhouse and raised in earthenware pots and 15-day old seedlings were treated with different salt dosages of 4,6 and 8 dS/m EC. Various physiological and biochemical parameters were studied under salt stress in all the four genotypes of chickpea.

The shoots as well as roots weight declined with increasing salt dosage which significantly affected the root to shoot ratio. The decline in weights was higher in DCP 92-3 and BG 267 as compared to CSG 9651 and CSG 8962 which showed significantly higher salinity tolerance. The kabuli genotype CSG 9651 had the highest dry matter accumulation. The nodule number, as well as nodule mass, increased with increase in salt concentration and were the highest under the highest saline concentration of 8 dS/m in both the tolerant genotypes, maximum being in CSG 9651. The salt sensitive genotypes revealed a significant decrease in both these parameters. The acetylene reduction activity {ARA} and the leghemoglobin content declined under salt stress in all the genotypes with least negative effects in the kabuli CSG 9651 and maximum negative effects in the desi DCP 92-3. The nodule PEPC activity also boosted with increasing salt concentration in both CSG 9651 and CSG 8961 whereas in the salt sensitive genotypes, an increase was observed only under the lowest salt concentration {4dS/m} and PEPC activity significantly decreased under high salt levels. A progressive degeneration in the chlorophyll pigments as well Rubisco activity was recorded in all the genotypes under salt stress. The deleterious effects of salinity had a direct effect on the crop productivity and there was a significant decrease in the various yield parameters like number of pods per plant, seed weight per plant, total dry matter accumulation, harvest index {HI} and the like. Maximum harvest index was observed in kabuli CSG 9651 and minimum in desi DCP 923. The kabuli genotype had significantly higher potential for salinity tolerance as compared to the desi.

Conclusions:
a) Important variabilities in terms of growth, dry matter accumulation, photosynthesis, nodule nitrogen and carbon fixation were observed amongst the different genotypes of chickpea.
b) The greater performance of symbiosis under saline conditions seemed to be determined mainly by the tolerance of the legume host plant.
c) In general, both kabuli genotypes seemed to have a better potential for salt tolerance as compared to the desi. Even the sensitive kabuli exhibited significantly higher salt resistance than the sensitive desi.
d) Tolerance to salinity seemed to directly related to a number of physiological and biochemical traits such as increased nodule number, nodule dry mass, stimulation in nodule PEPC activity, and higher Rubisco as well as nitrogenase activities in the tolerant kabuli and desi genotypes of chickpea.
e) The genotype having greatest capacity for nodulation performed best under stressed and unstressed conditions in terms of both nitrogen fixation and grain yield.
f) Understanding the mechanism that caused an increase in the nodule number and nodule dry mass under salt stress as well lead to boosted PEPC activity could help establish biochemical targets for breeding programs aimed at increasing salinity resistance in plants.

NODULE CO-OCCUPANCY OF *AGROBACTERIUM* AND *BRADYRHIZOBIUM* WITH POTENTIAL BENEFIT TO LEGUME HOST

Sohail Hameed*, Fathia Mubeen, Kauser A. Malik and Fauzia Y. Hafeez
National Institute for Biotechnology and Genetic Engineering (NIBGE)
P.O. Box 577, Jhang Road, Faisalabad, Pakistan. *e-mail: sohail@nibge.org

The study was carried out on rhizobial isolates, Ca-18 from chickpea, PS-1 and PS-2 from pea, Ma-8 from siratro, MN-S from mungbean and a reference strain TAL-102 from soybean. Plant infectivity test and acetylene reduction assay showed effective root nodule formation by all the isolates on their respective hosts except chickpea isolate Ca-18 that failed to infect its original host. All strains showed homology to a typical *Rhizobium* and *Bradyrhizobium* strains. Strain Ca-18 proved important for its significantly high phosphate solubilization (63.3 µg ml^{-1}) and indole acetic acid (35 µg ml^{-1}) producing ability. Random amplified polymorphic DNA (RAPD) analysis employed for the isolates discriminated the strains of *Rhizobium* and *Bradyrhizobium* in two different clusters and also discriminated the non-nodulating chickpea isolate Ca-18 from all the other nodulating rhizobial strains, showing least homology of 15 % and 18 % with *Rhizobium* and *Bradyrhizobium*, respectively and hence was not a (*Brady*)*rhizobium* strain. Partial 16S *rRNA* gene sequence analysis for strains MN-S, TAL-102 and Ca-18, showed 97 % homology among strain MN-S and TAL-102, confirming them both as strains of *B. japonicum* species. The non-infective isolate Ca-18 was 67 % different from the other two strains and was confirmed as an *Agrobacterium* strain (Hameed et al. 2004).

In a plant infectivity test conducted in growth pouches under controlled environmental conditions, nitrogen fixing nodules were formed in cowpea plants inoculated with *Bradyrhizobium* strain TAL-102 alone (13 nodules with 1.6 µ mol acetylene reduced h^{-1} mg^{-1} dry nodule wt), as well as in plants inoculated with Ca-18 and TAL-102 together (25 nodules with 6.8 µ mol acetylene reduced h^{-1} mg^{-1} dry nodule wt). However, no nodules or galls were formed in plants that were inoculated with Ca-18 alone. The root and nodule samples of all the treatments were also subjected to ultrastructure processing and Immunogold labeling. The results showed that, the *Bradyrhizobium* strain TAL-102 alone entered the central cortical region forming nodules. The Ca-18 cells when inoculated alone were localized only in the outer cortical cell layers of the cowpea roots, rather directly infecting the central cortical region of the roots. Immunogold labeling on root nodules formed by co-inoculation of TAL-102 and Ca-18 cells, showed the presence of both the strains within the central cortical region with gold labeling on only Ca-18 cells against which the antisera was used. This indicates that the *Agrobacterium* has the ability to enter the legume root cells alone through mutual recognition of the plant cell wall, but is unable to enter the central cortical region and cause nodulation. Fluorescent antibody staining /antisera cross reactivity tests using FAs of Ca-18 and TAL-102 against crushed nodule of cowpea further confirmed the co-occupancy of the two strains, as fluorescent staining in co-inoculated nodules was positive with both the antisera, but did not cross react with individual strains. The evidence of the presence of *Agrobacterium* in the infected region of the nodules have been provided by Lajudie et al. (1999) by isolating the strain from the infected region, but the evidence through ultrastructure localization is being reported here for the first time.

The study conclusively shows the significance of nodule cell co-occupancy by a non-rhizobial isolate with its potential benefit of phosphate solubilization and phytohormone production, the property that can be exploited for its practical application in large-scale biofertilizer production (Bashan Y and Holguin G 1997).

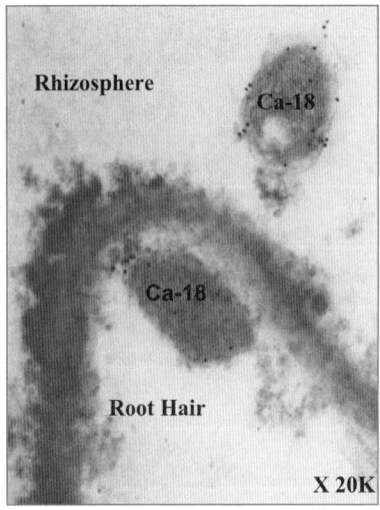

Immunogold labeled Ca-18 cells in rhizosphere & within cowpea root hair

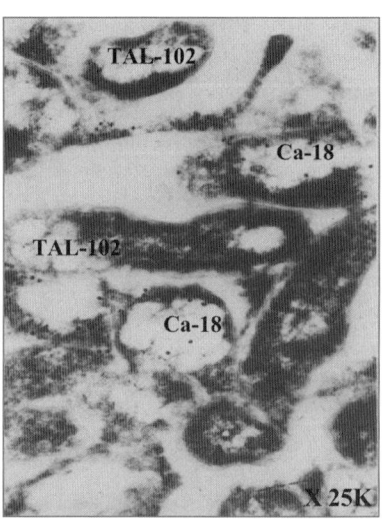

Co-occupancy of immunogold labeled *Agrobacterium* Ca-18 and unlabeled *Bradyrhizobium* TAL-102 in cowpea root nodule

References
Bashan Y and Holguin G (1997) Can. J. Microbial. 43, 103-121.
Hameed S et al. (2004) Biol. Fertil. Soil. 39(3), 179-185.
Lajudie P D et al. (1999) System. Appl. Microbiol. 22, 119-132.

SALT STRESS ADAPTATION IN ALFALFA BACTEROIDS : IMPORTANCE OF PROLINE BETAINE

Karine Mandon, Alexandre Boscari, Jean Charles Trinchant, Laurence Dupont, Geneviève Alloing, Didier Hérouart, and Daniel Le Rudulier
CNRS-INRA-Université de Nice-Sophia Antipolis, Centre INRA Agrobiotech, 400 route des Chappes, 06903 Sophia Antipolis Cedex, France

1. Introduction

Excessive salinity and drought are the most important environmental factors that greatly affect plant growth and productivity worldwide. During osmotic stress, plants induce processes that regulate the osmotic adjustment and maintain sufficient cell turgor for growth to proceed. Such adjustment requires the control of accumulation of organic solutes compartmented mainly in the cytoplasm, and termed osmolytes, compatible solutes, or osmoprotectants. These non-toxic, low-molecular weight molecules, raise the osmotic pressure and protect some macromolecular structures against denaturation. Proline (Pro) and glycine betaine (GB) are the most diversely nitrogenous osmolytes accumulated under osmotic stress conditions. However, if GB is produced in large quantity by several plant families (Chenopodiaceae, Amaranthaceae, and Gramineae), certain plants lack significant amounts of GB and produce other betaines, like proline betaine (PB), also known as N,N-dimethylproline, or stachydrine. PB is the main betaine in *Medicago sativa*, and other *Medicago* species including *M. truncatula*. PB is released by alfalfa seeds during germination, and is an inducer of nodulation (*nod*) genes in *Sinorhizobium meliloti*. Interestingly, *S. meliloti* transports this betaine by a high-affinity Na^+-coupled transporter, BetS (Boscari et al. 2002), and via an ABC (ATP-binding cassette) protein-dependent transporter, Hut (Boncompagni et al. 2000). In addition, *S. meliloti* uses PB as an energy source, and as an osmoprotectant. The aim of the present study was to gain an understanding of the effects of salinity on PB pools in alfalfa, particularly in nodule compartments, to analyze PB transport by purified symbiosomes and isolated bacteroids, and to investigate the effect of BetS overproduction in bacteroids.

2. Results

Salt stress was imposed on the plants 3 weeks after inoculation by adding 0.2 M NaCl to the growth medium. The leaf water potential decreased from -0.6 to -1.8 MPa during the first 2 weeks, and reached -2.1 MPa after 4 weeks. In parallel, a 40% and 60% reduction in acetylene reduction activity was noticed. After a 2-week delay, salt stress was followed by a strong increase in PB in shoots (x 10), roots (x 4), and nodules (x 8). Salinization also induced a rapid increase in Pro content. Based on a water volume of 0.85 ml/g fresh weight nodule tissue, the average PB and Pro concentrations in 4-week-stressed nodules were estimated at approximately 3.8 mM and 7.8 mM, respectively.

Indeed, salt stress resulted in a 4-fold increase in bacteroids after 2 weeks, and an 8-fold increase in cytosol and bacteroids after 4 weeks of salinization. The estimated PB concentration in bacteroids from 4-week-stressed nodules reached 7.4 mM compared to only 0.8 mM in control. PB was accumulated in bacteroids via transport from the cytosol through the peribacteroid membrane (PBM), and PB transport into symbiosomes was slightly stimulated during salinization. Pro level was also greatly enhanced in bacteroids, and after 4 weeks of salinization, its concentration was estimated at 11.8 mM, and only 0.8 mM in control. Since investigations with ^{14}C-Pro indicated that *S. meliloti* did not produce PB from Pro, in contrast to what was observed in alfalfa shoots, accumulation of PB in bacteroids is unlikely to be due to *de novo* biosynthesis. During symbiosome preparation, metabolites are more likely exchanged between the peribacteroid space (PBS) and the cytosol. Thus, realistic values of PB and Pro from the PBS cannot be obtained. Nevertheless, both compounds were identified in this compartment. Whatever the age of the nodules, the rate of PB and Pro uptake by symbiosomes was linearly enhanced at substrate concentrations increasing from 0.5 to 5 mM. There was no indication of a saturable carrier at these concentrations, which were physiologically relevant with PB and Pro concentrations determined in bacteroids from salt-stressed nodules.

In terms of bacteroid ion contents, 4 weeks of salinization induced a 2.2-fold increase in Na^+ level, a 3.2-fold decrease in the K^+ pool, while Cl^- level was not significantly modified. Such results suggest that the PBM is permeable to these ions when an excess is present within the cytosol. It has been accepted that symbiosomes behave as osmometers, and also that the PBS serves as the main compartment to accommodate changes in osmotic pressure. In our case, it seems reasonable to assume that accumulation of PB, Pro, and Na^+ in the PBS contributes to the increase in turgor pressure, which might be involved in the enlargement of the symbiosome volume observed on cross-nodule sections (Trinchant et al., 2004).

Since BetS is a major PB transporter in *S. meliloti*, the expression of *betS* was monitored *in situ* during the nodulation and nitrogen fixation process. The indigo coloration resulting from the expression of a *betS-lacZ* fusion was observed inside root hair infection threads, and the infection zone of young and mature nodules. The bacteroid Na^+-dependent PB transport activity was osmotically regulated, and strongly impaired in a *betS* mutant. Overproduction of *betS* induced an increase in PB content within bacteroids from nodules subjected to salinization. Simultaneously, acetylene reduction activity in 7-day-stressed nodules produced by the overexpressing strain was less affected by salt stress than in nodules obtained with the wild-type strain.

3. Conclusions

PB which is synthesized from Pro in alfalfa shoots is taken up by purified symbiosomes and free bacteroids. Salinization of the host plant strongly enhanced PB content in nodule cytosol and in bacteroids. As in free-living *S. meliloti* cells, BetS, a major PB transporter in bacteroids, could play a significant role in the initial phase of bacteroid adaptation when salt stress is applied to the plant host.

4. References

Boncompagni E et al. (2000) J. Bacteriol. 182, 3717-3725.
Boscari A et al. (2002) J. Bacteriol. 184, 2654-2663.
Trinchant JC et al. (2004) Plant Physiol. 135, 1583-1594.

ISOLATION OF GENES FOR SALT TOLERANCE FROM *SINORHIZOBIUM* LT11

W. Payakapong[1], P. Tittabutr[1], N. Teaumroong[1], D. Borthakur[2] and N. Boonkerd[1]
[1]School of Biotechnology, Suranaree University of Technology, Nakhon Ratchasima 30000, Thailand. [2]Department of Molecular Biosciences and Bioengineering, University of Hawaii, Honolulu, HI 96822, USA

Almost 40% of area over the world cannot be cultivated because of salinity problems. Salinity does reduce water available for plant growth resulting in decreased productivity of most crop plants. Some of leguminous plants can tolerate up to 0.1 M salt while *Rhizobium* strain that nodulate them are generally sensitive to salt. We have identified a *Sinorhizobium* strain LT11 from Thailand, which can tolerate up to 0.4 M salt and is effective in N_2 fixation. The objective of this work was to isolate genes for salt tolerance from this strain. The genomic DNA library of LT11 was transferred by conjugation to salt-sensitive rhizobial strain TAL1145 and the transconjugants were selected on YEM agar containing 0.1 M NaCl. Two hundred colonies that grew under salt were isolated and the recombinant cosmid clones were isolated from twenty-one of those colonies. These cosmid clones were classified into two groups after digestion with *Hind*III (Fig. 1). The analysis of sequence at the end of each fragments revealed interesting results. The first group of clones showed a relatively high homology with betaine aldehyde dehydrogenase cluster (*bet*A and *bet*B gene) of *S. meliloti*. The second group exhibited similarity with methylase, xanthine dehydrogenase, transcriptional regulator syrB (AraC family) and a conserved hypothetical protein (Fig. 2). These two groups of clones were introduced into *Bradyrhizobium japonicum* THA6 and the growth of the resulting strains was checked under different salt condition. Transconjugants showed better growth than THA6 wild type (Fig. 3).

Data reported here reveal that the betaine aldehyde dehydrogenase cluster is related to the salt tolerance property in *S. meliloti* and that the cosmid clones contains several genes similar to previously known ones. However, mutagenesis of the cosmids isolated in this study has to be done in order to identify which of the genes are involved in salt tolerance. In addition, we showed that the transfer of the cosmids carrying salt tolerance genes from *Sinorhizobium* LT11 into *B. japonicum* resulted in strains with improved tolerance to high salinity.

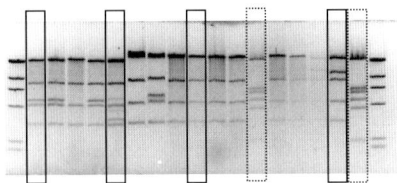

Fig. 1 Two different groups of DNA fragment pattern digested by HindIII cosmid clones. Dot square presented group 1. Solid square presented group 2.

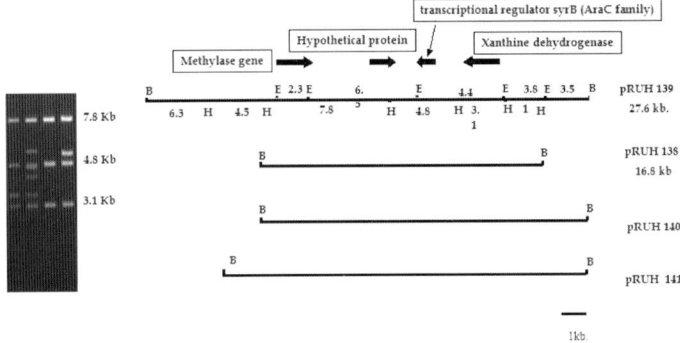

Fig. 2 Ethidium bromide strained agarose gel showing four overlapping salt tolerant cosmid clones of group 2 digested with HindIII and restriction map of these clones.

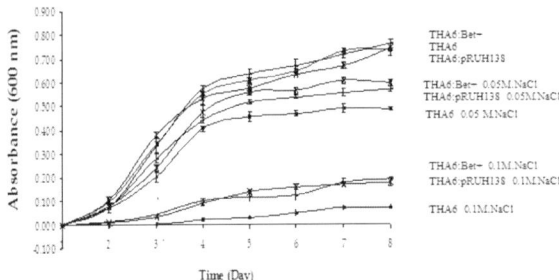

Fig. 3 Growth of THA6 and transconjugants in YEM liquid medium at different NaCl concentration.

References
Balassubramanian and Sinha (1976) Physiol. Plant. 36, 197-200.
Cordovilla, MP et al. (1994) J. Exp. Bot. 45, 1483-1488.
Rao, DLN, and Sharma, PC (1995) Indian. J. Exp. Biol. 33, 500-504.

REGULATION OF HEME AND IRON METABOLISM IN *BRADYRHIZOBIUM JAPONICUM*

Jianhua Yang, Yali Friedman and Mark R. O'Brian
Department of Biochemistry, State University of New York at Buffalo, Buffalo, New York 14214 USA

1. General Introduction. Heme is the prosthetic group or active moiety of proteins involved in many cellular processes. Heme is required for bacterial respiration to support the energy-intensive process of nitrogen fixation, and to detoxify reactive oxygen species generated by oxidative metabolism. It is also required for O_2 sensing to trigger bacteroid development as the active moiety of FixL (1). Finally, heme can serve as an iron source in rhizobia (2-4), as occurs for many bacterial pathogens of animals. Our group is interested in the regulation of heme metabolism in *Bradyrhizobium japonicum* and its integration with iron homeostasis. Heme biosynthesis is regulated by iron in *B. japonicum*, and this control is mediated by the iron response regulator (Irr). In turn, Irr is regulated by iron at both transcriptional and post-translational levels, which is the topic of this chapter. Regulation of *irr* by iron has revealed novel mechanisms of bacterial gene expression.

2. Irr mediates iron control of heme biosynthesis.
Ferrochelatase catalyzes the insertion of ferrous iron into protoporphyrin IX to form heme (protoheme) in the final step of the heme biosynthetic pathway. Iron can be a limiting nutrient for both prokaryotes and eukaryotes. The Irr protein from *B. japonicum* coordinates the heme biosynthetic pathway to prevent the accumulation of toxic porphyrin precursors under iron limitation (5). Loss of function of the *irr* gene is sufficient to uncouple the pathway from iron-dependent control as discerned by the accumulation of protoporphyrin under iron limitation. The Irr protein accumulates in cells under iron limitation, where it negatively regulates the pathway at *hemB*, the gene encoding the heme biosynthetic enzyme δ-aminolevulinic acid (ALA) dehydratase (5). Irr is a conditionally stable protein that degrades rapidly when cells are exposed to iron, allowing derepression of synthesis (6). This iron-dependent degradation is mediated by heme (6). Heme binds directly to Irr at a heme regulatory motif (HRM), which is necessary for its degradation. Accordingly, Irr persists in heme synthesis mutant strains

in the presence of iron, and a mutation in the HRM stabilizes Irr in the presence of the metal (6). Thus, heme is an effector molecule in Irr degradation that reflects the availability of iron for heme synthesis.

3. Irr interacts with ferrochelatase to mediate iron control of heme synthesis.
How can heme be a regulatory molecule in cells? In our view, a heme pool is untenable because heme is lipophilic, and would likely form adventitious associations in cells if it accumulated freely. Furthermore, heme is toxic because of its ability to catalyze the formation of reactive oxygen species. We demonstrated that Irr interacts directly with ferrochelatase, and responds to iron via the status of heme and protoporphyrin localized at the site of heme synthesis (7). Thus, heme can serve as a regulatory molecule without accumulating freely in cells. In the presence of iron, ferrochelatase inactivates Irr, followed by heme-dependent Irr degradation to derepress the pathway. Under iron limitation, protoporphyrin relieves the inhibition of Irr by ferrochelatase, probably by promoting protein dissociation, allowing genetic repression. Thus, metabolic control of the heme pathway involves a regulatory function of a biosynthetic enzyme to affect gene expression.

4. Two heme binding sites are involved in the regulated degradation of Irr.
Normal degradation of Irr involves heme binding to the heme regulatory motif (HRM) of Irr. We found that Irr confers iron-dependent instability on glutathione S-transferase (GST) when fused to it. Analysis of Irr-GST derivatives with C-terminal truncations of Irr implicate a second region necessary for degradation other than the HRM, and show that the HRM is not sufficient to confer instability on GST. The HRM-defective mutant IrrC29A degrades in the presence of iron, but much more slowly than the wild type protein. This slow turnover is heme-dependent as discerned by the stability of Irr in a heme-defective mutant strain. Whereas the HRM of purified recombinant Irr binds ferric (oxidized) heme, a second site that binds ferrous (reduced) heme was identified based on spectral analysis of truncation and substitution mutants. A mutant in which histidines 117 to 119 were changed to alanines abrogates ferrous, but not ferric, heme binding. Introduction of these substitutions in an Irr-GST fusion stabilizes the protein *in vivo* in the presence of iron. We conclude that normal iron-dependent Irr degradation involves two heme binding sites, and that both redox states of heme are required for rapid turnover.

5. *B. japonicum* Fur regulates the *irr* gene by binding to a novel site in the *irr* promoter.
Fur homologs are found throughout the eubacterial kingdom. A generalized model for Fur function posits that the protein binds ferrous iron when the metal is available, conferring the ability to recognize and bind to a Fur box DNA element to repress transcription of target genes (8). We cloned the *B. japonicum fur* gene and found that is necessary for normal *irr* mRNA accumulation (9,10). *B. japonicum* Fur (BjFur) recognizes a DNA element within the *irr* gene promoter that is dissimilar from the Fur box consensus sequence (11). The affinity of BjFur for the *irr* promoter was similar to that for the Fur box consensus, and thus, the defined DNA-protein interaction is

physiologically relevant. These observations suggest that a generalized model for Fur with respect to its target DNA needs to be expanded or modified.

Fur proteins are likely to be dimeric in solution (12,13), and the *Pseudomonas aeruginosa* protein was crystallized as a dimer (14). A model for Fur based on the crystal structure of the unbound protein depicts two dimers binding to target DNA. BjFur can occupy the *irr* promoter as a dimer and two dimers or a tetramer (11). BjFur was fully capable of repressing transcription initiation *in vitro* from a mutant *irr* promoter that allowed only dimer occupancy. Thus, the assumption that Fur requires two or more dimers to repress transcription may not be valid. Finally, our data show that Fur binding to a target promoter is sufficient to repress transcription.

6. *B. japonicum* Fur is an iron-responsive transcriptional repressor.
The discovery that Fur is a zinc metalloprotein and the use of surrogate metals for Fe^{2+} for *in vitro* studies question whether Fur is a direct iron sensor. We found that the affinity of *B. japonicum* (BjFur) for its target DNA increases 30-fold in the presence of metal, with a K_d value of about 2 nM (22). DNase I footprinting experiments showed that BjFur protects its binding site within the *irr* gene promoter in the presence of Fe^{2+}, but not in the absence of metal, showing that DNA binding is Fe^{2+}-dependent. *In vitro* transcription from the *irr* promoter using purified components was not inhibited by BjFur in the absence of metal, but BjFur repressed transcription in the presence of Fe^{2+} (22) Thus, BjFur is an iron-responsive transcriptional repressor *in vitro*. A regulatory Fe^{2+}-binding site (site 1) and a structural Zn^{2+}-binding site (site 2) implicated from the recent crystal structure of Fur from *Pseudomonas aeruginosa* are comprised of amino acids highly conserved in many Fur proteins (14), including BjFur. BjFur mutants containing substitutions in site 1 (BjFurS1) or site 2 (BjFurS2) bound DNA with high affinity and repressed transcription *in vitro* in an Fe^{2+}-dependent manner (22). Interestingly, only a single dimer of BjFurS2 occupied the *irr* promoter, whereas the wild type and BjFurS1 displayed one or two dimer occupancy. We suggest that the putative functions for metal-binding sites deduced from the structure of *P. aeruginosa* Fur cannot be extrapolated to bacterial Fur proteins as a whole.

7. Fur homologs in other rhizobia.
Fur homologs are found in the genomes of other rhizobia and in the α-Proteobacteria as a whole. A noticeable exception is the lack of a recognizable homolog in the genome of *Mesorhizobium loti*. Whereas Fur appears to be the only transcriptional regulator of iron homeostasis in the γ-Proteobacteria and some other bacteria, additional regulators are present in the rhizobia. The Irr protein described above is involved in ferric citrate transport (5), and perhaps heme transport (3), and homologs are found in most of the α-Proteobacteria. The RirA protein from *R. leguminosarum* mediates control of iron transport and siderophore synthesis, and homologs are found in *S. meliloti*, *Brucella* species and *Agrobacterium tumefaciens* (15). Interestingly, the Fur homolog of *S. meliloti* is adjacent to the *sitABCD* operon, which was shown recently to be involved in

manganese transport (16). Furthermore, three groups showed independently that the Fur homolog controls the *sitABCD* operon in *S. meliloti* (17,18) and *R. leguminosarum* (19), and it is a manganese-responsive regulator. Thus, whereas *B. japonicum* Fur is clearly involved in iron metabolism, the homologs in *S. meliloti* and *R. leguminosarum* have a different function, although they do show weak iron-responsiveness. Collectively, the studies of metal metabolism in rhizobia underscore a diversity of properties and functions of bacterial Fur homologs that cannot be fully appreciated from studying *E. coli* alone.

Deletions mutations of rhizobial Fur homologs have no obvious symbiotic phenotypes (9,20). However, a *B. japonicum* Fur mutant with multiple amino acid substitutions has a severe symbiotic phenotype (21). This mutant allele has a dominant-negative phenotype in partial diploid cells. These findings suggest a role for Fur in symbiosis

8. Acknowledgments.
Work from the O'Brian lab was supported by the National Institutes of Health (GM-067966), U.S. Department of Agriculture (2003-35319-13269) and the National Science Foundation (MCB-0089928).

9. References
1. Gilles-Gonzalez MA et al. (1991) Nature 350, 170-172.
2. Noya F et al. (1997) J. Bacteriol. 179, 3076-3078.
3. Nienaber A et al. (2001) Mol. Microbiol. 41, 787-800.
4. Wexler M et al. (2001) Mol. Microbiol. 41, 801-816.
5. Hamza I et al. (1998) J. Biol. Chem. 273, 21669-21674.
6. Qi Z et al. (1999) Proc. Natl. Acad. Sci. U.S.A. 96, 13056-13061.
7. Qi Z and O'Brian MR (2002) Mol. Cell 9, 155-162.
8. Escolar L et al. (1999) J Bacteriol 181, 6223-6229.
9. Hamza I et al. (1999) J. Bacteriol. 181, 5843-5846.
10. Hamza I et al. (2000) Microbiol. 146, 669-676.
11. Friedman YE and O'Brian MR (2003) J. Biol. Chem. 278, 38395-38401.
12. Coy M and Neilands JB (1991) Biochem. 30, 8201-8210.
13. Michaud-Soret I et al. (1997) FEBS Lett 413, 473-476.
14. Pohl E et al. (2003) Mol. Microbiol. 47, 903-915.
15. Todd JD et al. (2002) Microbiol. 148, 4059-4071.
16. Platero RA et al. (2003) FEMS Microbiol. Lett. 218, 65-70.
17. Platero R et al. (2004) Appl. Environ. Microbiol. 70, 4349-4355.
18. Chao TC et al. (2004) J. Bacteriol. 186, 3609-3620.
19. Diaz-Mireles E et al. (2004) Microbiol. 150, 1447-1456.
20. Wexler M. (2003) Microbiol. 149, 1357-1365.
21. Benson HP et al. (2004) J. Bacteriol.186, 1409-1414.
22. Friedman YE and O'Brian MR (2004) J. Biol. Chem. 279, 32100-32105.

RHIZOBIAL GENES ESSENTIAL FOR SALT TOLERANCE

Ju Quan Jiang, Wei Wei, Li Shi Xie, Lei Wang and Su Sheng Yang
Department of Microbiology, College of Biological Sciences, China Agricultural University, Beijing, China. E-mail: yangssh@cau.edu.cn

Salt-tolerance genes of *Sinorhizobium fredii* RT19 were identified by the construction and screening of a transposon Tn5-1063 library containing over 30,000 clones. Twenty-one salt-sensitive mutants were obtained and five different genes were identified by sequencing. Eight mutants were found with disruptions in the *phaA2* gene, while mutations in other genes were found in seven (*phaD2*), two (*phaF2*) and two (*phaG2*) mutants. A mutation in the *metH* gene was found in two of the salt sensitive strains.

Further sequence analysis showed these four genes, *phaA2, D2, F2* and *G2*, as well as other three genes *phaB2, C2* and *E2*, form a single transcription unit, which share the highest identity with *phaA2, B2, C2, D2, E2, F2* and *G2* of *Sinorhizobium meliloti* 1021 and also share very high similarity with the corresponding genes of the putative monovalent cation/H^+ antiporters of *Coxiella burnetii* RSA 49. And the *metH* gene shared the highest identity with the corresponding gene of *S. meliloti*, encoding 5' methyltetrahydrofolate homocysteine methyltransferase.

Growth experiments showed that *phaA2*, *phaD2*, *phaF2* and *phaG2* mutants were hypersensitive to Na^+/Li^+ and slightly sensitive to K^+ and not sensitive to sucrose and that *metH* mutants were highly sensitive to any of Na^+, Li^+, K^+ and sucrose. Complementary experiments showed that the *pha* gene cluster completely restored the growth of *pha* mutants even at the concentration of 0.6 mol l^{-1} NaCl.

Na^+ intracellular content measurements established that *phaA2*, *phaD2*, *phaF2* and *phaG2* are mainly involved in the Na^+ efflux in *S. fredii* RT19. Recovery of growth of the *metH* mutants incubated with different concentrations of NaCl could be obtained by the addition of methionine, choline and betaine, which showed that the *metH* gene is probably involved in osmoregulation in *S. fredii* RT19.

Monovalent cation/H^+ antiporter assay showed that wild type had very high $Na^+(Li^+)/H^+$ antiporter activity and the mutants lost this function, indicating that the protein PhaA2B2C2D2E2F2G2 shares $Na^+(Li^+)/H^+$ antiporter activity. And wild type showed higher K^+/H^+ antiporter activity than *pha* mutants, suggesting that the protein PhaA2B2C2D2E2F2G2 also functions as a K^+/H^+ antiporter.

A transposon Tn5-1063a mutagenesis library of *S. meliloti* 042BM was constructed and nine salt-sensitive mutants were isolated, which were unable to growth on FY plates containing 0.4 mol l^{-1} NaCl. Our interest is to provide information about the mechanism of salt tolerance in bacteria by studying the genes involved in salt tolerance. Here, eight different genes were identified.

In this study, most of mutants were more sensitive to salt than to glucose under the same osmotic stress level of 0.4 mol l^{-1}. Thus, these mutants should be classified as salt-sensitive rather than osmosensitive.

In strain W4, Tn5-1063 insertion site was located in *omp10*, encoding an outer membrane lipoprotein, which was an important cell wall element. Omp10, which has been identified as an immunogenic determinant in *Brucella*, is for the first time reported with regard to salt tolerance.

Strain W11 carried a Tn5-1063 mutation in *relA* gene, encoding the (p)ppGpp synthetase. (p)ppGpp is thought to be a stringent response factor and to be the positive regulator of *rpoS*, which encodes the stress-responsive RNA polymerase subunit σ^s in *E. coli*. Complementation test in *Escherichia coli relA* mutant indicated the gene has (p)ppGpp synthetase activity. The growth of *S. meliloti* 042BM *relA* mutant was partially restored by complementation on 0.2 mol l^{-1} and 0.4mol l^{-1} NaCl. Symbiosis defect was found in the *S. meliloti* 042BM *relA* mutant.

The *nuoL* gene, encoding NADH dehydrogenase I chain L transmembrane protein, was identified, which enforced the view of a Na^+/H^+ antiporter, which includes a NuoL subunit which has an important role in Na^+ resistance and energy metabolism. *nuoL* mutant could not grow in FY medium supplemented with 0.1 mol l^{-1} LiCl and was the most sensitive strain to salt in all mutants. Other genes included *greA* encoding a transcription cleavage factor, *gshB* encoding glutathione synthetase, a putative nuclease/helicase gene and two unknown genes.

In conclusion, the genes *phaA2, B2, C2, D2, E2, F2* and *G2* encode a monovalent cation/H^+ antiporter, which plays a crucial role on salt tolerance of *S. fredii* RT19. To our knowledge, this is the first report about the involvement of the gene *metH* in salt-tolerance. The adaptation of *S. meliloti* 042BM to salt is a complex multilevel regulatory process in which many different genes can be involved.

EFFECT OF LOW PH AND ALUMINUM ON THE NODULATION SIGNAL TRANSDUCTION IN MEDICAGO SATIVA [*]

Min Yang[1], Xiaofeng Li[1], Yongxiong Yu[2] and Minghua Gu[1]
[1]Faculty of Agri., Guangxi U, Daxue 100, Nanning, [2]Faculty of Zoological Sci., Southwest Agri. U, Chongqing, P.R.China

An estimated 40% of arable soils and 70% of non-arable soils of the world are acidic, and the growth of *Rhizobium spp.* and the infection of legume roots are restricted by low pH and Al toxicity in these soils. Nodulation of subterranean clover roots has been shown to be restricted at low pH and by the presence of Al (Kim et al, 1985). Expression of *nod* genes in *R. leguminosarum* biovar *trifolii* was affected at low pH and Al (Richardson et al. 1988). In present study, the effects of low pH and Al on the transduction of nodulation signal in alfalfa nitrogen-fixation system were investigated.

The alfalfa seedlings with second true leaf, inoculated with *Sinorhizobium meliloti* were grown in pHs (pH6.5 and pH4.5) and Al (0, 5, 10, 15 and 30 $\mu mol \cdot L^{-1}$ $AlCl_3$) solutions for 6 d, then changed to culture in sand with Fahraeus nitrogen-free nutrient solution (pH 6.5), and 30 d after inoculation, the nodules of plants were counted. Nodulation of alfalfa was restricted by low-pH and Al (Fig.1 and 2). The same results were obtained in the root elongation of alfalfa and the growth of *R. meliloti* in YM liquid medium.

Root hair deformation elicited by nod factor (NF) secreted from rhizobia is the primary morphological response of host to the rhizobia. Cultured in Fahraeus N-free nutrient solution with low pH and Al for 24 h, root hair deformation was shown to be inhibited in inoculated with *S. meliloti* (Fig.3 and 4) or the presence of NF while not inoculated with *S. meliloti* (Fig. 5 and 6), but there were no effects of low pH and Al on the root hair deformation in the absence of nodulation signal (neither in presence of NF nor inoculated with *S. Meliloti*), indicating that low pH and Al inhibited the transduction of nodulation signal. The result showed that low pH and Al depressed the nodulation of alfalfa, and it could be caused by the following factors: the first, low pH and Al inhibited the root elongation; the second, low pH and Al suppressed the growth of *S. Meliloti*; and the third, low pH and Al depressed the expression of *nod* genes

(Richardson AE,1988), and showed an inhibitory effect on the transduction of nodulation signal in symbiotic nitrogen-fixation system.

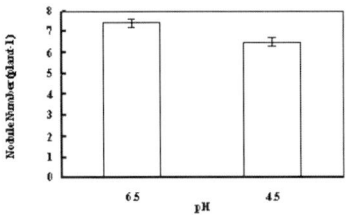

Fig.1 Effects of pHs on the nodulation of alfalfa inoculated with R. melilot.

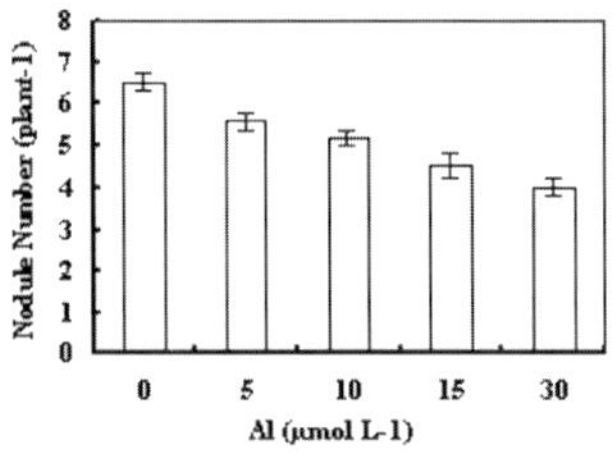

Fig.2 Effects of Al on the nodulation of alfalfa inoculated with R. melilot.

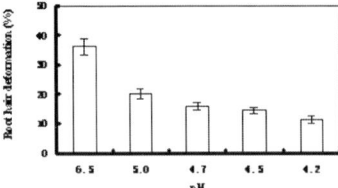
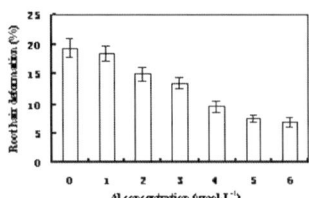

Fig.3 Effects of pHs on root hair deformation in alfalfa inoculated with R. melilot.

Fig.4 Effects of Al on root hair deformation in alfalfa inoculated with R. Meliloti.

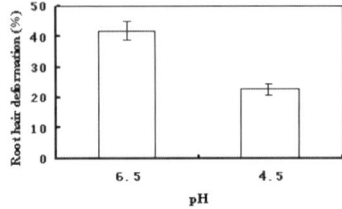
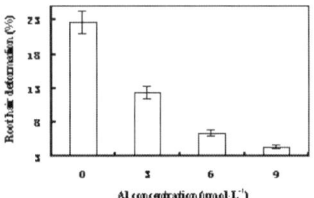

Fig.5 Effect of low pH on root hair deformation of alfalfa in the presence of NF while not inoculated with R. melilot.

Fig.6 Effects of Al on root hair deformation of alfalfa in the presence of NF while not inoculated with R. melilot.

This work was supported in part by 973 SKBRDP (Grant No. 2001CB108905) and 863 HTRDP (Grant No. 2001AA241164) and Chongqing Committee of Science and Technology Project (No. 7381)

References:
Kim MK et al. (1985a and b) *In* Proceeding of the XV International Grassland Congress. Science Council of Japan & Japanese Society of Grassland Science, Tochigi-ken, Japan, pp501-503.and 543-544.
Rhichardson AE et al. (1988) Appl. Environ. Microbiol.10, 2541-2548.

FURTHER INVESTIGATION OF THE ROLES OF POLY-3-HYDROXYBUTYRATE (PHB) AND GLYCOGEN IN *SINORHIZOBIUM MELILOTI-MEDICAGO* SP. SYMBIOSIS

Chunxia Wang[1], Marsha Saldanha[2], Xiaoyan Sheng[1], Kris Shelswell[2], Trevor C. Charles[2], and Bruno W. Sobral[1]
[1]Virginia Bioinformatics Institute, VA, USA; [2]Dept. Biol. University of Waterloo, Canada

Poly-3-hydroxybutyrate (PHB) and glycogen are major carbon storage compounds in *Sinorhizobium meliloti*. The roles of PHB and glycogen in rhizobia-legume symbiosis are not fully understood (1, 2, 3). Bacteroids with determinate nodules often accumulate high levels of PHB (up to 70% dry weight) (1). In contrast, bacteroids within indeterminate nodules do not accumulate PHB. In the symbiosis between *S. meliloti* and alfalfa, it has been reported that *phbC* mutants form bacteroids capable of fixing nitrogen as efficiently as the wild type (4, 5, 6), but are less competitive than wild type strain (5). There are two glycogen synthase-encoding genes in *S. meliloti*, *glgA1* in a cluster of other glycogen synthesis pathway genes on the chromosome, and *glgA2* on megaplasmid pSymb. Recently, it has been shown in *Rhizobium tropici* with *Phaseolus vulgaris* that there may be a link between glycogen synthase deficiency, decreased exopolysaccharide, and increased symbiotic performance (2). However, the reason for the increased symbiotic efficiency of the *glgA* mutant is uncertain (3). To determine the roles these compounds may play in the symbiotic process and in the overall physiology of the organism in the free-living and bacteroid states, mutants unable to synthesize PHB and/or glycogen were constructed. A *glgA1* mutation was constructed by in-frame deletion, preserving the expression of the downstream *pgm* gene, while a *glgA2* mutation was constructed by disruption of the gene with the Sp^r omega cassette (7). A pre-existing Tn5-generated mutation of the PHB synthase encoding gene *phbC* (8) was combined with *glgA1* and *glgA2* mutations to make all combinations of double mutants, and the triple mutant (Table 1). PHB was not detectable in free-living cells of any mutant containing the *phbC* mutation; glycogen was not detectable in any of the mutants containing the *glgA1* mutation. The production of PHB decreased significantly in the *glgA1* mutant (Rm11479), and the *glgA* double mutant (Rm11482). The production of glycogen increased significantly in the *phbC* mutant (Rm11105) in high carbon ratio media. Exopolysaccharide (EPS) was not detected in any of the mutants containing the *phbC* mutation, while the *glgA* double mutant (Rm11482) produced much more EPS in MOPS medium compared to the wild type (Rm1021), *glgA1*

(Rm11479), or *glgA2* (Rm11478) single mutant (Table 2). Symbiotic properties of these strains were investigated on *Medicago truncatula* and *M. sativa*. Our results indicated that the strains unable to synthesize PHB were still able to nodulate and fix nitrogen, but fewer nodules form and they do so at a reduced rate. Mutations in *glgA1* or *glgA2* resulted in increased plant dry weight of *M. truncatula*, but not *M. sativa*. Surprisingly, increased nitrogenase activities in the mutants were not correlated with increased plant dry weight, nodule number, or nodule dry weight. By further study of these mutants, a more complete understanding of the physiological roles of PHB and glycogen during the nodulation process will be obtained.

Table 1. Bacterial strains used in this study

Strain	Relevant characteristics	Reference
Rm1021	SU47 *str-21*, Sm^r	9
Rm11105	Rm1021 *phbC*::Tn5	10
Rm11478	Rm1021 *glgA2*::ΩSpSm	This study
Rm11479	Rm1021 *glgA1*ΔPstI	This study
Rm11480	Rm1021 *phbC*::Tn5, *glgA1*ΔPstI	This study
Rm11481	Rm1021 *phbC*::Tn5, *glgA2*::ΩSpSm	This study
Rm11482	Rm1021 *glgA1*ΔPstI, *glgA2*::ΩSpSm	This study
Rm11483	Rm1021 *glgA1*ΔPstI, *glgA2*::ΩSpSm, *phbC*::Tn5	This study

Table 2. Phenotype of the wild type and mutants

Strain	PHB (% cell dry wt)	Glycogen (ug/10^9 cells)			EPS/cell dry wt (g)		
	YEM	MOPS	YEM	TY	MOPS	YEM	TY
Rm1021	20.9c	51.8a	83.8b	-	1.6a	1.6	-
Rm11105	-	113.9c	104.1c	-	-	-	-
Rm11478	23.3c	50.1a	59.7a	-	1.6a	1.7	-
Rm11479	8.3a	-	-	-	1.3a	1.7	-
Rm11480	-	-	-	-	-	-	-
Rm11481	-	77.1b	ND	-	-	-	-
Rm11482	16.8b	-	-	-	2.2b	1.9	-
Rm11483	-	-	-	-	-	-	-

References
1. Mandon K et al. (1998) J. Bacteriol. 180, 5070-5076.
2. Marroqui S et al. (2001) J. Bacteriol. 183, 854-864.
3. Lodwig E and Poole P (2003) Critical Rev. in Plant Sci. 22 (1), 37-78.
4. Povolo S et al. (1994) Can. J. Microbiol. 40, 823-829.
5. Willis LB and Walker GC (1998) Can. J. Microbiol. 44, 554-564.
6. Aneja P and Charles TC (1999) J. Bacteriol. 181 (3), 849–857.
7. Frey J and Krisch HM (1985) Gene 36, 143-150.
8. Cai G et al (2000) J. Bacteriol 182, 2113-2118
9. Meade HM et al. (1982) J. Bacteriol. 147, 114-122.
10. Charles TC et al. (1997) Genetics 146, 1211-1220.

EFFECTIVENESS OF ACACIA RHIZOBIA

Kang Lihua, Jiang Yegen, Ma Haibin, Shang Junhong
Research Institute of Tropical Forestry, CAF, Guangzhou 510520, P. R. China

The broad aim of this research is to develop a collection of effective rhizobial strains which can be recommended as inoculants for a range of Acacia spp. and which are persistent in soil. The summary is: (1) Inoculation of Acacia with Rhizobium often lends to high levels N_2 fixation. Different Acacia species and different provenances within species respond differently to inoculation with the same sets of Rhizobium strains. (2)Chinese Rhizobium strain LL026 is highly effective in N_2 fixation for several Acacia provenances; Australian strain 1563 is also useful especially for A.mearnsii. (3)It is most improbable that a single Rhizobium strain highly effective for all Acacia provenances will even be found. (4)Peat is the best carrier for Acacia inoculant. (5)Acacia nodulation and N_2 fixation in the nursery are improved when clay soils are amended with organic matter. (6)Most Acacia rhizobia are sensitive to acidity, but strongly acid-tolerant strains do exist. (7)Calcium phosphate should not be used to correct acidity in nursery soil. (8)Soil rhizobia in Acacia plantations increase in number as time progresses. The conclusion is a collection of strains of *acacia rhizobia* has been obtained by isolation from Chinese soils and augmented with strains of Australian origin. Strains effective in fixing nitrogen with *Acacia mearnsii, A. implexa* and *A. melanoxylon* have been selected and their performance proven in lab, nursery and outplanting experiments. Effective, Long-life acacia inoculants have been prepared using finely-ground sterile peat as a carrier. The composition of a nursery soil suitable for the culture of vigorous, well- nodulated acacia seedlings has been defined. A nursery system that will grow vigorous, well-nodulated, mineral-N-independent acacia seedlings is an important finding. We established 6 field trials in 3 sites with a coverage of 4 hm2.

SCREENING FOR EFECTIVE SYMBIOTIC RHIZOBIAL STRAINS AND DETECTION OF NODULATION BY MOLECULAR MARKING IN ALFALFA

Dan-Ming Chen [1,2], Zhao-hai Zeng [3], Xin-hua Sui [1,2], Yue-gao Hu [3], Wen-xin Chen [1,2]
[1] College of Biological Sciences, China Agricultural University., [2] Key Lab. of Agro-Microbial Resource and Application, Ministry of Agri of China.
[3] College of Agronomy and Biotechnology, China Agricultural University.

Inoculation with rhizobia has been proven an effective way to increase alfalfa yield. There are big biodiversity of rhizobial resources, however, only few of them have been explored. In the present research, we selected for efficient rhizobial strains from our collections and tested the possibility of IGS-Restrictive Fragment Length Polymorphism (IGS-RFLP) as the method to determine the ratio of nodulation occupancy and competitiveness.

In the first experiment, alfalfa plants (cv. Vector) were inoculated with eighteen rhizobial strains (*Sinorhizobium meliloti*). After growing for 90 days, plants inoculated with strain CCBAU 30138, CCBAU 96077, CCBAU 96068 and CCBAU 75035 showed greater shoot dry weights, increased total nitrogen content and higher N_2-fixing efficiency.

The effect of the four strains was further demonstrated in a field experiment. It was found that the aboveground dry weight and protein content of inoculated alfalfa plants increased 11.9% and 16.3%, respectively on the average.

IGS-RFLP was used for determining the ratio of nodulation occupancy and competitiveness. When alfalfa was inoculated with CCBAU96068 and 96077, the ratio of nodulation occupancy reached 65% and 50% respectively.

It was concluded that CCBAU 30138, CCBAU 96077, CCBAU 96068 and CCBAU 75035 might be used as the effective inoculants for Vector cultivation in middle northern China. IGS-RFLP can be used to determine the ratio of nodulation occupancy and competitiveness.

Acknowledgements: Study made possible through Grant No.2003AA241150 from 863 of China. Joint contribution through Grant No 2001CB108905 from 973 of China.

REINOCULATION INCREASING SOYBEAN GRAIN YIELD IN BRAZIL

Mariangela Hungria[1,2], Rubens J. Campo[1], and Iêda C. Mendes[3].
[1]Embrapa Soja, Londrina, PR, Brazil; [2]CNPq, Brazil; [3]Embrapa Cerrados, Planaltina, DF, Brazil.

Soybean (*Glycine max*) was introduced into Brazil at the end of the 19th century, but commercial crop expansion started only in the 1960s, such that today the country is the second largest grain producer of this legume. Biological nitrogen fixation (BNF) has always been a significant concern and programs for both strain selection and identification of plant genotypes with superior symbiotic performance began in the late 1960s, and remain a major focus. Now, there are four strains officially recommended as inoculants in Brazil: *Bradyrhizobium elkanii* SEMIA 587 (from 1968-1975 and then since 1979) and 29W (= SEMIA 5019) (rsince 1979), and *B. japonicum* strains CPAC 7 (=SEMIA 5080) and CPAC 15 (=SEMIA 5079) (both since 1992). However, today most soils (>90%) cropped with soybean in Brazil have high populations of *Bradyrhizobium* (10^3 cells g^{-1} of soil or higher), established from previous inoculations, thus the need for reinoculation should be investigated. Field trials were performed in four Brazilian states, representing the main cropping areas. Inoculant treatments included single or combined recommended strains in a peat-based carrier, at a concentration of 10^8 cells g^{-1} of inoculant, applied at a rate of 500 g of inoculant 50 kg^{-1} seeds. Trials always included non-inoculated controls, with or without N-fertilizer (200 kg of N ha^{-1}, split at sowing and at flowering time).

29 field trials were performed in a 14-year period, and statistically significant increases were obtained due to reinoculation. Mean grain yield increases were 8%, representing a major contribution to the crop. Noteworthy are that yields higher than 4,000 kg ha^{-1} were obtained in BNF-dependent plants and that yield responses to inoculation occurred in a majority of soils having more than 10^3 indigenous rhizobia g^{-1} soil. Rates of BNF of up to 300 kg of N ha^{-1}, and providing up to 94% (estimated by the N-ureide technique) of total plant N were observed. Benefits from N release to the following crop were also observed (about 20-30 kg N ha^{-1}). Starter (30 kg N ha^{-1}) and pre- or post-flowering (50 kg N ha^{-1}) N-fertilization did not increase yield, so emphasizing biological efficiency. Partially supported by CNPq (PRONEX-41.96.0884.00 and 35216/1992-3).

PERFORMANCE OF A NEW INOCULATION TECHNOLOGY TO INCREASE SOYBEAN YIELD

Rubens J. Campo and Mariangela Hungria.
Embrapa Soja, CP 231, 86001-970, Londrina, PR; Brazil;

Peat became the standard carrier for legume seed inoculants around the world due to its excellent performance, protective character and low price. In Brazil, more than 95% of the inoculants commercialized up to 1997 were peat-based, but after 1997 several liquid inoculants were developed and released, now representing 50% of the market. Problems with the use of peat-based inoculants with sowing equipment, demand for seed treatment with fungicides, insecticides and micronutrients and demand for higher number of cells per seed to face soybean crop expansion in degraded soils stimulated the development of alternative and more effective inoculation technologies. This study aimed to compare the traditional seed inoculation methods, using peat- based or liquid inoculants, with the row inoculation method using liquid inoculants.

The experiments were set up in sites where soybean has been planted for several years showing a high-established *Bradyrhizobium* population ($>10^4$ cells/g soil). The trials were conducted for three years, following all technical recommendations and using a randomized complete block design with six replicates. The row application of liquid inoculants was done manually (two experiments), or mechanically (five experiments), just before dropping the seeds. The minimum concentration of cells was of 1×10^9/ ml of inoculant, and the amount of water plus inoculant applied in the row was around 50 l/ha. The results showed that nodulation (nodule number and dry weight), grain yield and N accumulated in grains were similar in seed or row inoculation, in soils with and without established population of *Bradyrhizobium*.

The seed inoculation method can be substituted by inoculation in the sowing row. However, the use of the row inoculation method requires six times more liquid inoculant than when applied on the seeds. The use of higher doses of inoculant in the row represents a strategy to avoid the toxicity of agrochemicals and micronutrients applied on the seeds.

SOILS CHARACTERISTICS AFFECTING BIOLOGICAL N FIXATION OF ORGANICALLY GROWN RED CLOVER IN FINLAND

Nykänen, Arja
MTT Agrifood Research Finland, 51900 Juva, Finland, Europe

Red clover (*Trifolium pratense*) is the most common forage legume in Finnish grasslands. It is cultivated as a mixture with grasses and cut twice during growing season. Amount of biologically fixed nitrogen is a key factor for N-balances in organic farming. Only a little research has been done on biological nitrogen fixation level in Finland

In this study we will study the variation of sward yield, its clover content and biological nitrogen fixation (BNF) of red clover. The red clover (*Trifolium pratense*) / timothy (*Phleum pratense*) / tall fescue (*Festuca arundinaceae*.) sward was established in 2003 with barley as nurse crop. A two hectares field under organic farming is situated in East-Finland (60°53'N 27°53'E). The variation of chemical (C, N, P, K, Mg, Ca, S, B, Co, Mo, Ni, Cu, Fe, Mn, Zn, pH), physical (earth worm burrows, root channels, macro pores, bulk density, total porosity) and microbiological characteristics (organic matter quality (POM), microbial biomas (C and N), soil respiration, netmineralization of N) of soil is examined as well as their influence on BNF and growth of sward. BNF will be measured with ^{15}N dilution technique and N difference technique.

Nutrient contents varied a lot inside the field, which soil type is considered as moraine containing silt, sand and clay. According to the Finnish soil classification, the status of these nutrients varied from 'poor' to 'good'. The dry matter yield of first cut varied from 2 530 to 10 870 kg ha^{-1} and second cut from 870 to 5 530 kg ha^{-1} in dry matter in 2004. The clover content varied 4-75 % and 1-73 % in dry matter respectively.

Very preliminary results show slight correlation between sward yield and some soil characteristics, as pH, potassium, manganese, cobolt, copper, molybdene and iron, show. Further studies will be carried out in becoming two years.

PRODUCTION AND APPLICATION OF RHIZOBIUM INOCULANT IN D. P. R. KOREA

Yong-Chan Kim, Yong-Nam Pak
Research Institute of Microbiology, Academy of Sciences, D. P. R. Korea

In recent years, the increasing demands for legume product have become the driving force of the active and productive investigation on rhizobium in DPRK, one of the original culture places of soybean. Here we report some results from the investigation on rhizobium inoculants (RI).

1. Production of the RI.
First, rhizobium strains were isolated from the nodules of the legume plants selected for good quality. Strains were treated by UV and 2, 4-dinitrophenyl hydrasin, and applied to legume plant. This manipulation was repeated three times and strains with high nitrogenase activity (more than 120 nmol C_2H_2/plant) were isolated. For field test, solid-phase fermentation (peat 40, fertile soil 20, soybean cake 5, superphosphate 0.1, $MgSO_4$ 0.05, moisture 60%) was used to prepare the large scale production of RI.

2. Approaches to improve the effect of RI.
The control of the nitrogen (N) and phosphorus (P) contents was very significant to improve the effect of RI. It was recommended that the N and P contents in soil should be controlled at 110~120 ppm and 100~110 ppm respectively to improve the effect of RI. Another experiment showed that the effect of potassium fertilizer on the effect of RI was not significant under our conditions. Since RI was affected by indigenous microbes, the application of seed dressing and 300~500 g/ha of base introduction were recommended. Co-applications with other effective soil microorganisms (AM fungi, *P. fluorescence*) were observed to improve the effect of RI.

3. PGPR effect of RI on non-legume plant.
R. leguminosarum was selected to investigate its PGPR effect on the yield of corn. RI presented PGPR effects not only by the N_2-fixing ability but also by the congregation of effective soil microorganism in the rhizosphere.

UNDERSTANDING THE GENETIC INSTABILITY IN *CICER ARIETINUM* ROOT NODULE BACTERIA

Saadia Naseem, Asma Aslam, Kausar A. Malik, [a] Fauzia Y. Hafeez, National Institute for Biotechnology and Genetic Engineering. P. O. Box. 577. Faisalabad, Pakistan.

The strains for inoculum production are selected for their superior broad host range, acid, temperature and salt tolerance characteristics. Three chickpea root nodule rhizobia Ca-34, Ca-BZ and Ca-1B that were selected for their capability of forming effective nodules on their host, lost their ability of nodulation after long term preservation and subsequent sub culturing. To understand this *in vitro* loss of nodulation ability different aspects of bacterial taxonomy including morphology, symbiosis, plasmid profile, analysis of symbiotic and ribosomal genes were assessed.

These rhizobia remained unable to induce nodulation or gall formation on their host plants. Analysis of plasmid profile showed one plasmid in Ca-34. No PCR amplification was detected with *nodA* and *nifH* primers suggesting that although the plasmid is present but it has lost the symbiotic genes or has undergone mutations. No plasmid was detected in Ca-BZ and Ca-1B. On the basis of partial 16S rRNA gene sequence anlysis Ca-34 was identified as *Ochrabactrum* sp. while Ca-BZ and Ca-1B were identified as *Rhizobium* and *Agrobacterium* respectively, although all were showing *Rhizobium* like characteristics when they were isolated. They have acquired inconsistently symbiotic genes or *Sym* plasmid during evolutionary process in soil before isolation that was lost during subsequent preservation. The study confirms that long-term preservation and extensive subculturing subject bacterial cultures to mutations leading to change their some characteristics effectively.

Reference:
Del Papa M F et al. (1999) Appl. Environ. Micorbiol. 65, 1420-1427.

POTENTIALS OF *BRADYRHIZOBIUM* SPECIES AS PROMOTERS OF PLANT GROWTH IN COTTON

[a] Fauzia Y. Hafeez, Nosheen Mushtaq, Sohail Hameed, Sumera Yasmin, and Kauser A. Malik.
National Institute for Biotechnology and Genetic Engineering (NIBGE).
P.O.Box. 577 Jhang Road, Faisalabad 38000, Pakistan

The beneficial effect of the symbiotic association between rhizobia and legumes is well known but the PGPR effects of rhizobia with non-legume is being intensively investigated.

The present study was conducted to screen the effective cotton (*Gossypium hirsutum*) associated plant growth promoting rhizobacteria (PGPR). The genetic relationship of twenty-four bacterial isolate was carried out by random amplified polymorphic DNA (RAPD), using 30 random primers. RAPD profiles of these isolates resulted in four clusters showing a considerable level of genetic diversity among these bacteria. The selected isolates S-1 and S-8 were studied for their ultra structure localization in the rhizosphere of cotton. The isolate S-8 was found below the epidermal layer in contrast to the isolate S-1 that resided only in the cervices present in characteristic foldings of cotton root cells. Moreover, localization was monitored to study its interaction with cotton using Immunogold labeling (Quadt and Kloepper 1996). Polyclonal antibodies raised against isolate S-8 showed an intensive gold labeling of *in vitro* grown culture and ultra thin sections of root tissue. Therefore, the isolate S-8 found to be a potent colonizer of cotton roots. The isolates S-1 and S-8 were identified as *Bacillus fusiformis and Bradyrhizobium* sp. respectively on the basis of partial 16S rRNA sequence analysis and *nodA* gene amplifications. Detailed studies using light and transmission electron microscopy combined with immunogold labeling suggested that *Bradyrhizobium* sp. S-8 is a novel strain regarding its high affinity towards cotton roots with potential benefit to the cotton crop by phytohormones production and nitrogen fixation (Hafeez et al. 2004). This indicates that specific bradyrhizobia have the potential to be used as PGPR for non-legumes.

References
1. Quadt-Hallmann A and Kloepper JW (1996) Can. J. Microbiol. 42, 1144-1155.
2. Hafeez YF et al. (2004) Australian J. Exp. Agri. 44, 617-622.

$^{15}N_2$ FIXATION AND TRANSLOCATION IN SUGARCANE (*Saccharum* sp.)

Takuji Ohyama[1], Atsushi Momose[2], Keiko Nishimura[1], Takahiro Hiyama[1], Noriko Ishizaki[1], Katsuya Kanbe[1], Kaushal Tewari[2], Norikuni Ohtake[1], Kuni Sueyoshi[1], Takashi Sato[3], Atsushi Sato[3], Yasuhiro Nakanishi[4], and Shoichiro Akao[5]
[1]Faculty Ag., Niigata Univ., Niigata 950-2181, Japan. [2]Grad. School Sci. and Tech., Niigata Univ., Niigata 950-2181, Japan. [3]Faculty Bioresource Science, Akita Prefectural Univ., Akita 010-0195, Japan. [4]Miyako Subtropical Exp. Farm, Tokyo Univ. Ag., Okinawa 906-0103, Japan. [5]Faculty Ag., Miyazaki Univ., Miyazaki 889-2192, Japan

Some diazotrophic endophytic bacteria, such as *Gluconacetobacter diazotriophicus* in sugarcane (*Saccharum* sp.), may significantly contribute N nutrition in the host plants. However, quantitative evaluation of the site of N_2 fixation and translocation of the fixed N has not been fully elucidated. ^{15}N tracer experiments were conducted on young sugarcane plants to investigate these questions. A young sugarcane seedling of about 30-cm height germinated from a cut stem was exposed to $^{15}N_2$ gas ($^{15}N_2$:Ar:O_2 = 24:56:20; 99.4 atom% ^{15}N) in a 500-mL plastic cylinder for 7 days, and the incorporation of ^{15}N into shoots, roots, and seed stem was analyzed after 3, 7, 14, and 21 days after $^{15}N_2$ treatment. After 3 days of $^{15}N_2$ feeding to the whole seedling, the ^{15}N abundance in the shoot, roots and the seed stem was 0.027 %, 2.221 % and 0.271 %, respectively. The roots showed the highest nitrogen fixation activity and the incorporation of ^{15}N in the shoot was very low. Due to the large amount of N in the seed stem, the distribution of total amount of ^{15}N among organs was higher in the seed stem (76 %) than in roots (17 %) and shoot (7 %) after 7 days of $^{15}N_2$ feedings. During the 3-week (1 week of $^{15}N_2$ feeding followed by 2 weeks of non-labeled condition) treatment period, about a half of non-labeled N, originating from the seed stem, was transported to the shoot and roots and utilized for their growth. However, the fixed ^{15}N either in roots or in the seed stem was not appreciably transported to the shoot.

In another experiment, sugarcane plants were harvested after 3 months cultivation after one day of $^{15}N_2$ feeding. Enrichment of ^{15}N could not be detected in the shoot. These results indicate that the endophytic diazotrophic bacteria in the roots and seed stem actively fix N_2 in young sugarcane seedling but the N fixed in early stage is not readily utilized for the shoot growth.

COMPARISON OF BACTERIAL ACTIVITIES INVOLVED IN NITROGEN CYCLING BETWEEN CONVENTIONAL RICE CULTIVATION AND THE SYSTEM OF RICE INTENSIFICATION (SRI)

Teaumroong Neung[a], Sooksa-nguan Thanwalee[a], Thies E. Janice[b] and Boonkerd Nantakorn[a]
[a]School of Biotechnology, Institute of Agricultural Technology, Suranaree University of Technology, Nakhon-Ratchasima 30000, Thailand, [b]Department of Crop and Soil Sciences, Cornell University, Ithaca, New York 14853 E-mail : neung@ccs.sut.ac.th

The system of rice intensification (SRI) is a novel system of rice cultivation that has been shown to increase yields up to 5-10 t ha^{-1} when compared with practices used in conventional rice cultivation. This study focused on the practices of intermittent irrigation during the vegetative growth phase and compost application in order to investigate the effects of rice cultivation systems on microbial processes in the nitrogen cycle including; N-mineralization, nitrification, denitrification and nitrogen fixation. Rice was grown in ChiangMai, Thailand during the dry season (February – June) 2003 by using four different management practices: (1) conventional system with compost, (2) conventional system without compost, (3) SRI system with compost, and (4) SRI system without compost. Rice yields obtained from the conventional systems were significantly higher than those obtained from the SRI systems. Nematodes were found during the intermittent irrigation in the vegetative growth phase and could be a factor in the observed yield reduction in SRI fields. Potentially mineralizable N differed significantly between soil depths, but not between rice cultivation systems, with the 0-10 cm depth having a higher N-mineralization rate. Nitrification rate was significantly higher in the SRI system, with and without compost addition. No differences in denitrification potential or nitrogen fixation were found between the cultivation systems or soil depths. Microbial community analysis using the Terminal Restriction Fragment Length Polymorphism (T-RFLP) technique revealed no significant differences in bacterial communities between the two cropping systems, but rather differences in relation to sampling time within the growing season.

Randriamiharisoa, R.P. (2002) Proceedings International Conference, Sanya, China, edited by N. Uphoff, et al., pp. 148-157.
Uphoff, N. (1999) Environment, Development and Sustainability 1, 297-313.

EFFECT OF INCREASING CORN YIELDS AND SAVING NITROGEN FERTILIZER BY INOCULATION OF *AZOSPIRILLUM BRASILENSE* YuMA

Tengyun Yao[1], Jinxiang Yu[1], Sanfeng Chen[2], Yanqi Chang[2], Jilun Li[2]
[1]Soil and Fertilizer Institute, Shanxi Academy of Agricultural Science, Taiyuan 030000, China
[2]State Key Laboratory for Agrobiotechnology and College of Biological Science, China Agricultural University, Beijing 100094, China

As a plant-associative bacterium, *Azospirillum* has been demonstrated the potential to promote plant growth and enhance the yield of crops in different soil and under different climatological conditions.

Field plant inoculating experiments with *Azospirillum brasilense* YuMA (*nifAc, draT -*) were carried out for three years in the sabulous soil in Shanxi province of northeast China. Under three different levels of urea (15, 12 and 9 kg/mu), the YuMA inoculants could increase corn yield remarkably, especially at 9 kg/mu of urea. The YuMA inoculants increased 9% and 31% of corn yields respectively compared with the Yu62 inoculants and no inoculants control while saving 40% ~ 20% of nitrogen fertilizer. It means that the genetically modified strains of *Azospirillum* can evidently promote the production of corn than the control and wide type.

Our results, based on three years' field experiments, showed that the *Azospirillum*-inoculated seed pieces fertilized with a low level of nitrogen gave higher yields than those obtained from non-inoculated or fully fertilized. The mode of action of *Azospirillum* proposed the possibility of more than one mechanisms being less significant when evaluated separated. But nitrogen fixation and hormonal activities are still considered to play the major roles in observed plant growth promoting effect. Other reports also indicated that *Azospirillum* can be used as a biocontrol agent, so it is a kind of multifunctional bacteria for sustainable agriculture.

Acknowledgements
This work was supported by the National Key Fundamental Research Program of China (973 Program, Grant No.2001CB108904).

ENDOPHYTIC NITROGEN-FIXING BACTERIA COLONIZATION INCREASES GROWTH AND NITROGEN ACCUMULATION IN RICE VARIETIES

Gui-Xiang Peng[1], Guo-Xia Zhang[2], Wei Hou[2], Hua-Rong Wang[2], Zhi-Yuan Tan[2*]
[1] College of Resource and Environment, South China Agricultural Uni., GuangZhou, 510642, P. R. China, [2] Lab of Molecular Genetics, College of Agriculture, South China Agricultural University, GuangZhou, 510642, P. R. China. *Corresponding author. E-mail: zytan@scau.edu.cn

Wild rice distributed in most of the areas in the world. Some of the endophytic nitrogen-fixing bacteria isolated from wild rice have been reported. We manage to isolate more endophytic nitrogen-fixing bacteria systematically from wild rice of *Oryza australiensis*, *O. grandiglumis*, *O. granulata*, *O. latifolia*, *O. longiglumis*, *O. longistaminata*, *O. malampuzhaensis*, *O. minuta*, *O. nivara*, *O. officinalis*, *O. punctata*, *O. rhizomatis*, *O. ridleyi*, *O. Rufipogon* and other gramineous plants by different culture methods. More than one hundred endophytic nitrogen-fixing strains have been obtained. Most of the strains showed to promote the rice growth and nitrogen accumulation. Some strains have a high nitrogenase activity tested by acetylene reduction methods. SDS whole-cell protein patterns and BOX fingerprinting were used as clustering methods. 16S rRNA gene sequencing analysis indicated that these strains belonged to *Azospirillum* sp., *Citrobacter* sp., *Pantoea* sp., *Azoarcus* sp., *Bacillus* sp., and perchlorate-reducing bacterium. More studies are continuing.

Acknowledgment
This research was supported by the National Natural Science Foundation of China (No. 30300001and 30470002) and by University Science Foundation grant from South China Agricultural University.

STUDY ON ASSOCIATIVE SYMBIOTIC NITROGEN BACTERIA IN RHIZOSPHERE OF OAT IN ALPINE REGION

Tuo Yao[1], Degang Zhang[1], Ruijun Long[1], Fauzia Y. Hafeez[2], Sumara Yasmin[2], Ghulam Rasul[2]
[1] Pratacultural College, Gansu Agricultural University, Lanzhou 730070 China; [2] National Institute for Biotechnology and Genetic Engineering, Faisalabad, Pakistan

Associative Symbiotic Nitrogen Bacteria (ASNB) strains from oat rhizosphere in Tibet Plateau were studied. The results obtained are as follows:

There are 8 ASNB strains isolated from root rhizosphere of oat. Nitrogenase activity is vary (112.5~1147.9 nmol C_2H_4/h/ml.) with strains, and only 2 strains, which nitrogenase activity is higher than 500 nmol C_2H_4/h/ml; 3 ASNB strains (*Azotobacter* sp. ChO4, *Azotobacter* sp. ChO5 and *Azospirillus lipoferum* ChO6) show phosphate solubilizing power (83.8~103.5 µg/ml); and all the ASNB strains show ability to produce auxin (IAA) although the ability is different (IAA concentration is 2.16~17.31 µg/ml) with the strains, and only 2 strains (*Pseudomonas* sp. ChO3 and *Azospirillus lipoferum* ChO6) show high IAA concentration (higher than 10µg/ml) detected by HPLC.

Effect of ASNB strains isolated from oat rhizosphere on growth of oat and quantification of nitrogen fixed were carried out by using ^{15}N Isotope Dilution technique in Hongland medium and in pot experiment as well. The results show that in hongland medium, most of strains are more benefit to oat growth although few strains have little influence on oat growth; %Ndfa (percentage nitrogen derived from atmosphere) and quantification of N fixed are distinct with strains, which is 13.78%~63.96% and 0.0653~0.3158 mg/plant respectively. But in pot experiment, %Ndfa (4.86%~27.68%) and quantification of N fixed (2.31~6.78 kg/hm^2) are much lower compared with in Hongland medium. Field experiments are in progress.

In the view of the above results, *Azospirillum lipoferum* ChO6, *Azotobacter* sp. ChO4, *Pseudomonas* sp. ChO3, *Azotobacter* sp. ChO5 and *Azotobacter* sp. ChO7 show higher potential as biofertilizer inoculums for oat.

THE PRESENCE OF THE SYMBIONT IN THE *AZOLLA-ANABAENA* SYMBIOSIS IS ESSENTIAL FOR THE ABILITY OF THE FERN TO UTILIZE COMBINED NITROGEN

D. Kaplan, M. Azeb, Y. Akkara, G. Granot, A. Nejidat, and Y. M. Heimer
The J. Blaustein Institute for Desert Research, Ben-Gurion University of the Negev, Sede Boqer campus 84990, Israel

In the *Azolla-Anabaena azollae* association, the fern provides the photo-assimilates while the N_2-fixing cyanobacterium supplies the entire N requirement (Peters and Meeks, 1989). Endophyte-free *Azolla* fern (EF), cured of its association, was dependent on combined N yet grew much slower than the association, displayed higher carbon (C) content, but lower N and protein contents (Kaplan and Peters, 1998), and developed a massive root system. We hypothesized that EF suffered from chronic N deficiency in the absence of the endophyte. Indeed, an *A. pinnata* association grown under a low-N_2 atmosphere develops symptoms of N deficiency, which are overcome by providing combined N to the medium (Table 1). Growth of EF was accelerated by replacing inorganic combined N with an organic N source (chasein hydrolysate + glutamine).

Table 1. Doubling time, biomass, and GS activity of association grown under ambient or low N_2, with or without inorganic N and of EF grown in air with added inorganic N.

	Growth Conditions and Plants				
	Association				EF
	Air	Air +NH_4NO_3	10% N_2	10% N_2+NH_4NO_3	Air + NH_4NO_3
DT	7.3± 0.4	7.7± 0.1	10.8± 1.7	8.21± 0.2	14.9± 1.1
DM	3.7± 0.1	3.7± 0.1	4.7± 1.0	3.4± 0.7	7.3± 0.1
N	6.2± 0.1	6.4± 0.1	1,4±0.02	5.2± 0.2	3.4± 0.1
MRL	11.6±3.8	8.4± 1.8	31.9±10.5	11.2± 4.6	27.8±6.1
GS	400±90	510± 55	75±60	295±120	30±10

DT: doubling time(days); DM: dry matter(%); N: nitrogen conten(%); MRL: mean root length; GS: glutamine syntetase activity (μmoles mgP min^{-1}).

The results support the hypothesis that the presence of the symbiont is essential for the fern's ability to uptake and assimilate combined nitrogen regardless of its source.
Peters GA, Meeks JC (1989) Ann. Rev. Plant Physiol. Plant Mol. Biol. 40, 193-210.
Kaplan D and Peters GA (1998) Symbiosis 24, 35-50.

MACROIDEA FOR RESEARCH ON BIOLOGIAL NITROGEN FIXATION AND MODE FOR ITS APPLICATION AND POPULARIZATION

Guo Yongjin[1,] Zhu Anni[1,] Ye Zengguang[2]
G-Laboratory, China[1]. New & Hope Star Bio-Technology Co Ltd., Beijing, China[2]

In the natural world, nitrogen-fixing microorganisms do not live in isolation to perform the function of fixing nitrogen. Simultaneously with being limited by various environmental factors, they and other colonies of micro-organisms coordinate each other for survival and reproduction in the process of the struggle between contradictions, the adaptive evolution and the unity of opposites. With regard to unicellular prokaryotes organism characterized mainly by extremely easy variation, the speed at which the human race researches them is far behind that of their variation.

In the course of the replacement of the traditional agriculture by the modern agriculture, the biological chain between the breeding industry and crop farming has been broken off. While obtaining enormous economic benefits, the modern agriculture has been confronted with such questions as environmental pollution, soil degradation and food safety. As to how to apply the technology of biological nitrogen fixation and build up a harmonious and balanced natural circulation bio-sphere, this thesis provides a practical example. A plant was established for treating faeces from 30,000 pigs on hand, producing 3,000 tons of organic fertilizer. According to repeated tests, the contents of N, P and K in the organic fertilizer are more than 17%. Many experiments on crops in the past two years have achieved good results. They depolluted the environment and provided material conditions for a sustainable development of the organic agriculture as well.

IMPROVEMENT OF SALT TOLERANCE OF *BRADYRHIZOBIUM JAPONICUM* BY EXPRESSION OF THE BETAINE TRANSPORTER BETS FROM *SINORHIZOBIUM MELILOTI*

Alexandre Boscari, Karine Mandon, Marie Christine Poggi, and Daniel Le Rudulier
CNRS-INRA-Université de Nice-Sophia Antipolis, Centre INRA Agrobiotech, 06903 Sophia Antipolis Cedex, France

Within the soil, rhizobia frequently encounter water stress and salinity that affect their growth and efficiency of nitrogen fixation. *Bradyrhizobium japonicum* is among the most salt sensitive species. In contrast to several other rhizobia, growth of *B. japonicum* at increased osmolarity cannot be restored by the osmoprotectant glycine betaine, GB. We have addressed the hypothesis that betaine transport through the Betaine Carnitine Choline Transporter system (BetS) from *Sinorhizobium meliloti* provides an alternative mechanism for *B. japonicum* to acquire betaines for osmoprotection. The *bets* gene, which encodes a high-affinity Na^+-dependent betaine symporter, was cloned into pVK100, and the resulting plasmid was transferred into *B. japonicum* USDA110. We found that BetS is functional in *B. japonicum* and regulated by the osmolarity, as observed in *S. meliloti* (Boscari et al., 2002). Salt-treated transformed cells accumulated large amount of GB, up to 240 mM at 150 mM NaCl, which was not catabolized. Such accumulation can be reversed through rapid efflux during osmotic downshock, indicating that *B. japonicum* possesses osmoregulated channels used for GB efflux during osmotic downshift. Growth of transformed cells at moderate NaCl concentration (80 mM) was obtained in the presence GB, whereas growth of USDA110 was totally abolished. Surprisingly, deleterious effect of 100 mM NaCl could not be overcome by GB, despite its accumulation. Growth experiments with a nonelectrolyte, mannitol, at a concentration osmotically equivalent to 100 mM NaCl, indicated that the inhibition essentially resulted from the toxic effect of ions. It is tempting to postulate that in the absence of potential gene-encoding Na^+/H^+ antiporters in *B. japonicum*, the expulsion of Na^+ from cells maintained at high NaCl concentration could not compensate Na^+ entry, which is amplified as a consequence of stimulated GB uptake through BetS (Boscari et al., 2004.)

Boscari A et al. (2002) J. Bacteriol. 184, 2654-2663.
Boscari A et al. (2004) Applied Environ. Microbiol. 70, 5916-5922.

EFFECT OF CADMIUM ON GROWTH AND NITROGEN FIXATION OF ALFALFA

En Ci, Ming Gao, Yong-xiong Yu
Southwest Agricultural University, Chongqing 400716, China

With the development of industry, Soil heavy metal pollution has become more and more serious, Cadmium is an important heavy metal that can pollute the environment, and plants are obviously affected with excessive cadmium accumulation (Barcelo T et al. 1986). Alfalfa is a legume forage plant, which distributes widely in the world. Rizobia-alfalfa symbiotic association is a high efficient N-fixing system. A study was conducted to evaluate the influence of seven levels of cadmium addition (0, 0.5, 1.0, 3.0, 5.0, 10 and 50 mg Cd/kg soil) on growth and nitrogen fixation of alfalfa by pot experiment The results showed that there was little effect of cadmium on seedling emergence ratio, aboveground biomass, height, and chlorophyll content of alfalfa in treatments with low cadmium addition in soil, and effect of cadmium on these indexes increased obviously in treatment with a certain high cadmium addition (>5.0 mg Cd/kg soil). Cadmium content of belowground part was higher than aboveground part, and cadmium content of different parts of alfalfa and available cadmium content in soil had a significant positive relation with cadmium addition in soil ($P<0.01$). The fresh weight of root nodule in treatments declined along with increase of cadmium addition in soil, and nitrogenase activity began to be restrained when 5.0 (mg/kg) of cadmium was added into the soil. When 3.0 (mg/kg) of cadmium was added into the soil, the growth of alfalfa was enhanced to a certain extent, and biomass and chlorophyll content were increased, and nitrogenase activity also got back.

References
Barcelo T, et al. (1986) J. Plant Physiol. 125, 17~25.

INVOLVEMENT OF STRESS, NITRIC OXIDE AND RHIZOBIAL SYMBIOSIS ON EXPRESSION OF GLOBIN GENE FAMILY IN LOTUS JAPONICUS

Yoshikazu Shimoda[1], Maki Nagata[1], Fumie Furuya[1], Akihiro Suzuki[1], Mikiko Abe[1], Shusei Sato[2], Tomohiko Kato[2], Satoshi Tabata[2], Shiro Higashi[1], Toshiki Uchiumi[1]
[1]Lab. Plant & Microbe, Dept. Chem. & BioSci., Kagoshima U., Japan.
[2]Kazusa DNA Res. Inst.

Globin gene family which consist of symbiotic and nonsymbiotic globin genes were identified in the genome of Lotus japonicus and characterized by their phylogenetic relation and expression profiles.

Three symbiotic globin genes (LjLb1,2,3) were expressed highly in root nodules and showed low level expression in young seedlings. Besides, transformed plants introduced LjLb3 promoter-GFP gene fusion complex showed early response to the inoculation of symbiotic rhizobium. From these results, we discussed the additional function and regulation of symbiotic globin genes.

Two distinct non-symbiotic globin genes (LjHb1,2) were also characterized based on their expressional analyses. Both genes were expressed in all tissues of mature plants, and enhanced expression was seen in root nodules. LjHb1 was strongly induced some stress conditions and by treatment with plant hormones, whereas LjHb2 was induced only by the application of sucrose. Moreover, LjHb1 was also induced transiently by the inoculation of symbiotic rhizobium and showed similar expression pattern to a candidate nitric oxide synthase gene. Using DA-FM as a detector of Nitric oxide, we indicated the expression of LjHb1 was linked to generation of nitric oxide and suggest the possibility that NO and globin might be involved in stress adaptation and early stage of the rhizobium-legume symbiosis.

Reference
T., Uchiumi, et al. (2002) Plant and Cell Physiol. 43(11), 1351-1358
Y., Shimoda, et al. (2005) Plant and Cell Physiol. (in press)

CHARACTERIZATION OF THE ALTERNATIVE PROTECTION PROGRAMS AGAINST OXIDATIVE-STRESS IN FREE LIFE AND SYMBIOSIS FROM *RHIZOBIUM ETLI*.

María del Carmen Vargas, Sergio Encarnación, María de Lourdes Girard, Agustín Reyes, Yolanda Mora and Jaime Mora.
Centro de Investigación sobre Fijación de Nitrógeno-UNAM. Av Universidad s/n. Apto. Postal 565-A. Cuernavaca, Mor. México.

R. etli CE3 has only one catalase, which is a bifunctional, inducible and heat-labile enzyme, which is encoded by the *katG* gene. This protein can be considered a marker of aerobic metabolism, because no detectible *katG* mRNA formation and KatG protein were detected in fermentative growth or in bacteroid conditions. *R. etli* catalase-peroxidase is fundamental for the direct detoxification of hydrogen peroxide, in free life whereas, in the nitrogen fixation process it shows a minor protective role. Additionally, protein profiles expressed under this metabolic condition indicate the existence in symbiosis of other compensatory mechanism, which could protect the cells against reactive oxygen species. Proteome analyses suggest that in symbiosis in the *katG*-mutant bacteroids, the presence a peroxiredoxin (*orf180*), in addition to new and over expressed proteins. Our analysis showed that the peroxiredoxin is encoded by a gene belonging to a symbiotic plasmid (pCFN42d), and preliminary results performed in symbiosis show that the *orf180* mutant strain is diminished in its capacity to fix nitrogen, indicating the crucial role that the peroxiredoxin plays during symbiotic process.

In *R.etli oxyR* gene is located on plasmid f (pCFN42f) contiguous to *katG* and transcribing in the opposite direction. The presence of the conserved cysteine residues C199 and C208 in OxyR protein, a potential helix-turn-helix DNA-binding motifs in the *oxyR* gene, besides the location of OxyR DNA binding site in the promoter region of *katG* gene, are structural evidences that indicate the regulatory role of the product of *oxyR* on catalase-peroxidase in *R. etli*. In this study the role of OxyR was demonstrated as an activator of the oxidative stress response in free life and a crucial factor for the establishment of an effective symbiosis with legume plants.

Part of this work was supported by CONACYT grant 40046-Z and DGAPA grants IN203003-3 and IX250004.

RHIZOBITOXINE BIOSYNTHESIS IN *BRADYRHIZOBIUM ELKANII* AND TRANSFER OF *RTX* GENES TO NON-PRODUCING RHIZOBIA

M. Sugawara[1], S. Okazaki[1], S. Nonaka[2], H. Ezura[2], and K. Minamisawa[1].
[1]Graduate School of Life Sciences, Tohoku University, Sendai, Japan.
[2]Gene Research Center, University of Tsukuba, Tsukuba, Japan.

Rhizobitoxine is synthesized by the root-nodule symbiont *Bradyrhizobium elkanii* and the plant pathogen *Burkholderia andropogonis*. Rhizobitoxine production by *B. elkanii* enhances nodulation and competitiveness, via the endogenous ethylene biosynthesis in host legumes, because ethylene acts as a plant hormone that restricts nodulation in many legumes. Although rhizobitoxine biosynthesis genes (*rtxACDEFG*) in *B. elkanii* USDA94 were identified, their function and rhizobitoxine biosynthetic pathway are poorly understood. Introducing genes for rhizobitoxine biosynthesis into rhizobia, which are unable to decrease ethylene levels in host plants, ought to enhance their symbiotic interactions with host plants. Our aims are: (i) to determine the rhizobitoxine biosynthetic pathway and function of unknown genes; and (ii) to construct rhizobitoxine-producing bacteria by *rtx* gene transfer into non-producing bacteria.

B. elkanii USDA94 cells fed with ^{15}N-amino-labeled glutamine produced ^{15}N-serinol (one mass unit higher than standard serinol) and 15N-dihydrorhizobitoxine (two mass units higher than the standard). In contrast, addition of 14N-glutamine and glutamine with 15N-amido-glutamine resulted in no mass increase. Thus, the glutamine amino-N is incorporated into serinol and a dihydrorhizobitoxine simultaneously. We constructed *rtxD* and *rtxA* mutants (Δ*rtxD*, Δ*rtxA*) and supplied them with 10 mM glutamine. Unlike wild type, glutamine addition did not enhance dihydrorhizobitoxine production in Δ*rtxD* cells. These results suggest that the *rtxD* gene product is likely involved in the condensation of serinol and glutamine with the *rtxA* gene product.

Also, we attempted to confer the rhizobitoxine-producing capacity on non-producing rhizobia. As a result, *R. radiobacter* (formerly *Agrobacterium tumefaciens*) C58C1RifR produced rhizobitoxine by the introduced *rtx* genes. This is the first report of the transfer of rhizobitoxine-producing capacity to non-producing rhizobia. Use of this strain is expected to enhance *Agrobacterium*-mediate genetic transformation in plants because ethylene reduces transformation efficiency.

ABOVE GOUND INDICATOR OF HERBICIDE INDUCED STRESS IN THE LEGUME-RHIZOBIA SYMBIOSIS

Elizabeth A Drew, Vadakattu VSR Gupta, David K Roget
CSIRO Land and Water, PMB2, Glen Osmond, 5064, South Australia
Email: elizabeth.drew@csiro.au

A lack of N benefits to cereal crops following grain legume production has been observed in dryland farming regions of southern Australia. Following applications of non-target herbicides (at recommended label rates) in grain legume crops it is not uncommon to see crop yellowing. The aim of this work was to determine if an above ground symptom (yellowing) of legume plants following applications of in-crop herbicides reflects effects on nodulation i.e. the legume-rhizobium symbiosis.

During 2003, a replicated field trial (plot size 25m x 1.6m) was sown to *Pisum sativum* cv. Parafield on a calcareous sandy loam at Waikerie, South Australia. Seed was inoculated with *Rhizobium leguminosarum* bv. viciae inoculum (BioCare Group E, SU303) prior to planting. The crop was left unsprayed (control) or sprayed at the 3-4 node crop stage with one of five herbicides at recommended label rates (Flumetsulam, Diflufenican, Fluazifop-P, Butroxydim+Fluazifop-P, Haloxyfop-R). Crop yellowing was measured using the Minolta SPAD 502 chlorophyll meter on the youngest fully extended leaf and the number of total and effective nodules (pink) per plant counted three weeks after herbicide application. With the exception of the herbicide mixture Butroxydim+Fluazifop-P, all herbicides (broadleaf and grass selective) caused significant yellowing ($P<0.001$) within three weeks of herbicide application. This trend was also apparent below ground with the same four herbicides causing significant ($P<0.01$) reductions in the number of effective nodules per plant (-23% for Haloxyfop to –44% for Flumetsulam) compared to the control. There was a significant ($P<0.001$) negative relationship ($R^2=0.57$) between leaf yellowing and the number of effective nodules present on peas three weeks after herbicide application. Crop yellowing or reduced photosynthesis is a recognized indicator of plant physiological stress. The legume-rhizobium symbiosis depends on the energy (C) supply from legume plant and therefore any reduction in photosynthesis has the potential to impact on the energy supply to nodules and affect the symbiosis. Thus, crop yellowing following herbicide applications can be indicative of below ground stress to the symbiosis.

CHARACTERIZATION OF PHENOL AND TRICHLOROETHENE DEGRADATION AND HEAVY METAL ABSORPTION BY RHIZOBIUM RALSTONIA TAIWANENSIS

Jui-Hsing Chou[1], Jo-Shu Chang[2], Chih-Hui Wu[2], Shu-Chen Chang[3], Wen-Ming Chen[1]
[1]Dept. Seafood Sci. Kaohsiung Marine Univ., Chinese Taiwan, [2]Dept.Chem. Engineering, Cheng Kung Univ., Chinese Taiwan, [3]Dept. Food Science, Tajen Inst. Technology, Chinese Taiwan

Ralstonia (*Wautersia*) *taiwanensis* is a root nodule bacterium originally isolated from *Mimosa* sp in southern Chinese Taiwan. Some strains of *R. taiwanensis* demonstrated the ability to grow on medium containing phenol as the sole carbon source. The dependence of phenol degradation rate on phenol concentration can be described by Haldane's model with a low K_S and an extremely high K_{SI}. *R. taiwanensis* TJ86 also achieved degradation for soil samples amended with phenol. Moreover, strain TJ86 cometabolically degraded trichloroethene (TCE) after being cultivated with media containing phenol as the carbon substrate. The sequence of large-subunit phenol hydroyxlase (LmPH) gene obtained from TJ86 displayed high homology to that of other phenol-utilizing bacteria. Results from kinetic and phylogenetic analysis suggest that strain TJ86 most likely belongs to group I phenol-degrading bacteria which were considered as efficient TCE degraders. Some strains of *R. taiwanensis* also demonstrated the ability to grow on medium containing heavy metals. The metal absorption can be described by Langmuir isotherm model. *M. pudica* plants inoculated with *R. taiwanensis* have higher Q_{max} (mg metal/g dry weight) than plans without inoculation and than the pure culture of *R. taiwanensis*. It is proposed that the symbiotic relationship between rhizobium *R. taiwanensis* and its host plant *Mimosa* sp. may have the potential for rhizoremediation of aquatic and soil environments contaminated by phenol and TCE and heavy metals.

Chen WM et al. (2004) Res. Microbiol.155, 672-680

COMPLEMENTATION AND FUNCTIONAL STUDY OF A PROLINE CATABOLISM MUTANT OF *SINORHIZOBIUM FREDII* HN01

Sheng Huang [1,2,3], Xue-Liang Bai [1,2,3], Qing-Sheng Ma [3], Xian-Lai Tang [3] and Bo Wu [1,2,3]
[1] Key Laboratory of Microbial and Plant Genetic Engineering, Ministry of Education. [2] Guangxi Key Laboratory of Subtropical Bioresources Conservation and Utilization. [3] College of Life Science and Technology, Guangxi University, Nanning 530005, P. R. China
For Correspondence: wubogx@gxu.edu.cn

Recent studies have focused on the possible role of amino acids, specifically proline, as a carbon source to support the growth of rhizobia during the colonization of the root, invasion and nodule formation. The work of Jiménez-Zurdo et al. (1995, 1997) and Tang et al. (2001) have suggested that the *putA* gene, which encodes proline dehydrogenase, was involved in root colonization, nodulation efficiency and competitiveness of the bacteria on plant roots. This study aims to demonstrate the relationship between proline catabolism and nodulation.

A mutant strain GXHN100, unable to catabolize praline, was screened from 6000 Tn5*gusA*5 random insertional mutants of *S. fredii* strain HN01. Sequencing analysis showed that an open reading frame, named *pmrA* (proline metabolic related), had a Tn5*gusA*5 insertion. A positive clone, named pGXHN300, was isolated from a gene library of *S. fredii* HN01 by colony *in situ* hybridization, and was subsequently introduced into GXHN100 to yield complemented strain GXHN300. Plant tests indicated that GXHN100 nodulated plants 2 days later and induced fewer nodules per plant than the wild type strain HN01. Moreover, the complemented strain GXHN300 could restore the nodulation ability of mutant to wild type strain. In 1:1 co-inoculation tests, nodule occupancy of HN01 and GXHN100 was 92.31% and 7.69% respectively. GXHN300 was as competitive as wild-type HN01. These indicated that mutation of *pmrA* gene of *S. fredii* HN01 led to a significant reduction in nodule competitiveness.

References:
Jiménez-Zurdo J et al. (1995) Mol. Plant-Microbe Interact. 8, 492 - 498.
Jiménez-Zurdo J et al. (1997) Mol. Microbiol. 23, 85 - 93.
Soto MJ et al. (2000) J. Bacteriol. 182, 1935 - 1941.
Tang XL et al. (2001) Agri. Sci. in China 9, 286 - 288.

Acknowledgements: Work supported by the 973 Project (No. 2001CB108901).

SECTION VI.

MOLECULAR ECOLOGY AND THE DISCOVERY OF NEW NITROGEN FIXERS

DIAZOTROPHIC ENDOPHYTES IN RICE: COLONIZATION AND NITROGEN FIXATION OF HERBASPIRILLUM AND CLOSTRIDIUM SPECIES

Kiwamu Minamisawa, Mu You, Tadashi Abe, Bin Ye, Asami Saito, Makoto Kawahara and Tadashi Sato
Graduate School of Life Sciences, Tohoku University, Katahira, Aoba-ku, Sendai 980-8577, Japan.

Nitrogen is one of the most important limiting nutrients for plant ecosystem production. Legumes acquire a symbiotic ability with rhizobia to fix atmospheric nitrogen during their evolution. As for non-leguminous plants, several diazotrophs in the rhizosphere and endosphere have been isolated and characterized such as *Klebsiella*, *Enterobacter*, *Alcaligenes*, *Azospirillum*, *Acetobacter*, *Azoarcus* and *Herbaspirillum*. However, it remains still unknown whether these bacteria contribute substantially to nitrogen economy of the gramineous plants (Reinhold-Hurek and Hurek 1998). Wild gramineous plants can often vegetate even in nitrogen-deficient soils, suggesting the presence and function of diazotrophic bacteria associated with the plants. Thus, we focused on diazotrophic bacteria in wild rice species and pioneer plants vegetated in lahar, and tried to isolate and characterize them.

1. Colonization of *Herbaspirillum* sp. in wild rice

Nitrogen-fixing bacteria were isolated on a modified Rennie (RMR) medium from stems of wild and cultivated rice species, which has been maintained in Japan. Based on 16S rDNA sequences, the diazotrophic isolates were phylogenetically close to four genera: *Herbaspirillum*, *Ideonella*, *Enterobacter* and *Azospirillum* (Elbeltagy et al. 2000, 2001). Phenotypic properties and signature sequences of 16S rDNA indicated that three isolates (B65, B501 and B512) were *Herbaspirillum* sp. To examine whether *Herbaspirillum* sp. B501 isolated from wild rice, *Oryza officinalis*, endophytically colonizes rice plants, *gfp* gene encoding green fluorescent protein (GFP) was introduced into the bacteria. Observations by fluorescence stereomicroscopy showed that the GFP-tagged bacteria colonized shoots and seeds of aseptically grown seedlings of the original wild rice after seed inoculation (Elbeltagy et al. 2001). Conversely, no GFP fluorescence was observed for shoots of cultivated rice *O. sativa*. Observations by fluorescence and electron microscopy revealed that *Herbaspirillum* sp. B501 mainly colonized intercellular space in the leaves of wild rice. Colony counts of surface-sterilized rice seedlings inoculated with the GFP-tagged bacteria indicated significantly more bacterial populations inside the original and other wild rice species than in cultivated rice varieties.

2. Nitrogen fixation of *Herbaspirillum* sp. in wild rice

After inoculation of *Herbaspirillum* sp. B501, *in planta*-nitrogen fixation in young seedlings of wild rice *O. officinalis* was significantly detected with the acetylene reduction and $^{15}N_2$ gas incorporation assays. Thus, *Herbaspirillum* sp. B501 is a diazotrophic endophyte compatible with wild rice, particularly *O. officinalis*. Moreover, we examined whether the transcript of nitrogenase gene (*nifH*) is detectable in *Oryza officinalis* W0012 seedlings inoculated with *Herbaspirillum* sp. B501*gfp*1. Total RNA extracted from the inoculated plants was subjected to *nifH* and *gfp* RT-PCR. The transcript *nifH* and *gfp* were successfully detected in the plants according to the acetylene-reducing activity. Interestingly, the level of *nifH* transcript was markedly fluctuated according to daily rhythm (light-dark cycles for the plants) (You et al. Unpublished results). This result suggested that nitrogen fixation of endophytes in plants fluctuated based on physiological states of the host plants.

We carefully evaluated their nitrogen fixation in *O. officinalis* is until a ripening stage (110 days) by ^{15}N-isotope dilution method. *Herbaspirillum* sp. B501*gfp*1 has colonized in stem, leaf sheath and root at more than 10^4 CFU/ g fresh weight until ripening stage. The nitrogen fixed by the bacteria was account for small fractions (ca. 0.5%) of total nitrogen even in the compatible host plants. Data also suggested ambiguous nitrogen sources (ca. 6.4%) except for ^{15}N-NH_4NO_3 (99.45 atom%) in nutrient solution for wild rice.

3. Anaerobic nitrogen-fixing consortium consisting of clostridia and non-diazotrophs

The most-probable-number (MPN) method, using RMR medium, enabled to detect significant numbers of nitrogen-fixing bacteria (10^3-10^5) in surface-sterilized stems of wild rice and in pioneer plants such as *Miscanthus sinensis* and *Saccharum spontaneum*. However, we failed to isolate the nitrogen-fixing bacteria by normal procedures using RMR medium. We therefore checked if the nitrogen-fixing activity could be due to several species of microbes. We identified nitrogen-fixing anaerobic clostridia together with diverse facultative or aerobic bacteria such as *Bacillus* and *Enterobacter* spp. A representative strain B901-1b of the clostridia grew, produced molecular hydrogen and showed acetylene reduction activity under oxygen concentrations lower than 0.4% (v/v). Further analysis showed that the clostridia were able to grow and fix nitrogen, after the accompanying bacteria eliminated molecular oxygen by respiration. We observed that *Clostridium* sp. B901-1b when maintained in the mixed culture with the accompanying bacteria, *Enterobacter* sp. B901-2, under the air remained viable (Minamisawa et al. 2004). The interpretation of these findings is as follows: (1) The nitrogen fixers appear as unculturable, because they are anaerobic diazotrophs. *In situ* they live in association with non-diazotrophic facultative and aerobic bacteria and are more difficult to isolate and count. (2) Obligate anaerobic bacteria such as clostridia exist as bacterial endophytes. It is worth noting, that to date little attention has not yet been paid to anaerobes such as *Clostridium* species inside plants. In contrast, the main emphasis on diazotrophic endophytes was on aerobic and facultative anaerobic bacteria (Reinhold-Hurek B. and Hurek T. 1998),

4. Phylogeny and properties of the anaerobic nitrogen-fixing consortia (ANFICO)

In order to examine whether the anaerobic consortia are widespread among gramineous plants, we have isolated anaerobic nitrogen-fixing consortia from various plants including *Miscanthus sinensis*, *Saccharum spontaneum*, wild rice species, and sugarcane. *Clostridium* spp. are classified into 17 clusters (Collins et al 1994). All of anaerobic nitrogen-fixing isolates fell into clusters I and XIV and they could be subdivided into five groups (groups I, II, III, IV and V) (Fig. 1). The non-diazotrophic accompanying bacteria were phylogenetically dispersed to the beta-, gamma-Proteobacteria and to High CG- and Low GC- of the Gram-positive lineage. Although these clostridia showed nitrogen-fixing activity (acetylene reduction activity, ARA) in co-culture with the non-diazotroph strains, their ability to reduce acetylene in pure culture under microaerobic or anaerobic conditions appears as group-specific (Fig. 1). In pure culture *Clostridium* isolates from group IV showed some tolerance to O_2, but clostridia from the other groups displayed ARA only under anaerobic conditions. Group II clostridia never expressed nitrogen-fixing activity in pure culture and they required some heat stable compounds provided by the accompanying bacteria (Minamisawa et al. 2004).

5. Population and distribution of *Clostridium* species in plants

From the ecological point of view, we want to address the population of clostridia in gramineous plants and their contribution to nitrogen economy in plants. To determine clostridial populations in the grass *Miscanthus sinensis*, we designed group-specific primers of 16 rDNA, and developed a MPN-TRFLP (terminal restriction fragment polymorphism) method. Clostridia were detected in strongly surface-sterilized leaves, stems, and roots of the plant at approximately 10^4 to 10^5 cells/g fresh weight; they made up a high proportion of N_2-fixing bacterial populations, as determined by MPN counts associated with acetylene reduction assay. Phylogenetic grouping by MPN-TRFLP analysis revealed that the clostridial populations belonged to group II in cluster XIVa and groups IV and V in cluster I. This result was supported by culture-independent TRFLP analysis using direct DNA extraction from the plants. When phylogenetic populations were compared between *M. sinensis* and the soil around the plants, group II clostridia were found to exist exclusively in *M. sinensis* (Miyamoto T. et al. 2004). When the culture-independent TRFLP analysis using direct DNA extraction was extended to several crops, group II clostrdia was frequently detected in stems and leaves of rice, corn, tobacco and sugarcane.

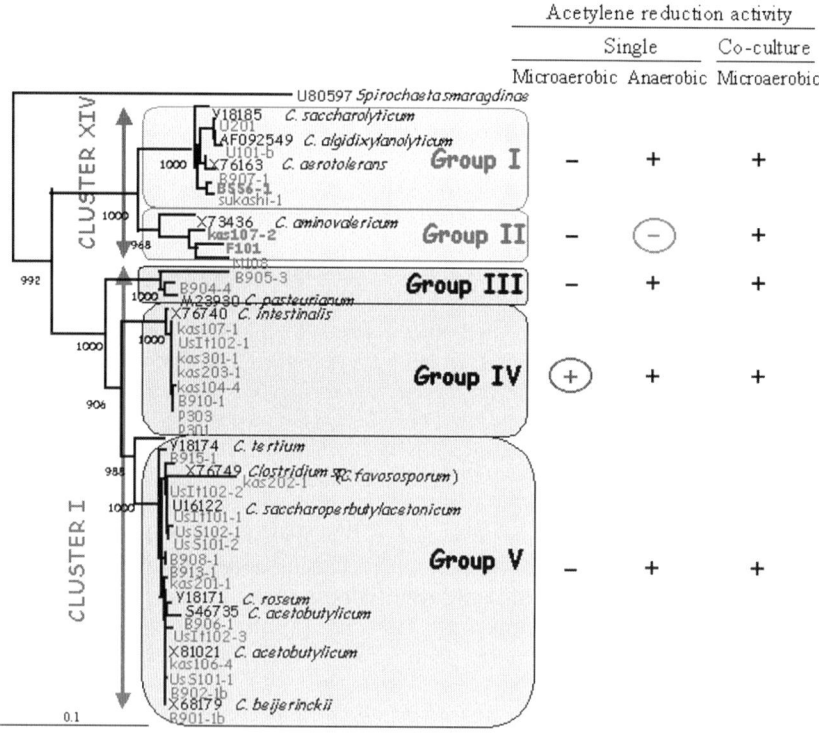

Fig.1. Phylogenetic positions of clostridia from gramineous plants and acetylene reduction activity in culture as associated with accompanying non-diazotrophic bacteria under different O2 regimes (adapted Minamisawa et al. 2004)

6. Functions of endophytic clostridia

A last question arose as to whether the consortia expressed nitrogen fixation in plants. For this purpose an experimental system to re-inoculate the consortia in plants was developed. We first examined colonization of the anaerobic consortia collection in *Miscanthus sinensis*, a pioneer plant in Japan. When *Enterobacter* sp. B901-2 was used as the accompanying non-diazotroph, colonization levels were dependent on *Clostridium* stains, which ranged from 10^1 to 10^5 cells/g plant fresh weight. In particular, two strains Kas107-1 (group IV) and Kas201-1 (group V) showed the highest colonization in *Miscanthus sinensis*. ARA and *nifH* expression of the inoculated young plants was assayed, using the same methodology as for the *Herbaspirillum* works. So far, ANFICO-inoculated plants (*Miscanthus sinensis*) basically showed no nitrogen fixation in the dark without carbon source supply. When carbon sources were supplied from root of the inoculated plants, we detected ARA and *nifH* expression of endophytic clostridia, suggesting the potential of *in planta*-nitrogen fixation of clostridia (Saito et al. Unpublished results).

We investigated another function of ANFICO including clostridia. ANFICO-inoculated grass *Miscanthus sinensis* showed more tolerance to salinity stress than uninoculated plants. Microscopic observations using an immunostaining technique revealed endophytic colonization by the clostridia and the maintenance of a substantial clostridial population until the end of plant cultivation. We conclude that endophytic *Clostridium* sp. Kas201-1 together with *Enterobacter* sp. B901-2 enhanced salinity tolerance of the host plant.

7. Acknowledgements
This work was supported in part by Promotion of Basic Research Activities for Innovative Biosciences (PROBRAIN), Bio-oriented Technology Research Advancement Institute.

8. References
Collins et al. (1994) Int. J. Syst. Bacteriol. 44, 812-826.
Elbertagy A. et al.(2000) Soil Sci. Plant Nutr. 46. 617-629.
Elbertagy A. et al. (2001) Appl. Environ. Microbiol. 67, 5285-5293.
Minamisawa K. et al. (2004) Appl. Environ. Microbiol. 70, 3096-3102.
Miyamoto T. et al. (2004) Appl. Environ. Microbiol. 70(11) (in press)
Reinhold-Hurek B. and Hurek T. (1998) Trends Microbiol. 6, 139-144.r

NEW PERSPECTIVES OF GRASS-ASSOCIATED NITROGEN FIXATION

B. Reinhold-Hurek, F. Battistoni, A. Krause, A. Ramakumar, F. Friedrich,
A. Sarkar, M. Böhm, S.Gemmer, L.Miché, L. Zhang, and T. Hurek
University Bremen, Laboratory of General Microbiology, Faculty of Biology
and Chemistry, P.O. Box 33 04 40, D-28334 Bremen, Germany

In agriculture, nitrogen is one of the most widely used fertilizers with a still increasing global input. In natural ecosystems, the availability of nitrogen often limits plant growth. It also impacts productivity and the species composition of plant communities and ecosystem processes at all scales. The biological reaction counterbalancing the loss of N from soils or ecosystems is biological nitrogen fixation. This process, unique to *Bacteria* and *Archaea* is estimated to contribute globally 200-300 million tons of fixed N per year; about 90-130 Tg N per year for terrestrial systems (Galloway et al., 1995) and about 100-200 Tg N per year for marine systems (Karl et al., 2002).

One of the best-studied interactions between microbes and eukaryotes, in which the eukayotic partner directly profits from N_2 fixation, is the root nodule symbiosis between rhizobia and legumes. However, the most important crops worldwide, wheat, rice and maize, belong to the *Poaceae*, which do not naturally form these specialized symbiotic structures. It has been shown that some grass-related crops, such as certain Brazilian sugar cane cultivars can derive a substantial part of the plant nitrogen from biological nitrogen fixation in natural soils without any inoculation (Boddey, 1995; Lima et al., 1987). In most cases, many different diazotrophic bacteria can be isolated from roots of grasses or cereals (Engelhard et al., 2000; Reinhold et al., 1986; Reinhold-Hurek et al., 1993b; Stoltzfus et al., 1997). This makes it difficult to determine the microbial diazotrophic partner(s) that actually contribute the fixed nitrogen. Among the various colonization sites, such as rhizosphere soil, rhizoplane and root interior, much of the current research on grass-associated diazotrophs focuses on endophytic bacteria (James & Olivares, 1998; Reinhold-Hurek & Hurek, 1998). Endophytic microorganisms multiply and spread inside plants without causing damage to the host or conferring an ecological threat to the plant (Quispel, 1992). Although they invade plants, they cannot be regarded as typical pathogens or endosymbionts. Crucial for understanding the functions of diazotrophic grass endophytes is the long-standing question of whether the host plants profit from nitrogen fixation. This has recently been demonstrated for *Azoarcus* sp. BH72 and its host Kallar grass (*Leptochloa fusca* L. Kunth). The comparison of wild type and a *nif* mutant showed a significant gain of plant nitrogen after inoculation with the wild type; total N-balance and natural ^{15}N abundance corroborated that fixed nitrogen was contributed. Moreover, *nifH*-mRNA of strain BH72 was found to be predominant in plant roots (Hurek et al., 2002). Thus, grass endophytes are able to supply fixed nitrogen to the plant as has also been demonstrated for *Gluconacetobacter diazotrophicus* and sugar cane (Sevilla et al., 2001). This makes the *Azoarcus* sp.-grass system an highly interesting model system for a novel type of plant-microbe interactions.

1. The *Azoarcus* sp. genome project
In cooperation with Alf Pühler's Biefeld Genomics group (BIG, University Bielefeld, Germany) and funded by the BMBF Germany in the GenoMik framework (Hurek and Reinhold-Hurek, 2003), the genome of *Azoarcus* sp. strain BH72 was estimated to be 4.6 MB in size by the thermal renaturation method (Reinhold-Hurek *et al*, 1993), which is in good agreement with 4.37 MB after sequencing and assembly of shotgun clones and closure of all gaps. Neither pulse-field electrophoresis nor Eckhard gels revealed evidence for several chromosomes or extrachromosomal elements (unpublished observations), which also agrees with results of the sequencing project. In order to allow an independent analysis of the genome structure and to facilitate the assembly of contigs and gap closure, we constructed a BAC library of total DNA of *Azoarcus* sp. BH72 (Stella Reamon-Büttner and B. Reinhold-Hurek, unpublished). Screening by Southern-hybridization with known BH72 genes, end-sequencing of BAC-clones, and comparison of *in silico* and *in gel* restriction patterns and clone size allowed to arrange an aligned BAC-library of the genome (with two gaps), which corroborated the genome assembly. The results of the gene annotations will be given in the near future.

2. Response of rice towards endophytic infection with *Azoarcus* sp. strain BH72
To study the interaction of diazotrophic endophytes and rice in more detail, a proteomic approach for roots of rice seedlings was established. Seedlings inoculated with bacteria were grown in quartz sand for two weeks, total proteins extracted and compared by two-dimensional gel electrophoresis. Jasmonic acid, known to be involved in the response of rice, e.g., to stress, wounding and pathogens, induced or elevated the expression of numerous proteins in uninoculated plants. Inoculation with strain BH72 led to a largely different protein pattern with only few overlapping proteins upregulated under both conditions in a compatible (well-colonized) interaction. This suggests that endophytic colonization by strain BH72 does not elicit a typical stress- or pathogen-related response in rice roots. However, the plant response visible in the proteome was dependent on close physical contact with the bacteria. Mutants with a decreased ability to colonize the root interior or even the root surface induced successively less changes in the rice root proteome.

3. Culture-independent detection of diversity and activity of diazotrophs *in situ*
It is important to stress here that *Azoarcus* spp. may occur in an "unculturable" state in nature, where they are highly active but defy attempts of cultivation. It is a common

observation in microbial ecological studies that the majority of bacteria present are as yet uncultivated. Cultivation-independent methods to specifically study diazotrophs are based on the structural genes for nitrogenase, the key enzyme complex for nitrogen fixation. The *nifH/anfH/vnfH* genes (mostly summarized as *nifH* gene throughout the text) encoding the nitrogenase iron protein are evolutionarily conserved and highly valuable for phylogenetic analysis and for detection and identification of diazotrophs by cultivation-independent methods (Hurek et al., 1997).

When a novel microhabitat is analyzed, molecular ecological studies are mostly based on sequence analysis of *nifH* fragments. After amplification by PCR from environmental samples, *nifH* fragments are cloned, and representative single clones are analyzed by sequencing and phylogenetic analysis of the deduced protein sequence. Such studies have revealed a wide range of as yet uncultured bacteria in rice roots and other habitats (Engelhard et al., 2000; Hurek et al., 1997; Ueda et al., 1995; Zehr et al., 1998). For a more rapid assessment of *nifH* biodiversity than sequence analysis, techniques such as denaturing gradient gel electrophoresis (DGGE) (Lovell et al., 2000), PCR-restriction fragment length polymorphism (RFLP) (Poly et al., 2001), and fluorescently labeled terminal restriction fragment length polymorphism (T-RFLP) (Widmer et al., 1999) have been applied. PCR-products are amplified with primers of which one is fluorescently labeled, digested by an restriction endonuclease, and the polymorphic digestion products are separated, e.g., by an automated sequencer. T-RFLP takes advantage of high sensitivity and high resolution by combination of PCR and automated sequencing technologies to separate the polymorphic fragment, and has been reported to have a slightly higher resolution than DGGE. We applied this method to analyze the effect of application of nitrogen fertilization, environmental conditions or *Oryza* species on the root-associated *nifH* gene pool. Especially nitrogen fertilization had a dramatic effect on the population; within 15 d after application, the diversity changed, apparently being lowered. Another new development is the application of DNA microarrays, "phylochips". They use oligonucleotides designed to specifically detect single *nifH* genes or groups of genes, which are covalently bound to a glass surface and hybridized to a fluorescing probe, e.g., generated by *nifH* –targeted PCR from environmental nucleic acids. We have currently developed a *nifH*-phylochip based on short oligonucleotides, which is applicable to environmental samples, but it does not yet cover all diazotrophic diversity.

In order to study not only presence, but also activity of bacteria by culture-independent methods, one promising approach is to analyze mRNA. Bacterial mRNA usually undergoes a rapid turnover and is thus not very stable, allowing to estimate rapid changes in the response of bacteria to environmental changes. Recently, a method was developed by T. Hurek to extract sufficiently stable mRNA from soil-grown rice roots and to amplify it by reverse transcriptase (RT)-PCR (Hurek et al., 2002). In order to be able to compare bacterial *nifH* expression in different environments and to evaluate the expression level, a most-probable-number (MPN) RT-PCR was carried out. When instead of dilutions of DNA, dilutions of RNA are used for RT-PCR, mRNA-levels can be estimated in comparison to cultivated bacteria (Hurek et al., 2002). This method is currently used by us for screening of different rice species and cultivars for the capacity to support high levels of bacterial nitrogen-fixation activity.

Boddey RM (1995). Crit. Rev. Plant Sci. 14, 263-279.
Galloway JN et al. (1995) Global Biogeochem. Cycles 9, 235-252.
Karl D et al. (2002) Biogeochemistry 57/58, 47-98.
Lima E et al. (1987) Soil Biol. Biochem. 19, 165-170.
Engelhard M et al. (2000) Environ. Microbiol. 2, 131-141.
Hurek T et al. (1997) J. Bacteriol. 179, 4172-4178.
Hurek T et al. (2002) Mol. Plant-Microb. Interact. 15, 233-242.
Hurek T and Reinhold-Hurek B (2003) J. Biotechnol. 106, 169-178.
James EK and Olivares FL (1998) Crit. Rev. Plant Sci. 17, 77-119.
Lovell CR et al. (2000) Appl. Environ. Microbiol. 66, 3814-3822.
Poly F (2001) Appl. Environ. Microbiol. 67, 2255-2262.
Quispel A (1992) in DPS Verma (Ed.), Molecular signals in plant-microbe communications, pp. 471-491. Boca Raton, FL: CRC Press, Inc.
Reinhold B et al. (1986) Appl. Environ. Microbiol. 52, 520-526.
Reinhold-Hurek B and Hurek T (1998) Trends Microbiol. 6, 139-144.
Reinhold-Hurek B et al. (1993) Int. J. Syst. Bacteriol. 43, 574-584.
Sevilla M et al. (2001) Mol. Plant-Microb. Interact. 14, 358-366.
Stoltzfus JR et al. (1997) Plant Soil 194, 25-36.
Tan Z et al.(2003) Environ. Microbiol. 2, 1009-1015.
Ueda S et al. (1995) J. Bacteriol. 177, 1414-1417.
Widmer FB et al. (1999) Appl. Environ. Microbiol. 65, 374-380.
Zehr JP et al. (1998) Appl. Environ. Microbiol. 64, 3444-3450.

ELECTRON DONATION AND ALTERNATIVE NITROGENASES IN CYANOBACTERIAL DINITROGEN FIXATION

Hermann Bothe[1] and Gudrun Boison[2]
[1]Botanisches Institut, University of Cologne, D-50923 Köln, Germany.
[2]Department of Marine Microbiology, Netherlands Institute of Ecology NIOO-KNAW, P.O. Box 140, NL-4400 AC Yerseke, The Netherlands.

1. Electron donation to nitrogenase in cyanobacteria

Dinitrogen fixation requires ATP and either ferredoxin or flavodoxin in the reduced form as the carrier transferring electrons directly to nitrogenase. Cyanobacterial heterocysts contain a special type of ferredoxin, fdxH (Masepohl et al. 1997). Under iron-limited growth, ferredoxin is substituted by the FMN-containing flavodoxin, as shown a long time ago (Smillie, 1965; Trebst and Bothe, 1966). The main pathway for electron generation is from glucose-6-P via glucose-6-P dehydrogenase, NADPH, NADPH:ferredoxin oxidoreductase (=FNR), to ferredoxin (flavodoxin). However, mutants defective in any of the proteins involved are not negative for N_2 fixation (Masepohl et al., 1997). Thus, reducing equivalents for N_2 fixation may be generated also by other pathways, e.g., from NADH and NADH dehydrogenase via photosystem I, from H_2 via hydrogenase or from pyruvate cleaved in the so-called pyruvate clastic reaction (Bothe and Neuer, 1988).

Despite contrary statements (Bauer et al., 1993), flavodoxin effectively transfers electrons to nitrogenase when properly reduced to the fully reduced state (Bothe et al., 1971). Flavodoxin exists in three redox states, the oxidized-, semiquinone- and fully reduced forms. Only the redox couple, fully reduced/semiquinone state, with an E_o' of about -500 mV, is able to effectively transfer electrons to nitrogenase. The generation of fully reduced flavodoxin by NAD(P)H ($E_o' = -320$ mV for NAD(P)H/NAD(P)$^+$) can be achieved only to a very limited extent due to thermodynamics. However, the situation is more favourable from pyruvate ($E_o' \sim -550$ mV for the cleavage of pyruvate and coenzyme A to acetylcoenzyme A, CO_2 and 2 electrons) or from H_2 ($E_o' = -420$ mV for $H_2/2H^+$).

2. The pyruvate:ferredoxin (flavodoxin) oxidoreductase in cyanobacteria

A somewhat controversial issue also exists about the cyanobacterial pyruvate:ferredoxin (flavodoxin) oxidoreductase (PFO) catalyzing the cleavage of pyruvate and coenzyme A to the reaction products just mentioned. The enzyme was found early on in the cyanobacterium *Anabaena variabilis* (Leach and Carr, 1971). Subsequently, our laboratory characterized PFO of *Anabaena cylindrica* biochemically in much detail (Neuer and Bothe, 1982). Its activities were determined by measuring: the reduction of methyl viologen (a chemical substitute for ferredoxin), the formation of CO_2, the synthesis of acetohydroxamate formed from acetylcoenzyme A; or the formation of pyruvate from CO_2, reduced ferredoxin and acetylcoenzyme A generated from acetylphosphate. This latter pyruvate synthesis is indicative of the occurrence of PFO in an organism, since thermodynamic reasons forbid such formation by the pyruvate dehydrogenase complex. However, these data were controversial.

In 1993, two groups, independently of each other, published sequences of cyanobacterial *nifJ* encoding PFOs from either *Anabaena* PCC 7120 (Bauer et al., 1993) or *A. variabilis* (Schmitz et al., 1993). The enzyme from *Anabaena* PCC7120 (Bauer et al., 1993) was expressed only under iron-limited growth. The sequenced parts of *nifJ* from both organisms had only sequence homologies of around 75%, whereas the sequence similarities among these strains usually amounted to at least 95%. Thus, it was suspected that the organisms contain two different *nifJ* genes. The genome-sequencing projects then revealed that the unicellular *Synechocystis* PCC6803 contains one *nifJ* gene and, in addition, genes coding for phosphotransacetylase and acetate kinase. Assuming the expression of these genes, this cyanobacterium would have, surprisingly, the possibility to generate ATP also fermentatively even under autotrophic growth. The genome sequencing of the heterocystous *Anabaena* PCC 7120, indeed, revealed two different PFO sequences. The gene characterized by our group (Schmitz et al., 1993) is constitutively expressed, even under saturating amounts of Fe in the medium, and is, in this respect, different from the one described by the Chicago group (Bauer et al., 1993). The sequence analysis showed that all the cyanobacterial PFOs cluster with genes from strict anaerobes, like *Clostridium* or *Desulfovibrio*, or with those which are expressed only under O_2-exclusion as for PFO from *E. coli* (Schmitz et al., 2001). However, as shown by the *lux* reporter system, the gene from the unicellular *Synechococcus* PCC 7942 is expressed in cells grown both aerobically and in Fe-replete medium (Schmitz et al., 2001). The role of PFO in cyanobacteria remains to be elucidated.

3. Hydrogenases in dinitrogen fixation
Reducing equivalents for N_2 fixation may also come from the recycling of H_2. The gas is formed, for example, as a by-product in parallel with ammonia synthesis in nitrogenase catalysis. Cyanobacteria possess two types of Ni-hydrogenases. One enzyme, the uptake hydrogenase, encoded by *hupS* and *hupL*, occurs in heterocysts and is apparently functionally associated with cyanobacterial nitrogenase, as in many other non-cyanobacteria. Sequencing revealed that the other enzyme, the bidirectional hydrogenase, is an $NAD(P)^+$ reducing enzyme (Schmitz et al., 1995). It is encoded by the five genes, *hoxE, F, U, Y,* and *H*. The products of *hoxEFU* catalyze electron transfer to $NAD(P)^+$ ("diaphorase" part of the protein). These proteins show strong sequence similarities to the NADH- and FMN-binding subunits of respiratory complex I. It was,

therefore, postulated that HoxEFU serve as components of both bidirectional hydrogenase and NADH:Q oxidoreductase (Schmitz and Bothe, 1996). However, *Nostoc* PCC 73102 apparently does not possess any of the *hox* genes but respires with similar rates as other cyanobacteria (Boison et al., 1999). Thus, the diaphorase part of bidirectional hydrogenase is not involved in respiration, at least not in this cyanobacterium and several others sequenced in the meantime. Possibly, the enzyme serves as a valve to dispose of excess reductant at high light intensities (Appel and Schulz, 1998). A role for this enzyme in dinitrogen fixation has been ruled out several times; recently, by comparing nitrogenase activities of hydrogenase single and double mutants of *Anabaena* PCC7120 (Matsukawa et al., 2002). However, the bidirectional hydrogenase is not ubiquitous in cyanobacteria, and a common ecological and/or physiological feature of those strains lacking the enzyme has not yet been figured out.

The *hoxE* gene is not present in the well-characterized NAD^+-reducing, soluble hydrogenase from *Ralstonia eutropha* (Friedrich and Schwartz, 1993). The role of this cyanobacterial gene product, e.g., linkage of the complex to the membrane, remains to be elucidated. The cellular distribution of the bidirectional hydrogenase could be species specific in cyanobacteria. In *Synechococcus* PCC6301, the enzyme might reside at the cytoplasmic membrane (Kentemich et al., 1989) or be located at the thylakoid membrane in *Synechocystis* PCC 6803 (Appel et al., 2000) or to be evenly distributed in the cytoplasm with enrichment at the thylakoid membrane in *A. variabilis* (Serebriakova et al., 1994). The cellular localisation of this enzyme needs to be reassessed.

4. Nitrogenase types in cyanobacteria
In addition to the conventional Mo-nitrogenase, organisms can possess a Vanadium or an Fe-only containing enzyme as originally shown for *Azotobacter vinelandii* (Bishop et al., 1980; Bishop and Joerger, 1990). With cyanobacteria, the occurrence of V-nitrogenase was inferred first from physiological evidences in *A. variabilis* (Kentemich et al., 1988). The enzyme was then molecular biologically characterized in detail by T. Thiel (1993). In addition, cyanobacteria contain a second type of Mo-nitrogenase as was described simultaneously by two different groups (Schrautemeier et al., 1995; Thiel et al., 1995). The long known, classical nitrogenase resides in heterocysts, whereas the second Mo-enzyme is expressed in vegetative cells under anaerobic conditions or under low O_2 tensions. The two Mo-nitrogenases are encoded by different gene sets. The second Mo-nitrogenase has only been described for *A. variabilis* and resembles, by its expression under anaerobic conditions, the enzyme from the filamentous, non-heterocystous *Plectonema boryanum*. Physiologically and biochemically, the enzyme has not extensively been studied. Thus, details of its catalytic properties (H_2-evolving capability, quotient between N_2 fixation and H_2 formation, reduction of C_2H_2 beyond C_2H_4 partly to C_2H_6) remain to be elucidated.

5. The occurrence of alternative nitrogenases in Nature
Alternative nitrogenases are known only from laboratory cultures. Apart from a single report, where PCR-products of alternative nitrogenases were found in sea-water

sediments (Loveless et al., 1999), no other ecological studies assessing the distribution of alternative nitrogenases in natural habitats are known to us.

This laboratory screened several, intuitively unusual appearing habitats for the expression of alternative nitrogenases. Samples from the environments were first tested for the formation of C_2H_6, which is indicative for the occurrence of an alternative nitrogenase. The assay turned out not to be sensitive enough to assess potential alternative-nitrogenase expression with certainty. Oligonucleotide primers specific for alternative nitrogenase genes were then developed and these allowed us to amplify segments of all nitrogenase types.

In the course of the study, one habitat, the Sachsenstein at Bad Sachsa at the South ridge of the German Harz mountains, turned out as being particularly interesting (Boison et al., 2004). This 40-m high cliff consists of more than 90% gypsum. The surface of the cliff is composed of shards, which can easily be split off by hand and broken into small pieces. The fissures beneath the shards contain a distinct blue-greenish coloured cyanobacterial layer. 16S-rDNA sequencing and microscopic examination showed that this layer is composed of more than 90% of the unicellular cyanobacterium *Chroococcidiopsis*, which forms quadrangular packages of 16 cells or multiples thereof. Sequencing of the PCR-product of a *nifH* segment showed that *Chroococcidiopsis* possesses the second Mo-nitrogenase, as described for the vegetative cells of *A. variabilis*. A phylogenetic tree of known data and sequences from GenBank revealed that the second Mo-nitrogenase is more widely distributed among cyanobacteria than known hitherto. In laboratory cultures of *Chroococcidiopsis*, C_2H_2 reduction requires rather low light intensities and low O_2 concentrations to proceed. The shards from the gypsum rock were assayed for C_2H_2 reduction within 5 h after removal from their natural location. The reaction then proceeds in air but is dependent on a sufficient supply of water. The conditions which allow N_2-fixation to proceed in air beneath the shards remain to be elucidated.

From an evolutionary point of view, *Chroococcidiopsis* is an interesting organism. The first cyanobacteria of 3×10^9 years ago have a similar appearance in fossil records as the recent *Chroococcidiopsis*, and sequence data indicate that a *Chroococcidiopsis*-related cyanobacterium could be ancestor of heterocystous forms (Fewer et al., 2002). Nowadays, *Chroococcidiopsis* may only survive at ecological niches where the competition is low. It has been described for the Inselbergs of South Africa and Venezuela, for the Antarctic at sites where the ice melts for 3-4 months a year, and for salt brines in Israel.

The search for a V- or Fe-only nitrogenase in the gypsum rock at Bad Sachsa was unsuccessful as yet. A PCR-product from DNA isolated from the blue-greenish layer beneath the shards showed high sequence identities to that of a V-nitrogenase, but the PCR-product could not yet be assigned to any of the isolates. A screening program of our laboratory cultures revealed that *Anabaena azotica* has genes coding for a V-nitrogenase. Detailed data will be published elsewhere. The *A. azotica* culture was kindly supplied to us by the former co-worker Dr. Dai Heping from Wuhan, P. R. China. It occurs in Chinese water rice fields and starts blooming when the temperature rises to such an extent that the fern *Azolla* dies (Li, 1981). *A. azotica* seems to be worth for detailed investigations at its natural habitats.

6. References

Appel J and Schulz R (1998) J. Photochem. Photobiol. 47, 1-11.
Appel J et al. (2000) Arch. Microbiol. 173, 333-338.
Bauer CC et al. (1993) Proc. Natl. Acad. Sci. USA 90, 8812-8816.
Bishop PE et al. (1980) Proc. Natl. Acad. Sci. USA 77, 7342-7346.
Bishop PE and Joerger RD (1990) Ann. Rev. Plant Phys. Plant Mol. Biol. 44, 109-125.
Boison G et al. (1999) FEMS Microbiol. Lett. 174, 159-164.
Boison G et al. (2004) Appl. Environ. Microbiol., in press.
Bothe H et al. (1971) in Flavins and Flavoproteins, p. 211-226, H. Kamin (ed), University Park Press, Baltimore.
Bothe H and Neuer G. (1988) Methods Enzymol. 167, 496-501.
Fewer D et al. (2002) Mol. Phylogen. Evol. 22, 82-90.
Friedrich B and Schwartz (1993) Ann. Rev. Microbiol. 47, 351-383.
Kentemich T et al. (1988) FEMS Microbiol. Lett. 51, 19-24.
Kentemich T et al. (1989) Z. Naturforsch. 44c, 384-391.
Leach CK and Carr NG (1971) Biochim. Biophys. Acta 245, 165-174.
Li S (1981) Acta Hydrobiol. Sinica 7, 417-423.
Loveless TM et al. (1999) Appl. Environ. Microbiol. 65, 4223-4226.
Masepohl B et al. (1997) Mol. Gen. Genetics 253, 770-776.
Matsukawa H et al. (2002) Appl. Microbiol. Biotechnol. 58, 618-624.
Neuer G and Bothe H (1982) Biochim. Biophys. Acta 716, 358-365.
Smillie RM (1965) Plant Physiol. 40, 1124-1128.
Schmitz O et al. (1993) Arch. Microbiol. 160, 62-67.
Schmitz O et al. (1995) Eur. J. Biochem. 233, 266-276.
Schmitz O and Bothe H (1996) Naturwissenschaften 83, 525-527.
Schmitz O et al. (2001) FEMS Microb. Lett. 195, 97-102.
Schmitz O et al. (2002) Biochim. Biophys. Acta 1554, 66-74.
Schrautemeier et al. (1995) Mol. Microbiol. 18, 357-359.
Serebriakova NA et al. (1994) Arch. Microbiol. 161, 140-144.
Thiel T (1993) J. Bacteriol. 175, 6276-6286.
Thiel T et al. (1995) Proc. Natl. Acad. Sci. USA 92, 9358-9362.
Trebst A and Bothe H (1966) Ber. Dtsch. Bot. Ges. 79, 44-47.

ROLE OF UNICELLULAR DIAZOTROPHS IN OCEANIC NITROGEN FIXATION

Joseph P. Montoya
School of Biology, Georgia Institute of Technology, Atlanta GA 30332, USA

Diazotrophs play an important role in supporting biological production in the open ocean. A variety of marine prokaryotes are capable of fixing N_2 and the colonial cyanobacterium, *Trichodesmium*, has long been viewed as the major contributor to oceanic N_2-fixation because of its large size and its tendency to form dense surface aggregations (Capone et al., 1997). *Trichodesmium* can clearly make a significant contribution to the marine nitrogen budget and has been the subject of a great deal of research at sea (e.g., Capone et al., 1997) and, more recently, in laboratory culture (e.g., Ohki et al., 1986). These efforts have led to significant advances in our understanding of the environmental factors and physiological processes that regulate N_2-fixation by *Trichodesmium*, and have added substantially to our knowledge of the spatial and temporal distribution of *Trichodesmium* and its contribution to the nitrogen cycle.

The number and known diversity of marine N_2-fixers has grown in recent years with the application of culture-independent molecular techniques for identifying and characterizing microbes in natural environments (e.g., Zehr et al., 1998; Zehr et al., 2001), but the impact of their activity is only now becoming clear (Montoya et al., 2004). Molecular techniques have demon-strated that a variety of small, unicellular diazotrophs are broadly distributed in oceanic systems (Figure 1). The recent discovery of unicellular diazotrophs at the Hawaii Ocean Timeseries Station ALOHA in the North Pacific Subtropical Gyre (see http://www.soest.hawaii.edu/HOT_WOCE/intro.html for station details) is a particularly interesting example (Zehr et al., 2001). N_2 fixation by these unicells is potentially a major source of new nitrogen to support biological production in oligotrophic waters and has been completely ignored in past efforts to construct nitrogen budgets for the open ocean.

The small size and relatively low numeric abundance of unicellular diazotrophs together pose significant challenges in any attempt to quantify their contribution to the oceanic nitrogen budget. The classic experimental approach for measuring N_2-fixation activity, the acetylene-reduction (AR) assay, has a limit of detection that reflects the volumetric concentration of diazotrophs in a sample. Although the AR assay is inherently sensitive, it is not optimal for quantifying the rate of N_2-fixation by small, rare cells. Preconcentration of samples before the start of the acetylene-reduction assay may allow detection of activity in dilute suspensions (Falcon et al., 2004), but the extensive handling involved in the concentration steps makes it difficult to extrapolate such measurements to the water column.

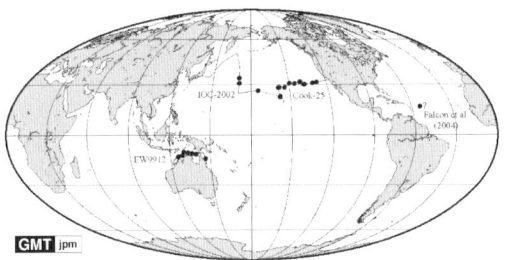

Figure 1. Distribution of observations of unicellular diazotroph activity or occurrence. Significant rates of N_2 fixation were detected at stations shown for cruises Cook-25 and EW9912 (Montoya et al., 2004). Unicellular diazotroph *nifH* sequences were detected at the marked stations for cruise IOC20032. Falcon et al. (2004) report measurable rates of N_2 fixation by unicellular diazotrophs at unspecified locations in the subtropical Atlantic.

The use of stable isotope tracers in quantifying the rate of N-cycle processes is now well-established in marine science and the increasing availability of highly sensitive isotope ratio mass spectrometers has brought this technique into broad use in studies of N_2-fixation in oligotrophic waters (e.g., Dore et al., 2002; Mills et al., 2004). In contrast to the AR assay, the limit of detection of the ^{15}N tracer method reflects the analytical precision of the mass spectrometer as well as the dilution of tracer by the standing stock of biomass, some of which is nondiazotrophic. As a result, the limit of detection of the ^{15}N tracer method varies inversely with biomass concentration (Montoya et al., 1996), making it ideal for studies in oligotrophic waters, particularly since the experiments can be conducted with minimal handling and disruption of the system. An additional benefit of the stable isotope tracer approach is that rates of N_2 fixation and CO_2 fixation can be measured simultaneously by amending an incubation bottle with both $^{15}N_2$ and $[^{13}C]$-HCO_3^-. Following incubation, the abundance of both ^{15}N and ^{13}C can be measured in a single sample by continuous-flow isotope ratio mass spectrometry, ensuring that both rates are measured in the same population.

We have used this experimental approach to study the role of diazotrophy in supporting primary production at Station ALOHA, the long-term biogeo-chemical timeseries site north of Oahu. Our experiments at ALOHA have allowed us to explore the patterns of N_2 fixation and the role of potential environmental factors in controlling the rate of N_2

fixation by small diazotrophs. Our experiments ran for as long as 36–48 h after the addition of tracer, with incubation end points typically occurring at 4–10 h intervals spanning the light/dark cycle. With this approach, we were able to measure integrated rates of activity extending from the start of an experiment (typically late afternoon of day 1) to the time of filtration, and to resolve the gross pattern of diel activity.

Results from two of our experiments are shown in Figure 2. In both cases, we found no evidence for diel phasing of N_2 fixation or for any treatment effect associated with addition of PO_4^{3-}. Instead, our measured rates are rather uniform through each of the two experiments, though our first time-point, taken near dawn after the first night of the experiment, appears to be an outlier in each case. Initially, we anticipated differences between experiments ending at dawn and at dusk that would reflect different rates of activity during the light and dark portions of the diel cycle. Clearly, this isn't the case in our data, which imply continuous N_2-fixation activity through both day and night by unicellular diazotrophs in these waters.

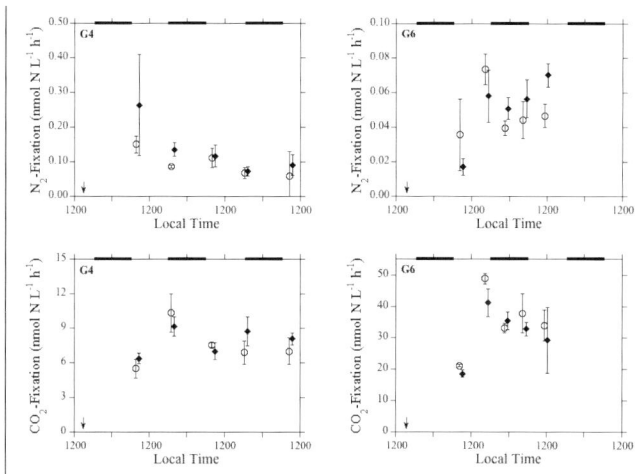

Figure 2. Rates of N_2 fixation (upper panels) and CO_2 fixation (lower panels) in experiments carried out with water from Station ALOHA (Experiment G4, 10–13 Aug 2001) or just outside Kaneohe Bay, Oahu (Experiment G6, 17–19 Feb 2002). In each panel, the start of the experiment is marked with a small arrow. Symbols represents the mean of three replicate incubations and error bars show the standard deviation of the replicate analyses. Open circles, controls ; filled diamonds, bottles amended with 1 μM PO_4^{3-}. Control and treatment symbols are offset slightly in time for clarity. The CO_2-fixation rates shown are normalized to the total light period between the start of the experiment and the end point of an individual incubation.

In contrast, we found clear evidence for diel phasing of CO_2-fixation activity in our incubations, with notably lower integrated rates measured in incubations ending at dawn than at dusk. The rates shown in Figure 2 have been normalized to the total length of the light period between the start of the incubation and the end-point in each case to remove the effect of the dark period on the calculated rates. When treated in this way, our CO_2-fixation data show rather constant rates through each time-series with the exception of the first time point, which again tended to be an outlier from the general trend. In any case, our CO_2-fixation measurements imply that bottle and handling effects did not have a large impact on the overall activity of the plankton in our incubation bottles.

In dual-isotope (^{15}N and ^{13}C) experiments of this sort, the ratio of specific N- and C-fixation rates (h^{-1}) can be used as an index to the relative abundance and activity of diazotrophs. For balanced growth of a pure population of diazotrophic cyanobacteria, the ratio of specific fixation rates should be 1:1. In our experiments, the ratio of specific rates was never lower than ca. 40:1 and ranged as high as 400:1. Our experiments were long enough to integrate over one or more full day-night cycles and so should provide a good estimate of the net C and N dynamics of the system. High specific C:N fixation ratios therefore imply that community carbon is turning over much more rapidly than communitiy nitrogen. Since diazotrophs make up only a small portion of the population, this isn't surprising and we can use the specific fixation ratio as a rough index to the relative abundance of diazotrophs in the community. In this context, our range of specific rate ratios (from 40:1 to 400:1) implies that diazotrophs comprised between 2.5% (40:1 ratio) and 0.25% (400:1 ratio) of the metabolically active plankton.

Although we have done our most detailed experiments at Station ALOHA, we have measured substantially higher rates of N_2 fixation by unicells elsewhere in the Pacific (Montoya et al., 2004). On one recent cruise from Honolulu to San Diego (T. Villareal, Chief Scientist), we carried out N_2-fixation experiments using water from the mixed layer and/or the pigment maximum (Figure 3). When integrated through the water column, our data imply that N_2 fixation by small diazotrophs contributes an average of 500 µmol N m^{-2} d^{-1} to the upper water column, which is comparable to the rate of vertical injection of subsurface NO_3^- into the mixed layer (Allen et al., 1996; Dore et al., 2002; Lee et al., 2002). This cruise provided the first evidence that N_2 fixation by unicellular diazotrophs is wide-spread in the North Pacific and can make a significant contribution to the oceanic N budget on a basin scale.

Our focused experiments at Station ALOHA and our broad scale surveys both show that unicellular diazotrophs fix nitrogen at significant rates in oligotrophic waters of the Pacific. These small, broadly distributed diazotrophs may be the most important single source of new nitrogen in the oligotrophic ocean and may make a globally significant contribution to oceanic new production (Montoya et al., 2004).

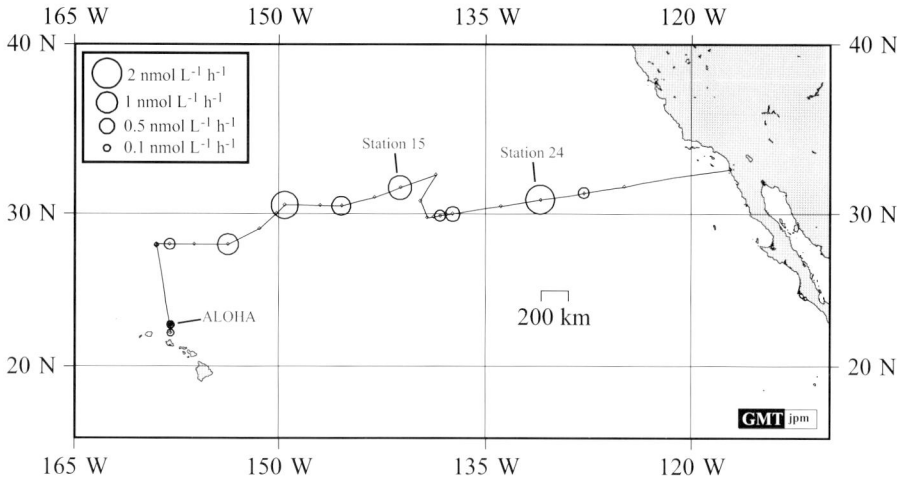

Figure 3. Volumetric rate of N_2 fixation by small diazotrophs at the Hawaii Ocean Timeseries (HOT) Station ALOHA and during cruise Cook-25 (Jun - Jul 2002). The area of each circle is directly proportional to the rate measured in $^{15}N_2$ tracer experiments (nmol L^{-1} h^{-1}). Stations where no rate measurements were carried out are marked with small diamonds.

Acknowledgments

I thank Dave Karl and the entire HOT team for support at Station ALOHA and Tracy Villareal for making shiptime available to us. I am grateful to Carrie Holl, Jason Landrum, and Poneh Davoodi for technical support. This work was supported by NSF grants OCE-99-77528 and OCE-04-25583.

References

Allen CB et al. (1996) Deep-Sea Res. I 43, 917-936.
Capone DG et al. (1997) Science 276, 1221-1229.
Dore JE et al. (2002) Limnol. Oceanogr. 47, 1595-1607.
Falcon LI et al. (2004) Appl. Env. Microbiol. 70, 765-770.
Lee K et al. (2002) Geophys. Res. Lett. 29, doi: 10.1029/2001GL014198.
Mills MM et al. (2004) 429, 292-294.
Montoya JP et al. (1996) Appl. Environ. Microbiol. 62, 986-993.
Montoya JP et al. (2004) Nature 430, 1027-1031.
Ohki K et al. (1986) Mar. Biol. 91, 9-13.
Zehr JP et al. (1998) Appl. Environ. Microbiol. 64, 3444-3450.
Zehr JP et al. (2001) Nature 412, 635-638.

NEW NITROGEN-FIXING MICROORGANISMS FROM THE OCEANS: BIOLOGICAL ASPECTS AND GLOBAL IMPLICATIONS

Jonathan P. Zehr[1], Barbara Methe[2] and Rachel Foster[1]
[1]Department of Ocean Sciences, E&MS A438, 1156 High Street, University of California, Santa Cruz, CA 95064, USA. [2]The Institute for Genomic Research, 9712 Medical Center Drive, Rockville, MD 20850, USA

The open ocean environment is typified by low concentrations of inorganic nutrients and trace elements, and most turnover rates are high. Nitrogen fixation has only recently been recognized as a quantitatively important process in the open ocean (12,16). Since concentrations of ammonium and nitrate are typically in the low nanomolar range, N_2-fixing species should have an ecological advantage in the open ocean. For most of the last four decades, it has been assumed that the major, and perhaps the only, nitrogen fixers were the free-living filamentous, colony-forming, non-heterocystous cyanobacteria of the genera, *Trichodesmium* (6), and symbiotic heterocystous cyanobacteria (*Richelia*) (20), which live in association with certain genera of diatoms. A little more than half of the global biological nitrogen fixation occurs in the ocean (141 Tg N yr^{-1} in the pelagic zone alone; 10,11), of which 10% (16) to over 50% (10) has been ascribed to *Trichodesmium* (85 Tg N yr^{-1}).

Several years ago, we detected nitrogenase (specifically *nifH*) genes in the Pacific Ocean subtropical gyre as well as in the Atlantic Ocean, which were not derived from either *Trichodesmium* or *Richelia*. The phylogenetic analysis of these *nifH* genes suggested that some of the organisms were unicellular cyanobacteria because the sequences clustered with those of *Cyanothece*, *Gloeothece* and *Synechococcus* (renamed *Cyanothece*) (Figure 1). Microscopic observations confirmed their presence in the plankton (22). Some isolates of unicellular cyanobacteria had previously been recovered from the marine environment, but were not believed to be abundant in the oceans. The unicellular cyanobacteria are much more difficult to observe than the colony-forming *Trichodesmium* or the symbionts in diatoms. These include *Cyanothece* sp. ATCC 51142, *Synechococcus* sp. BG 043511, and *Crocosphaera* (marine *Synechocystis*) sp. WH501. The strain of *Cyanothece*, which has been the most studied (strain ATCC 51142), was isolated from an inter-tidal area in the Gulf of Mexico. The "*Synechococcus*" appears to be in the *Cyanothece* group and is reported to have been isolated from shallow waters in the Bahamas (21). True marine *Synechococcus* are

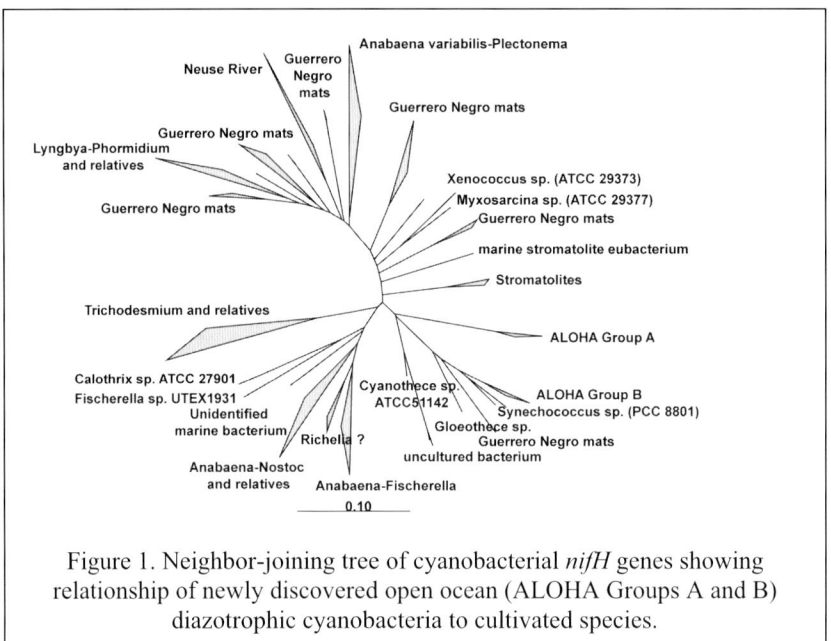

Figure 1. Neighbor-joining tree of cyanobacterial *nifH* genes showing relationship of newly discovered open ocean (ALOHA Groups A and B) diazotrophic cyanobacteria to cultivated species.

much smaller and do not fix N_2. The one diazotrophic strain, called *Synechococcus* RF-1 that was isolated from a rice paddy field, is now named *Cyanothece* in the Pasteur Collection. *Crocosphaera* sp. WH8501 is the most oceanic of the isolates, and was originally isolated in the Atlantic Ocean off the coast of Brazil, but has now also been recovered from the Pacific Ocean (22). A draft sequence of the genome of *Crocosphaera* has been completed and the sequencing of the *Cyanothece* genome is underway.

All of these isolates have similar 16S rRNA sequences (approximately 95% identical) and all are several micrometers in diameter. There are at least two major groups of diazotrophic unicellular cyanobacteria found in the open ocean on the basis of *nifH* sequences. One group is most closely related to *Cyanothece* sp. ATCC 41152 (Group A) and the other to *Synechococcus* RF-1 (Group B, the WH8501-like cyanobacteria). Representatives of the Group A oceanic type have yet to be cultivated. Group B *nifH* genes have been reported from the Atlantic Ocean (7) and the Arabian Sea (15). Our group has detected both Group A and B *nifH* phylotypes in both the Atlantic and Pacific Oceans, and they are similar in abundance to *Trichodesmium*.

Recently, along a cruise transect in the subtropical Pacific, unicellular cyanobacteria were shown to be an important source of newly fixed nitrogen (17,22; Montoya, this volume). We have used cultivation-independent approaches to identify and quantify nitrogen-fixing microorganisms in the open ocean. Results of real-time (quantitative) PCR indicate that the unicellular cyanobacteria can be more abundant than the previously recognized cyanobacteria, including *Trichodesmium* (3). The uncultivated Group A phylotype is generally the most abundant of the two unicellular groups.

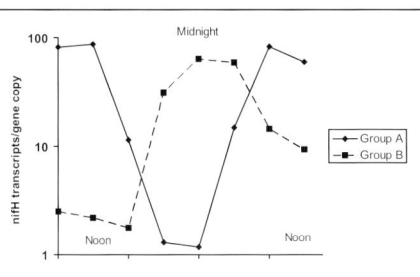

Figure 2. *nifH* gene expression in natural populations of unicellular cyanobacteria from the Pacific Ocean near Kaneohe Bay, Hawai'i.

The diazotrophic unicellular cyanobacteria are interesting from biological, ecological and physiological perspectives. It is not yet known how these microorganisms compete in the environment and whether or not they have different temporal (seasonal) and spatial (vertical and horizontal) distributions. They have been observed to fluctuate in relative abundance (M. Church, C. Short and J. Zehr, unpub. data). Intriguingly, the two groups have very different *nif*-gene expression patterns. The daily phasing of *nif*-gene expression and nitrogen-fixation activity has been documented in unicellular cyanobacteria (*Synechococcus* RF-1, ref. 13 and *Cyanothece*, ref. 5) as well as *Trichodesmium* (2). The Group A and Group B cyanobacteria have distinct daily rhythms (4) that have been detected in the environment using quantitative reverse-transcriptase PCR (QRT-PCR) (Figure 2). The rhythms are consistent between samples collected at different places and different times. The Group B RF-1 like sequence expresses nitrogenase during the night (4,7), which is consistent with a pattern of nitrogen fixation at night usually found in nonheterocyst-forming cyanobacteria. The Group A expression cycle is exactly the opposite with highest expression during the day. Daytime N_2 fixation is usually found in heterocyst-forming cyanobacteria, but is also characteristic of *Trichodesmium*. This is intriguing, as it suggests that the Group A unicellular cyanobacteria may be fixing nitrogen aerobically, simultaneous with oxygenic photosynthesis (4). The possibility for such behavior has been demonstrated in cultures of unicellular cyanobacteria under constant illumination (9) or at high growth rate in light-dark cycles. This group of cyanobacteria may provide a model system for understanding aerobic nitrogen fixation because it is a much simpler organism than the filamentous and colony-forming *Trichodesmium*.

The unicellular cyanobacteria, *Trichodesmium* and *Richelia*, are competing N_2-fixing microorganisms in a relatively homogenous environment that provides little opportunity for spatial separation of niches, except for the vertical gradients in light and nutrients. The "paradox of the plankton" has yet to be solved for phytoplankton: why are there so many species when the best competitor should out-compete them all (14,18)? This quandary equally applies to the diazotrophs of the open ocean. Clues to how these diazotrophs compete reside within the genotype and phenotype. Comparison of draft genome sequences of *Trichodesmium* sp. IMS 101 and *Crocosphaera* sp. WH8501 show that both genomes are relatively large (7.8 Mb and 6.3 Mb, respctively). The *Crocosphaera* genome is almost twice the size of the non-N_2-fixing unicellular cyanobacteria, *Synechocystis* sp. PCC6803 or *Gloeobacter violaceus*, which have similar morphology but genome sizes of only 3.5 and 4.7 Mb, respectively. The *Crocosphaera* genome has more genes that are closely related to *Synechocystis* sp. 6803 than does *Trichodesmium* (1436 vs. 360 best Kegg hits, respectively, http://genome.ornl.gov/microbial), which may be indicative of the unicellular lineage. Both *Crocosphaera* and *Trichodesmium* have high numbers of hits to *Anabaena* genes (2357 vs. 2095, respectively). Intriguingly, there are more signal transduction proteins in the *Crocosphaera* genome (243) than in the *Trichodesmium* genome (181), even though *Trichodesmium* is also a free-living planktonic organism and is filamentous. The *Trichodesmium* genome is about 24% larger than the *Crocosphaera* genome, but there is about 34% more signal transduction genes in the unicellular *Crocosphaera* genome. Only 89 of the 243 signal transduction genes in *Crocosphaera* are recognized in a reciprocal BLAST between the *Trichodesmium* and *Crocosphaera* genomes, suggesting that there may be unique sensory and signal transduction pathways. Similarly, there are more transcription-related proteins in the *Crocosphaera* genome. It is possible that some of the unicellular cyanobacteria form symbiotic associations with protists, and this could be reflected in genome size. Unicellular phycoerythrin-containing cells have been observed in association with dinoflagellates, tintinnids and other organisms, and have 16S rRNA sequences that are closely related to the *Cyanothece* group (1,8).

Our understanding of the magnitude and role of N_2 fixation in the open ocean is constrained by our lack of understanding at the fundamental biological levels as well as the ecosystem biogeochemical dynamics. Our best estimates of N_2 fixation in the ocean come from our knowledge of *Trichodesmium*, which is easily observed, collected and even visualized from space (19). The unicellular cyanobacteria are microscopic and, therefore, not as easily observed or isolated. Less is known about what constrains their distribution, including interactions with light, temperature or nutrients and trace elements (e.g., phosphorus and iron). Furthermore, there are hints of more diazotrophs, including symbioses with single-celled eukaryotic algae, that have yet to be identified (8). Genomic information will guide experimentation and development of even remote instrumentation for examining the distributions and factors constraining N_2 fixation in the sea.

Acknowledgments
This contribution is dedicated to Dr. John Gallon for his many contributions and his inspiration to many of us in the field. I am grateful to Matt Church and Cindy Short for providing data and to Elizabeth Mondragon and Charles Park for their technical support. These sequence data were produced by the US Department of Energy Joint Genome Institute http://www.jgi.doe.gov/. Research has been supported by NSF OCE-9977460 and OCE-0131762, and the Gordon and Betty Moore Foundation.

References
1. Carpenter EJ and Foster RA (2002) in Cyanobacteria in Symbiosis, pp. 11-18. Rai AN et al. (Eds.) Dordrech: Kluwer Academic Publishers.
2. Chen Y-B et al. (1998) J. Bacteriol. 180, 3598-3605.
3. Church MJ et al. (2004) Aquat. Micr. Ecol., in press.
4. Church MJ et al. Appl. Environ. Microbiol., submitted.
5. Colon-Lopez M et al. (1997) J. Bacteriol. 179, 4319-4327.
6. Dugdale RC et al. (1961) Deep-Sea Res. 7, 298-300.
7. Falcon LI et al. (2004) Appl. Environ. Microbiol. 70, 765-770.
8. Foster RA (2004) Ph. D. dissertation. Stony Brook, NY: Stony Book University.
9. Gallon JR (2001) Plant and Soil 230, 39-48.
10. Galloway J et al. (2004) Biogeochemistry, in press.
11. Galloway JN (2003) in Treatise on Geochemistry, pp. 557-583. Schesinger W (Ed.) San Francisco: Elsevier Health Sciences.
12. Gruber N and Sarmiento JL (1997) Global Biogeochem. Cycles 11, 235-266.
13. Huang T-C et al. (1988) FEMS Microbiol. Lett. 50, 127-130.
14. Hutchinson GE (1961) Amer. Nat. 95, 137-145.
15. Mazard SL et al. (2004) Appl. Environ. Microbiol., in press.
16. Michaels AF et al. (1996) Biogeochemistry 35, 181-226.
17. Montoya JP et al. (2004) Nature 430, 1027-1032.
18. Siegel DA (1998) Limnol. Oceanogr. 43, 1133-1146.
19. Subramaniam A et al. (1999) Limnol. Oceanogr. 44, 6078-6617.
20. Venrick EL (1974) Limnol. Oceanogr. 19, 437-445.
21. Waterbury JB and Rippka R (1989) in Bergey's Manual of Systematic Bacteriology Volume 3. Staley JT (Ed.) Baltimore: Williams & Wilkins.
22. Zehr JP et al. (2001) Nature 412, 635-638.

BIODIVERSITY AND PHYLOGENY OF RHIZOBIAL GERMPLASM IN CHINA

Wen Xin Chen[1], En Tao Wang[1,2] and Wen Feng Chen[1]
[1]Key Laboratory of Agro-Microbial Resources and Application, Ministry of Agriculture/Department of Microbiology, College of Biological Sciences, China Agricultural University, 100094, Beijing, China. [2]Escuela Nacional de Ciencias Biológicas, Instituto Politécnico Nacional, México D. F., 11340, México

The nitrogen-fixing interaction of symbiotic rhizobia with leguminous plants makes rhizobia the most valuable microbial resources in the natural and agricultural ecosystems. The existence of diverse rhizobia guarantees the successful colonization of leguminous plants in different environments. The diverse geographic conditions in the vast territory of China offer great opportunities for the divergent evolution of plants and microorganisms. A total of 1,485 leguminous species within 172 genera have been recorded in China. It can be estimated that there should be diverse rhizobial populations associated with these legume plants in distinct geographic regions. Since 1973, we have collected and researched the rhizobial germplasm in China and obtained some success as described below.

1. Survey of nodulation of leguminous plants and collection of rhizobial germplasm.
Our thirty-year research program into the nodulation of legumes and their rhizobia has involved cooperating with more than 100 people from 20 institutions in a combined effort across 700 counties over 32 provinces/cities in China. So far, we have investigated 600 leguminous species within more than 100 genera, in which the nodulation of more than 300 species had not been recorded in the earlier list in the "*Leguminosae*" edited by Allen and Allen (1981).

So far, we have obtained 7,000 nodule specimens and 6,600 rhizobial isolates. Among them, 1,500 rhizobial strains have been characterized, based on 100 phenotypic features and these data were stored in our database of rhizobia. From the characterization, we have found many strains that possess high tolerance traits. Some strains tolerate pH 3.8

or pH 12, others tolerate 6.0% NaCl, and others could grow at 40°C or 4 °C. These strains have been used as inoculants in China's western regions in a government-supported national reforestation and restoration program. Our rhizobial strains have had a significant impact on the growth of leguminous plants; for example, increasing biomass by 30% or more.

2. Rhizobial diversity in China
We used polyphasic taxonomic methods to analyse the phenotypic and genotypic characters of Chinese rhizobial isolates, to group the bacterial strains into specific levels, and to reveal the diversity of rhizobia. Our results have enabled us to describe and publish 2 new genera and 15 new species and 9 defined species will be published

(1) Genus *Sinorhizobium* (Chen et al., 1988; emended by de Lajudie et al., 1994).
This genus name was first proposed on the basis of numerical taxonomic results of the fast-growing rhizobia isolated from soybeans grown in China (Chen et al., 1988). There were only *S. fredii* and *S. xinjiangense* in this genus at that time and all the strains in these two species were found only in China. So, we named them *Sinorhizobium.* Later, it was emended to include more species (de Lajudie et al., 1994). So far, 10 species have been described in this genus, in which 3 species or new combinations, *S. fredii, S. xinjiangense,* and *S. kummerowia,* were described based upon our work. The species *S. meliloti (*Yan et al., 2000*)* and *S. saheli* (Yao et al., 2001) were also found in China.

(2) Genus *Mesorhizobium* (Jarvis et al., 1997).
We suggested this genus name to accommodate the mono-phylogenic group represented by *M. loti, M. huakuii* and *M. tianshanense,* when we described the last one as a new species (Chen et al., 1995). This name was published formally in 1997 and presently 10 species are included in this genus. Among them, 5 were proposed following our research in China; *M. huakuii* (Chen *et al.*1991) for rhizobia of *Astragalus sinicus. M. tianshanense* (Chen *et al.* 1995) for a group of isolates from *Glycine max* and other plants. *M. amorphae* (Wang *et al.* 1999) for isolates from *Amorphae* spp., and *M. temperatum* and *M. septentrionale* (Gao *et al.*, 2004) for isolates from *Astragalus adsugens.*

(3) Genus *Rhizobium* (Frank, 1889).
In this genus, 12 species have been reported and 5 were isolated only in China. They are *R. mongolense* (van Berkum *et al.,* 1996), and *R. yanglingense, R. indigoferae, R. hainanense* and *R. loessense* reported by our group. We also identified *R. leguminosarum* in China. Recently, we have classified 3 new specific groups: the 1st was from common bean, the 2nd came from *Trifolium* spp. in the south of China, and the 3rd was from *Dalbergia* spp.

(4) Genus *Bradyrhizobium* (Jordan, 1982).
Of the 4 species defined in this genus, 3 have been found in China. They are *B. liaoningense* (Xu *et al.,* 1995), *B. yuanmingense* (Yao *et al.,* 2001), and *B. japonicum.*

The first two were found only in China. *B. elkanii* was reported for rhizobia associated with soybeans in the USA. Recently, 5 new species in this genus have been found in our laboratory. Two of them are isolates from *Phaseolus vulgaris* and *Cassia tora*, the other two from *Pueraria* spp., and one is from *Cercis rasemosa* and *Amorpha fruticose* (unpublished).

These genera and species all belong to the alpha-proteobacteria class. Recently, some new spp. of *Burkholderia* and *Ralstonia* that are capable of nodulating legumes were described in the beta-proteobacteria subgroup. *Ralstonia taiwanensis* and *Bulkholderia caribensis* were found to be widespread in Taiwan, China (W. M. Chen *et al.*, 2003). We also found one strain of rhizobia isolated from *Mimosa* sp. to belong to the genus *Bulkholderia*.

Chinese rhizobial resources have enriched the taxonomic system of rhizobia. These results also showed that the diversity of rhizobia is far from completely discovered and that the rhizobia in temperate regions seemed more diverse than in tropical regions.

Recently, there has been a proposal concerned the amalgamation of *Rhizobium*, *Allorhizobium* and *Agrobacterium* and another concerning the nomenclature of *Sinorhizobium* and *Ensifer*. These proposals have led us to think about the definition of the rhizobial genus and its nomenclature. Firstly, we agree that the definition of a genus can be based only upon the phylogeny of 16S rRNA and the phenotypic distinction can be used as a reference. Secondly, both the modern research results and the history of literature should be respected when a nomenclature change is made.

In the case of *Agrobacterium* and *Rhizobium* and of *Sinorhizobium* and *Ensifer*, we think it is better to keep the symbiotic species in the genera *Rhizobium* and *Sinorhizobium*, respectively. The phyto-pathogenic species can be maintained in the genus Agrobacterium and the nonsymbiotic species related to *Sinorhizobium* can be put in the genus *Ensifer*. The nomenclature of these genera has been used for a long time without causing problems for people in understanding their phylogenetic positions.

3. Diversity of rhizobia-legume symbiosis and geographic distribution
Specificity of host plants for the rhizobia has been recognized for a long time. But, in 2000, Perret reported the molecular basis of symbiotic promiscuity. We also have found that the plant may change its symbiotic partners in different geographic zones. For example, soybean is an important crop that originated in China, where it has been cultivated for more than 5000 years, and has now been introduced to other countries across the world. We found that at least 7 species within the genera *Bradyrhizobium*, *Mesorhizobium* and *Sinorhizobium* have been defined in the rhizobia associated with soybean. Of these 7 species, only *B. elkanii* strains were not found in China. The Chinese species are predominant in different geographic regions. *B. japonicum* is distributed across most areas of China. In the Northeastern region, *Bradyrhizobium* species is the main microsymbiont of soybean. In Xinjiang, in the Northwest region, no

Bradyrhizobium strains from soybean have been reported so far. The species *M. tianshanense* and *S. xinjiangense* associated with soybean were found only in Xinjiang. *S. fredii* was predominant in the Central and North of China.

With the common bean, different species of rhizobia are associated with different countries and with different regions in China. *R. etli* was isolated from the USA, *R. tropici* from Mexico, *R. gallicum* and *R. giadinii* both from France. In China, we have not yet found these species. In China's Southwest region, we identified the strains isolated from common bean as *R. leguminosarum,* in Central China, they were *B. japonicum* and another two new species in *Bradyrhizobium* (unpublished).

We identified the strains nodulated with *Lespedez* isolated from the USA as *B. japonicum* and *B. elkanii,* but from North China *as S. saheli* and *B. yuanmingense.* Rhizobia associated with clover trefoil isolated from some Chinese regions have also been analysed by us. We found that some of them isolated from Central China belong to *Bradyrhizobium* and *Mesorhizobium* spp.*,* from Southwest China belong to *Rhizobium* sp. and *Mesorhizobium,* and from Northwest China belong to *R. leguminosarum* (unpublished).

We could not yet point out the factors that determine the association of a legume species with a certain rhizobial species in different geographic regions. However, some reports indicated that the soil pH affects rhizobial populations (Anyango *et al.*, 1995). Our findings showed that the adaptation of rhizobia to local environments should be considered in the selection and utilization of rhizobial inoculants.

Our results also indicated that many legume plants in the same geographic region had one rhizobial species in common. For example, the slow-growing rhizobia isolated from 18 species in 8 genera of legumes in Hainan province, South China, were identified as *B. japonicum* and the fast growers isolated from 12 species in 9 genera were *R. hainanense* (Gao *et al.* 1994; Chen *et al.*1997). While in Xinjiang, in the Northwest region, rhizobia isolated from 8 species in 6 genera of legumes were *M. tianshanense* (Chen *et al.* 1995). All of these results showed that the symbiotic relationship between the rhizobia and the legumes is diversity. These results also confirmed the existence of coevolution between the host legumes and their microsymbionts.

4. Conclusions
Based upon the research on rhizobia in China, the following conclusions could be drawn.
1) From different geographic regions of China a large number of rhizobial strains with different characters have been collected and conserved to provide a big pool of germplasms and rhizobial genes.
2) In China's vast territory there exists a large diversity of rhizobia with many species and genera in *Rhizobiales.* It has much enlarged the rhizobial taxonomic system.

3) The symbiotic diversity among legumes and rhizobia is demonstrated in two ways. First, by the fact that single legume species may be nodulated by several different rhizobial genera and species in different geographic regions of China. And, second, by single rhizobial species being able to nodulate several legume species, even several genera, in some Chinese geographic regions.

5. References

Allen ON and Allen EK (1981) The *Leguminosae*: Madison. University of Wisconsin Press, WI, USA.
Anyango B et al. (1995) Appl. Environ. Microbiol. 61, 416-421.
Chen WM et al. (2003) J. Bacteriol. 185, 7266-7272.
Chen WX et al. (1988) Int. J. Syst. Bacteriol. 38, 392-397.
Chen WX et al. (1991) Int. J. Syst. Bacteriol. 41, 275-280.
Chen WX et al. (1995) Int. J. Syst. Bacteriol. 45, 153-159.
Chen WX et al. (1997) Int. J. Syst. Bacteriol. 47, 870-873.
de Lajudie et al. (1994) Int. J. Syst. Bacteriol. 44, 715-733.
Gao JL et al. (2004) Int. J. Syst. Evol. Microbiol., in press.
Jordan (1982) Int. J. Syst. Bacteriol. 32, 136-139.
Tan ZY et al. (2001) Int. J. Syst. Evol. Microbiol. 51, 909-914.
Van Berkum et al. (1998) Int. J. Syst. Bacteriol., 1998, 48, 13-22.
Vandamme et al. (2002) Syst. Appl. Microbiol. 25, 507-512.
Wang ET et al. (1999) Int. J. Syst. Bacteriol. 49, 51-65.
Wei GH et al. (2003) Int. J. Syst. Evol. Microbiol. 53, 1575-1583.
Xu LM et al. (1995) Int. J. Syst. Bacteriol. 45, 706-711.
Yan AM et al. (2000) Int. J. Syst. Evol. Microbiol. 50, 1887-1891.
Yao ZY et al. (2002) Int. J. Syst. Evol. Microbiol. 52, 2219-2230.

THE SPECIES PARADIGM IN BACTERIOLOGY: FROM A CROSS-DISCIPLINARY SPECIES CONCEPT TO A NEW PROKARYOTIC SPECIES DEFINITION (WITH EMPHASIS ON RHIZOBIA)

Kristina Lindström[1], Zewdu Terefework[1], Petri Auvinen[2], Lars Paulin[2], Stefan Weidner[3], Anke Becker[3], Silvia Rüberg[3], and Helge Gyllenberg[1]
[1]Department of Applied Chemistry and Microbiology and [2]Institute of Biotechnology, University of Helsinki, Finland. [3] Lehrstuhl für Genetik, Universität Bielefeld, 33594 Bielefeld, Germany.

The prokaryotic species suffers from a lack of proper conceptualisation. The reason might be that the species is man-made. We need the species to have structures to better understand the living world and to be able to communicate about the organisms we use. The organisms themselves neither need the concepts nor the names we put on them.

1. Philosophical background

The roots of the species definition can be traced back to Plato's conceptual idea - behind everything is a never changing, temporary, independent idea (in our case taxonomic category). Aristotle's "Definitio, Genus, Differentia, Species" can be interpreted as Genus meaning the part of Definitio, which describes (or defines) the general, whereas Species describes the special, giving in our case rise to two taxonomic categories. Linnaeus defines genera, which are divided into a natural (divine) system with ranks above and below the genus level, e.g., several ranked taxonomic categories.

Mayr (1957) lists three philosophical concepts applicable to the species rank: (i) the *typological concept* related to Plato's idea concept, static and stressing differences between species; (ii) the *non-dimensional concept*, which is relational and based on distinction – either the presence or absence of interbreeding (reproductive gap); and (iii) the *multidimensional concept*, involving gene flow among interbreeding populations in a multidimensional system. Mayr ends up supporting a synthetic species concept, the *biological concept*, which is a combination of the two latter basic concepts. Species are groups of actually or potentially interbreeding natural populations, which are reproductively isolated from other such groups.

2. New species concepts for prokaryotes

Mayden (1997) evaluated twenty-two species concepts applicable to prokaryotes, using four criteria: theoretical significance, generality, operationality, and applicability. Based on these criteria, he makes the following suggestion. The *primary concept* is the evolutionary species concept, ESC, originally suggested by Simpson (1961). To be fully implemented, ESC must be supplemented with more operational, accessory notions of biological diversity or *secondary concepts*. ESC and secondary species concepts applied together reveal species diversity. The ESC differs from Platos's static idea concept, but is in harmony with Aristotle's more dynamic thoughts about everything being subject to change; material and form constantly moving towards a final purpose.

So far, the primary species concept for prokaryotes was really just that organisms belonging to the same species should be sufficiently similar, whereas those belonging to separate species should be sufficiently different. The secondary concepts were merely technical advice about which methods to use in order to sort the organisms and, as a guideline, a 70% or higher DNA:DNA homology among organisms within a species. In addition, a distinctive phenotypic character was to be given to define species. This kind of species concepts and definitions might have very little to do with biology.

An aspect of micro-organisms that is often neglected in current taxonomic discussions is the fact that bacteria are not individuals but clones. A clone is a collection of identical cells, but we should rather talk about populations. A prokaryotic population is a mixture of genetically diverging clonal cell lines on which natural selection acts, and a taxonomic category consists of evolving populations, not individual cells.

The primary concept ESC incorporates the biology and the phylogeny of evolving organisms, giving the theoretical framework. If lateral gene transfer occurs within a taxonomic category, it adds to its coherence, if it occurs between taxa, the category diverges and may give rise to new taxa. If the result is a new species, we talk about speciation. The secondary concepts provide the practical or applied definitions, guidelines or tools to obtain a clear picture of what can be accepted as a species. We propose that the following concepts are used in prokaryotic taxonomy:

PRIMARY CONCEPT
- The evolutionary species concept (discipline: biology)

SECONDARY CONCEPTS
- Maximisation of information content (discipline: bioinformatics)
- A species is an artefact of a community of practice (discipline: cognitive sciences, social learning) (Wenger, 1998)

3. Taxonomy and genomics
A modern taxonomy should rely on the best methods and knowledge available. With over one hundred prokaryotic genomes fully sequenced, we are facing a paradigm shift in microbiology. Can taxonomy gain from this?

According to current knowledge, the bacterial genome can be seen as consisting of the core genome and the accessory genome. The core genome encodes housekeeping or necessary functions for the microbe to survive. The accessory genome is composed of genes needed for niche adaptation, e.g., symbiotic genes and stress-adaptation genes in rhizobia. Phylogenetic studies of genes representing these genome compartments are necessary to increase our understanding of the evolving rhizobial genomes. Whereas the core genome is by definition necessary for each microbe, the accessory genome may or may not carry certain genes. Thus, the composition of the genome – the presence and absence of genes and the order of genes – becomes important to determine in modern taxonomy. Currently, this scientific field is moving very fast and we can gain from methods and results obtained with other groups of organisms (Figure 1). In the BACDIVERS EU consortium, we address these questions and we have started to use the *Sinorhizobium meliloti* 1021 microarray for genome comparisons.

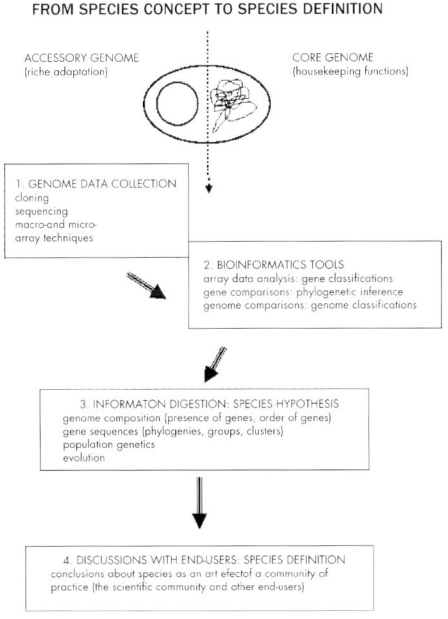

Figure 1. A flow-chart describing taxonomic work as outlined in the text.

4. Maximization of information content
In creating taxonomic categories, the maximization of information is a necessary goal. Because taxonomic categories describe diversity, and diversity can be defined as the amount of distribution of information in a community, this is an additional argument for a purposeful taxonomy. A maximal information content also attaches predictability of the behaviour and features of taxonomic categories (including the species).

5. A species as an artefact of a community of practice
One of the new secondary species concepts is the involvement of end-users as a taxonomic method. Who are the end-users of bacterial names? Those are the people that need and use the name labels that taxonomist put on the bacteria. Named bacterial species are useful, even necessary, concepts for the communication within most fields of applied bacteriology. Therefore, species definitions and descriptions as well as nomenclature should be left to experts on the biology and taxonomy of organisms in the specific fields in interaction with other end-users. This would, of course, imply that different lines of approach may be followed in different applied branches and some confusion from the general point of view. Instead, the scientists and the practitioners in the respective branches of applied bacteriology would know what they are speaking about. In social cognitive sciences, "communities of practice" are distinguished as important environments of learning and knowledge. A community of practice develops the artefacts suitable for the activities of that community. A community of practice might be a group of scientists working with the same group of organisms and the bacterial species might the artefact primarily defined to suit the needs of that community.

In the rhizobial field, we have many questions that need to be discussed. We are not served by a taxonomy for the taxonomists, but we need to create a taxonomy that most of us can be happy with. We have listed a few topics that could be discussed in the rhizobium community:
- *What kind of information should there be in the species name?*
- *Should we have a taxonomy for taxonomists or for end users?*
- *What shall we call rhizobia – those bacteria with nod genes?*
- *Is "beta-rhizobia" OK?*
- *Agrobacterium vs. Rhizobium*
- *Should rhizobial species names be accompanied by a biovariety designation?*
- *What about non-nodulating rhizobia?*

6. References
Mayden RL (1997) in MF Claridge et al. (Ed.) Species. The units of biodiversity. Chapman & Hall, London, UK, p. 31-42.
Mayr E (1957) in E Mayr (ed.) The species problem. The American Association for the Advancement of Science, Washington D.C., p. 1-22.
Wenger W (1998) Communities of practice: Learning, meaning and identity. Cambridge University Press, Cambridge, UK.

DIRECT AMPLIFICATION OF RHIZOBIAL *nodC* SEQUENCES FROM SOIL TOTAL DNA AND COMPARISON TO *nodC* DIVERSITY OF ROOT NODULE ISOLATES

Sarita Sarita[1], Parveen K. Sharma[1], Ursula B. Priefer[2] and Juergen Prell[2]
[1]CCS Haryana Agricultural University, Hisar, India. [2]RWTH Aachen, Institute for Botany, Section of Soil Ecology, Aachen, Germany

The idea of using PCR to selectively amplify sequences of partial bacterial communities related to either different phylogenetical origins or special physiological properties is widespread nowadays. These PCR approaches can be based on 16S rDNA target sequences or on conserved genes that are related to physiological functions, which lack a phylogenetical origin. The first study using a common rhizobial *nod* gene as a target for a PCR approach was published by Zézé et al. (2001). This group investigated the diversity of a fraction of the rhizobial community by the amplification of *R. leguminosarum* bv. *trifolii* and bv. *viciae nodD* sequences from soil total DNA, followed by RFLP analysis. The resulting *nodD* RFLP genotypes were compared to genotypes obtained from strains isolated from *Trifolium repens* nodules. The common rhizobial *nod* genes are responsible for the synthesis of the Nod-factor core structure, which is similar in all rhizobial species investigated until now. The idea of using a common *nod* gene as a target to amplify sequences, which represent a greater part or maybe the whole rhizobial community, consisting of competitive and non-competitive strains, is to gain new insights into rhizobial diversity. A fast way to screen rhizobial diversity at any location to check any aspect related to the rhizobial community is of upcoming interest. For example, a group-specific PCR approach could be used to investigate competition of rhizobia for permissive plant species, where selection of potential symbionts is more difficult.

In the study presented here (Sarita et al., 2004), about 427 *nodC* RFLP profiles were analysed. There were 209 from rhizobial nodule isolates that originated from six different trap plant species from soil of a chickpea and a wheat field site in India and 218 PCR clones that resulted from PCR products directly amplified from total DNA of the same soil samples, following different PCR protocols. PCR products were obtained from all rhizobial strains isolated from trap plants and from some reference strains routinely used in the lab, which served as an additional control for the suitability of the primer system. With the restriction enzymes *Msp*I and *Rsa*I, the isolates produced 15 different *nodC* RFLP genotypes from four rhizobial genera, whereas the PCR clones generated 24 different genotypes.

The community DNA PCR was able to amplify 6 out of 8 genotypes from the chickpea and 7 out of 14 genotypes from the wheat field site. Additionally, 6 and 10 new *nodC* genotypes could be identified. These genotypes were all related to species that could have been potentially trapped by one of our plants, which makes us speculate that they belong to non-competitive strains. Phylotype richness estimation showed that the wheat field datasets were under represented. The clone library sizes were too small for a clear evaluation. This shows that rarefaction analysis during the screening of the libraries is essential for a detailed validation.

The detection limit of our PCR was determined to be around 50 nodule-forming-units/g of soil in the case of *Mesorhizobium mediterraneum* and *Sinorhizobium meliloti* genotypes. *Bradyrhizobium* genotypes were difficult to isolate from sand mesocosms inoculated with about 10^5 cells/g of sand. NGR234-like genotypes were not recaptured from soil samples and mesocosms. This led us to conclude that the capability of this (and maybe also other) group-specific primer approaches depends on the detection limit of each single genotype. Such detailed validations of other group-specific PCRs are in general lacking (e.g., *nifH*, *nirS*, *nirK*, *amoA*, etc.).

References
Zézé A et al. (2001) Environ Microbiol 3, 363-370.
Sarita S et al. (2004) FEMS Microbiol. Ecol. (submitted for publication).

RECENT STUDIES ON THE *RHIZOBIUM*-CEREAL ASSOCIATION

Frank B. Dazzo[1], Youssef G. Yanni[2], Rizk Rizk[2], M. Zidan[2], Abu-Bakr M. Gomaa[3], Andrea Squartini[4], Yu-Xiang Jing[5], Feng Chi[5], and Shi-Hua Shen[5]
[1]Dept. Microbiology, Mich. State Univ., East Lansing, MI USA; [2]Sakha Agric. Res. Station, Kafr El-Sheikh, Egypt; [3]Agric. Microbiol. Dept., Nat. Res. Centre, Cairo, Egypt; [4]Dipt. Biotecnologie Agrarie, Univ. Padova, Padova, Italy; [5]Inst. Botany, Chinese Acad. Sciences, Beijing, China

Studies conducted worldwide have validated the natural, endophytic association of rhizobia with cereals (e.g., rice, wheat, barley, wild rice, maize, sorghum, millet) in rotation with legumes, and its strain / variety specificity in promoting vegetative growth, grain yield and agronomic fertilizer N-use efficiency. *Rhizobium leguminosarum* bv. trifolii is the dominant species of rhizobia capable of forming an endophytic association with rice and wheat in the Egyptian Nile delta. A total of 21 out of 23 field inoculation trials conducted so far in the Nile delta have indicated that these cereal crops can benefit from inoculation with selected rhizobial endophytes, based on their direct ability to promote cereal growth independent of BNF, plus the newly fixed nitrogen provided by BNF with the clover rotation. The natural spatial distribution of a high-performing, biofertilizer-candidate strain indigenous to the Nile delta is being mapped by a combination of microscopy, image analysis, and geostatistics in crop fields at the Km-scale relevant to the rice farmer and on rice roots at the µm scale relevant to the microbe. For these geostatistical studies, new indices of local abundance have been introduced as the Z variate for both spatial scales over the sampled domains.

Other related studies have been done to gain a better understanding of the route(s) used by rhizobia to enter, disseminate and colonize the interior of rice plants. For these studies, a plasmid conferring stable, constitutive expression of *gfp* and antibiotic resistance markers was introduced into various wildtype rhizobia (*Sinorhizobium meliloti*, *R. leguminosarum* bv. viciae, *Azorhizobium caulinodans, Mesorhizobium haukuii*). Populations of these marked rhizobia associated with various rice tissues were analyzed by a combination of fluorescence confocal microscopy, CMEIAS *in situ* image analysis, and viable plate counts on media containing appropriate antibiotics. The results indicated a dynamic infection process in rice, beginning with preferential colonization and entry of the bacteria at lateral root emergence, followed by their endophytic ascending migration into and growth within aerial plant tissues that include the stem base, leaf sheath and leaves of rice. Thus, the endophytic rhizobia-cereal association should be considered as far more invasive than previously thought, therefore heightening its interest as an experimental research model of plant-microbe association and its potential value for exploitation in sustainable agriculture to produce the world's most important cereal crops for the future.

ASCENDING MIGRATION OF ENDOPHYTIC RHIZOBIA FROM ROOTS TO LEAVES INSIDE RICE PLANTS

Feng Chi[1], Shi-Hua Shen[1], Hai-Ping Cheng[2], Yu-Xiang Jing[1] and F. B. Dazzo[3]
[1]Key Laboratory of Photosynthesis & Environmental Molecular Physiology, Institute of Botany, Chinese Academy of Sciences, Beijing 10093, China;
[2]Lehman College, the City University of New York, New York 10468, USA;
[3]Dept. of Microbiology & Molecular Genetics, Michigan State University, East Lansing, Michigan 48824, USA

Introduction
Rhizobium is the most thoroughly studied diazotroph due to its ability to induce nitrogen-fixing root nodules on leguminous plants and provide them with nitrogen nutrient (1). Recently many research studies have shown that rhizobia can also colonize the root interior of nonlegumes (2-6). *R. leguminosarum* bv. trifolii can endophytically colonize rice roots, promoting the plant's growth and grain yield under lab and field conditions (2, 5). In spite of the widespread occurrence of natural endophytic rhizobia-cereal associations, much remains unknown about its infection and colonization processes. Here we examined the infection process throughout the growth cycle of the rice plant.

Materials and methods
Five rhizobial strains (*Sinorhizobium meliloti* 1021, *Azorhizobium canlinodans* ORS 571, *Sinorhizobium meliloti* USDA 1002, *Rhizobium leguminosarum* USDA 2370, *Mesorhizobium huakui* 93) were tagged with a constitutively expressed *gfp* gene and inoculated on rice plants in gnotobiotic and open potted soil cultures. At various time points, plants were harvested to examine the colonization, dispersion and growth dynamics of the rhizobial reporter strains by computer-assisted microscopy using laser scanning confocal microscopy and CMEIAS image analysis (7) and by viable plating on TY medium containing appropriate antibiotics.

Results
Rhizobia not only colonized rice roots by crack entry between displaced epidermal cells at lateral root emergence, but also ascended into the leaf sheath above the stem base and within leaves. The fluorescent bacteria spread to and located within intercellular spaces and plant cell interiors in neighboring regions of epidermal, cortical and vascular root tissues, and also disseminated upward to arenchyma and vascular tissue within leaf sheaths and leaves. The *gfp*-tagged bacteria were isolated from the various surface-sterilized tissues and verified that they were fluorescent, resistant to the appropriate antibiotics, and were able to nodulate their respective legume host under gnotobiotic conditions.

The local abundance of fluorescent bacteria was quantitatively analyzed by *in situ*

CMEIAS image analysis of confocal optisections from infected tissues. These analyses showed that the *gfp*-tagged bacteria covered 3.66% to 20.13% of the local sample area examined and accumulated to local population densities ranging from 2.61 x 10^8 to 9.03 x 10^{10} bacteria per cm^3 of infected plant tissue at various plant locations.

Viable plating studies confirmed the ascending migration of endophytic rhizobia from roots up into above-ground rice tissues. The population dynamics of the five rhizobial test strains displayed similar tendencies to transiently increase in density, followed by a slow decline within these tissues. The long-term persistence of the endophytic rhizobia within rice tissues was indicated by reisolating all five test strains from surface-sterilized leaf sheaths and leaves of rice plants at the mature heading stage 3 months after inoculation and growth in potted soil. Factors affecting the magnitude of these dynamic population kinetics *in planta* would include an expanding growth of plant tissue inaccessible to the rhizobia, and survival / (in)compatibility of the endophytic rhizobia in response to various stresses concurrent with loss of culturability.

The growth-regulating phytohormones indole-acetic acid (IAA) and gibberellins (GA) were measured within roots and above-ground tissues of rice plants 20 days after their roots were inoculated with *S. meliloti* 1021 or *A. caulinodans* ORS571. Levels of GA_3 and IAA within above-ground tissues of inoculated plants were elevated above those within uninoculated control plants. The local density of the endophytic rhizobia that migrated up into the leaf sheath above the roots was analyzed by CMEIAS geostatistics, and the best fit semivariogram autocorrelation model was used to produce a kriging interpolation map that depicts the predicted gradients of bioactive metabolites produced by these endophytic rhizobia *in situ*.

Concluding Statements
Various rhizobia are not only able to colonize the rice root interior, but also have the ability to migrate upward into the stem base, leaf sheaths, and leaves in a dynamic infection process where the rhizobia grow to high local population density. These bacteria remain metabolically active during rice development from the vegetative to the reproductive stages, even though their colony forming ability eventually declines with time. This intimate interaction results in elevated levels of phytohormones in the aerial tissues of rice. This new finding indicates that the natural, endophytic *Rhizobium*-rice association is far more invasive than previously thought. Thus, its relevance as an important experimental research model of beneficial plant-bacteria interactions is heightened, as is its potential value for exploitation in sustainable agriculture needed to produce the world's most important cereal crops for the future.

References
Long, S. R. (2001) *Plant Physiol.* 125, 69-72.
Yanni Y. G. *et al.* (1997) *Plant Soil* 194, 99-114.
Biswas J. C. *et al.* (2000) *Agron. J.* 92, 880-886.
Chaintreuil C. *et al.* (2000) *Appl. Environ. Microbiol.* 66, 5437-5447.
Yanni Y. G. *et al.* (2001) *Austr. J. Plant Physiol.* 62, 845-870
Gutierrez-Zamora M., Martinez-Romero E. (2001) *J. Biotechnol.* 1, 117-126.
Dazzo F. *et al.* (2004) http://cme.msu.edu/cmeias/ &
http://lter.kbs.msu.edu/Meetings/2004ASM/Abstracts/Dazzo.htm

THE ECOLOGY OF *AZORHIZOBIUM CAULINODANS* ORS571 COLONIZING ON ROOT SURFACES OF *ARABIDOPSIS THALIANA*

Taichiro Iki, Toshihiro Aono, Oyaizu Hiroshi
University of Tokyo, 1-1-1Yayoi Bunkyo-ku, Tokyo Japan

Azorhizobium caulinodans ORS571 induces root and stem nodules on a tropical legume, *Sesbania rostrata*. Inside the nodules, ORS571 invades host cells intracellularly and differentiates into bacteroids, which fix atmospheric nitrogen efficiently. In general, Rhizobia fix nitrogen only in symbiosis, but ORS571 has a unique ability to grow on nitrogen-free media using atmospheric nitrogen as a sole nitrogen source. This particular feature is of importance if the strain can associate to cereals and supply nitrogen to the host plant. Colonization of rice, wheat, and *Arabidopsis thaliana* by *A. caulinodans* ORS571 has been reported. Early research focused on the intercellular colonization or root tissues, especially xylem colonization, because xylem elements are considered to be appropriate for nitrogen fixation. ORS571 may be able to exchange metabolic compounds including fixed nitrogen directly with the host plant. The oxygen concentration gets lower within the xylem and plant tissues than that of the root surface environment. The low oxygen concentration is considered to be suitable for nitrogenase activity. This paper focused on the colonization of the root surfaceI *A.thaliana*, and on its effect on the host plant.

Materials and Methods
Arabidopsis thaliana (col) was grown on agar. *A. caulinodans* ORS571 (WT) and its GFP expressing strain (pHC60) were used for inoculation. The base components of agar media were half strength, nitrogen free MS media The NH_4NO_3 concentration was changed (0mM, 1mM, and 10mM). The media contained 2% of sucrose for plant growth, and ORS571 cannot use sucrose for their carbon source, therefore we changed bacterial carbon source by adding succinate di-sodium (0% or 0.05%). ORS571 WT and pHC60 was incubated in YEB media at 37 °C, for 24hours. Then the bacterial solution was resuspended with distilled water and the bacterial concentration was modified as OD600 = 0.1. The resuspended solution was used for inoculation and 10ul was added on the whole root, when the root length was about 1cm. Petri dishes were directly put on

the stage and observed with a light and a fluorescent microscopy. When we tried to observe individual ORS571 colonizing on root surfaces, it was necessary to observe the root in high magnification, and in this case, we extracted the root samples from the surface of the media.

Results
On the third day after inoculation, the seedlings of *A.thaliana* formed mucilage like structure around their root surfaces and ORS571 actively swarmed inside it. On the 7th day after inoculation, ORS571 formed novel sphere colonies on primary root surfaces, which was surrounded with plant mucilage like structure. The observed root surface structures, such as mucilage like sheath and bacterial sphere colonies were fragile and easily damaged by physical attacks like sticking with a fine needle.

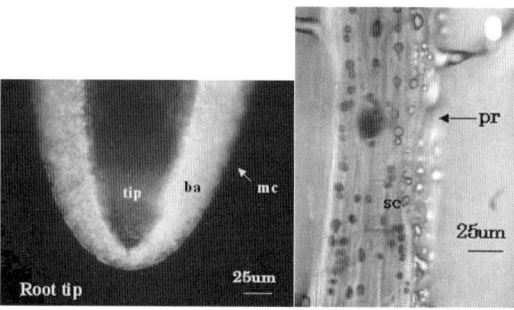

Figure 1: Colonization of A. thalina with ORS571. Left:ORS571 was swarming inside mucilage like structure at the primary root tip; Right:ORS571 formed novel sphere colonies on primary root surfaces.

Next, we investigated how the sphere colonies originate and develop. The colony sizes on the middle site of the primary root were larger than that on the tip. Plants were co-inoculated with WT and pHC60 in equal amounts. It was found that individual sphere colonies start from a single cell attachment on the primary root tip. This led us to propose a model for root surface colonization by ORS571: i) inoculated ORS571 swarms actively inside mucilage like structure around root surfaces of *A.thaliana*; ii). some of the swarming bacteria attach the tip of the primary root; iii) then, one attached cell grows and develops a sphere colony.

Further investigation revealed a negative correlation between the size of sphere colonies and the growth promotion effect on the plant. This was tentatively explained as follows. The inoculated bacteria use succinate as carbon source. The sphere colony developed on root surfaces. Too much bacteria grows on root surfaces damages the root. Then, the root growth is regulated.

Observation of the bacteria on the root surfaces was done at the highest magnification with fluorescent microscopy. ORS571 usually forms rod shape incubating in YEB or MMO medium; but it was observed that most of the colonizing bacteria formed branched Y-shape. Furthermore, some of them were greatly enlarged. This is reminiscent of the differentiation of some rhizobia into bacteroids, forming branched Y-shape. It will be of interest to analyze the correlation between Y-shaped cell existing ratio and *nifH* expression.

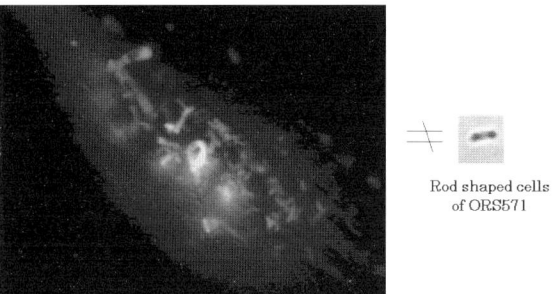

Figure 2: Y-shaped cells on primary root surfaces.

Then, we measured Y-shape ratio of ORS571 colonizing on root surfaces of *A.thaliana*. Y-shape ratio means Y-shaped cells existing ratio on root surfaces. The inoculated plant roots were extracted from agar media, and soaked into diluted tween20. After 60 minutes of sonication, the bacterial solution was observed with light microscopy. Total cells in one sight and Y-shaped cells were counted. By integrating the Y-shape ratio and nifH expression, the positive correlation was confirmed.

Conclusions
We accumulate the important knowledge on root surface colonization.
1: Inside fragile mucilage like structure, ORS571 forms sphere colonies from a single cell attachment.
2: Some of the root surface cells differentiate into Y-shaped cells like bacteroids observed in symbiosis.
3: The root surface colonization probably has some effects on the host growth as well as the intercellular colonization.
4: ORS571 has both positive and negative effect on the inoculated host. The appropriate use on agriculture is important.

References
E.C.Cocking (2003) Plant and Soil. 252, 169-175.
Oke and S.R.Long (1999) Curr. Opinion Microbiol; 2, 641-646

EVOLUTIONARY IMPLICATION OF NITROGENASE-LIKE PROTEINS IN THE PLANT KINGDOM AND PROSPECTS FOR *NIF* GENE TRANSFER IN MODEL EUKARYOTES

Q. Cheng[1,3], J. Yang[3], A. Day[2], M. Dowson-Day[1] and R. Dixon[1]
[1]Nitrogen Fixation Laboratory, John Innes Centre, Norwich NR4 7UH, UK. [2]School of Biological Sciences, University of Manchester, Manchester M13 9PT, UK. [3]Department of Plant Sciences, University of Cambridge, Downing Street, Cambridge CB2 3EA, UK.

Over the past two decades, strategies for the engineering of nitrogen fixation (*nif*) genes into higher plants plastids have been proposed, but the O_2 sensitivity of the nitrogenase component proteins has been considered to be a major problem because nitrogen fixation is not compatible with photosynthetic O_2 evolution (1). We have used homologous recombination events to replace precisely the entire *chlL* coding region in the *C. reinhardtii* plastome with the eubacterial genes *nifH* (Figure 1). Surprisingly, *nifH* can partially restore the capacity for chlorophyll biosynthesis in the dark. Recent *in vitro* reconstitution of DPOR with purified BchL, BchN-BchB subunits (2) also indicate the structural and mechanistic similarity between DPOR and nitrogenase. Since *K. peumoniae nifH* can complementarily replace the function of *C. reinhardtii chlL*, by analogy ChlL might also function in ATP-coupled electron transfer to the other components of the light-independent protochlorophyllide reductase (DPOR) encoded by ChlN and ChlB, which have sequence similarities with NifD and NifK (3). However, the details of the subunit structures, biochemical properties and requirements for biosynthesis of DPOR are as yet largely unknown. It has been proposed that the "chlorophyll iron proteins" evolved from the nitrogenase Fe protein as a consequence of a gene duplication event (4). Our results may have provided *in vivo* evidence that these two proteins are similar in function as well as structure.

The ability of NifH to function partially in the dark-dependent chlorophyll biosynthesis pathway raises a number of questions concerning the requirements for the biosynthesis and maintenance of activity of the Fe protein within the chloroplast environment. The Fe protein is the most O_2-sensitive component of nitrogenase and its ability to replace ChlL partially suggests that O_2-sensitive enzymes may function in chloroplasts when *C. reinhardtii* is grown in the dark. This is perhaps not surprising since hydrogenase, another O_2-sensitive enzyme located in the chloroplast, is active in dark-grown Chlamydomonas or in anaerobically adapted cells grown in the light (5). A further requirement for Fe protein activity is a suitable low potential electron donor. A variety of ferredoxins and flavodoxins can function with nitrogenase (6) and, since reducing equivalents are generally abundant in the chloroplast, this is not likely to present a problem.

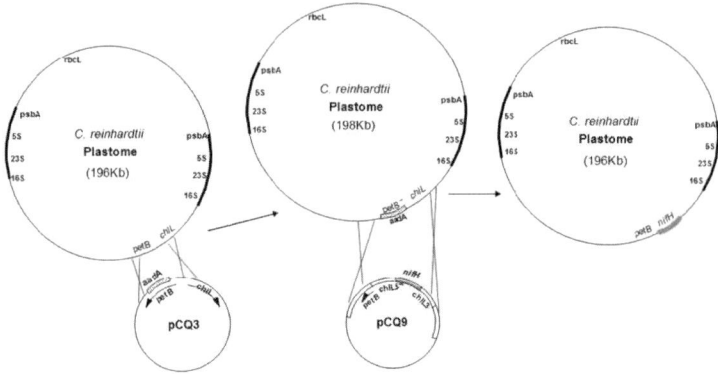

Figure 1. Schematic diagrams of constructs and two-step chloroplast transformation. Two-step chloroplast transformation via homologous recombination by bombardment vector pCQ3 to obtain *petB* mutant which was used as a recipient for the secondary transformation by delivering vector pCQ9 bearing *nifH* gene to obtain *C. reinhardtii nifH* transplastomic line.

Assuming that a [4Fe-4S] cluster is required for NifH and ChlL function in the protochlorophyllide reductase complex, the biosynthesis of this cluster is presumably achieved by ancillary proteins. In diazotrophs, the products of two genes, *nifS* and *nifU*, are necessary for Fe-protein activity, being required for the mobilisation of sulphur and iron for Fe-S cluster formation (7). However, homologues of these genes have been found in non-diazotrophic eubacteria, as well as in yeast and humans, suggesting that they may provide a ubiquitous pathway for Fe-S cluster assembly (8). The product of a third gene, *nifM*, is required for activation and stability of the Fe protein in nitrogen-fixing organisms. When *nifH* is expressed in either *E. coli* or yeast in the absence of *nifM*, a very low level of dimeric Fe protein is synthesized (9). NifM is suggested to have a chaperone-like role in maintaining the apo-Fe protein in the correct conformation to accept the [4Fe-4S] cluster and the carboxyl-terminal region of the protein shares homology with peptidyl-proline cis/trans isomerases (10). The structural similarities between the Fe protein and the 'chlorophyll iron proteins' suggests that an equivalent of *nifM* must be present in organisms which contain *chlL*. Since homologues of *nifU*, *nifS* and *nifM* are not present in the liverwort chloroplast genome, we have searched updated Chlamydomonas genome database and identified *nifU*, *nifS* of *nifM* equivalents, respectively, and are currently locating their mutants in order to investigate relevant phenotypes.

Our next approaches will be the replacement of the putative *nifDK*-like DPOR components, the *chlN* and *chlB* genes, with those of Mo-nitrogenase structural *nifDK* genes. Based on a similar assumption that the ChlNB complex may resemble the NifDK complex and harbour a similar metal scaffold, which is provided by yet unknown DPOR biogenesis proteins in the chloroplast, the expectation is to alter the enzymatic DPOR structure towards a functional nitrogenase *in vivo*.

Additionally, such research may also open avenues for studying evolutionary relationships among the light-independent protochlorophyllide reductase and the highly evolved light-dependent protochlorophyllide reductase (LPOR) and, moreover, the non-existent (so far) light-dependent nitrogenase, which we may be able to designed (11).

References
1. Dixon R et al. (1997) *Plant and Soil* 194, 193-203.
2. Fujita Y and Bauer CE (2000) *J. Biol. Chem.* 275, 23583-23588.
3. Fujita Y (1996) *Plant Cell Physiol.* 37, 411-421.
4. Burke DH et al. (1993) *Proc. Natl. Acad. Sci. USA* 90, 7134-7138.
5. Happe T et al. (1994) *Eur. J. Biochem.* 222, 769-774.
6. Buckel W et al. (2004) *Curr Opin Chem Biol.* 8, 462-467.
7. Dos Santo PC et al. (2004) *J. Biol. Chem.* 279, 19705-19711.
8. Frazzon J et al. (2002) *Biochem. Soc. Trans.* 30, 680-685.
9. Paul W and Merrick M (1989) *Eur. J. Biochem.* 178, 675-682.
10. Gavini N and Pulukat L (2002) in *Nitrogen Fixation: Global Perspectives*. Finan T et al. (Eds) Wallingford, Oxford: CABI International, pp. 228-232.
11. Yang J and Cheng Q (2004) *Plant Biology* 6, 537-544.

NOVEL *MIMOSA*-NODULATING STRAINS OF *BURKHOLDERIA* FROM SOUTH AMERICA

Euan K. James[1], Wen-Ming Chen[2], Sergio M. De Faria[3], Jean L. Simões-Araùjo[3], Roseangela Straliotto[3], Rosa M. Pitard[3], Jui-Hsing Chou[2], Yi-Ju Chou[2], Edmundo Barrios[4], Alan R. Prescott[1], Janet I. Sprent[1], J. Peter W. Young[5]
[1]School of Life Sciences, University of Dundee, Dundee DD1 5EH, UK.
[2]Laboratory of Microbiology, Department of Seafood Science, National Kaohsiung Marine University, Kaohsiung City 811, Chinese Taiwan.
[3]EMBRAPA-Agrobiologia, km 47, Seropédica, 23851-970, RJ, Brazil.
[4]Tropical Soil Biology and Fertility (TSBF) Institute of CIAT, Centro Internacional de Agricultura Tropical (CIAT), A.A. 6713, Cali, Colombia.
[5]Department of Biology 3, University of York, P.O. Box 373, York YO10 5YW, UK.

It is now well established that members of the β-proteobacteria, particularly strains of *Burkholderia* and *Ralstonia* (*Wautersia*), can be isolated from nodules. However, only *Wautersia taiwanensis* has actually been confirmed to form genuinely symbiotic N_2-fixing nodules with legumes, in this case with *Mimosa* spp. (Chen et al., 2003). Although symbiotic genes (*nifH*, *nodA*) are found in the *Burkholderia* strains, STM678 and STM815, so far there is very little physiological and structural evidence as to their symbiotic nature, and they have been shown only to form ineffective nodules on the promiscuous legume, *Macroptilium atropurpureum* (Moulin et al., 2001). In this study, we show that several strains isolated from *Mimosa* spp. native to South America are *Burkholderia* and that at least five of them can effectively nodulate their legume hosts.

Twenty *Mimosa*-nodulating strains from Brazil and Venezuela were shown via comparisons of their 16S rDNA sequences to be in the genus *Burkholderia* (Figure 1). Four clusters could be discerned. Cluster A consisted of six strains isolated from *M. pellita* (syn. *M. pigra*) nodules in Venezuela (MAP3-1, MAP3-2, MAP3-3, MAP3-4, MAP3-5, MAP3-6), as well as the Brazilian strains, Br3454 (*M. scabrella*), Br3461 (*M. bimucronata*), Br3464 (*M. flocculosa*), Br3467 (*M. pellita*) and Br3470 (*M. bimucronata*). Cluster B consisted of Br3429, Br3432 (*M. acutistipula*), Br3437 (*M. scabrella*), Br3466 (*M. tenuiflora*), and *B. tuberum* STM678 (*Aspalathus carnosa*). Cluster C consisted of Br3405, Br3407 (*M. caesalpiniaefolia*), Br3446 (*M. laticifera*), *B. caribensis* TJ182 and *B. phymatum* STM815 (*Machaerium lunatum*), and Cluster D of Br3462 (*M. flocculosa*) and Br3469 (*M. camporum*).

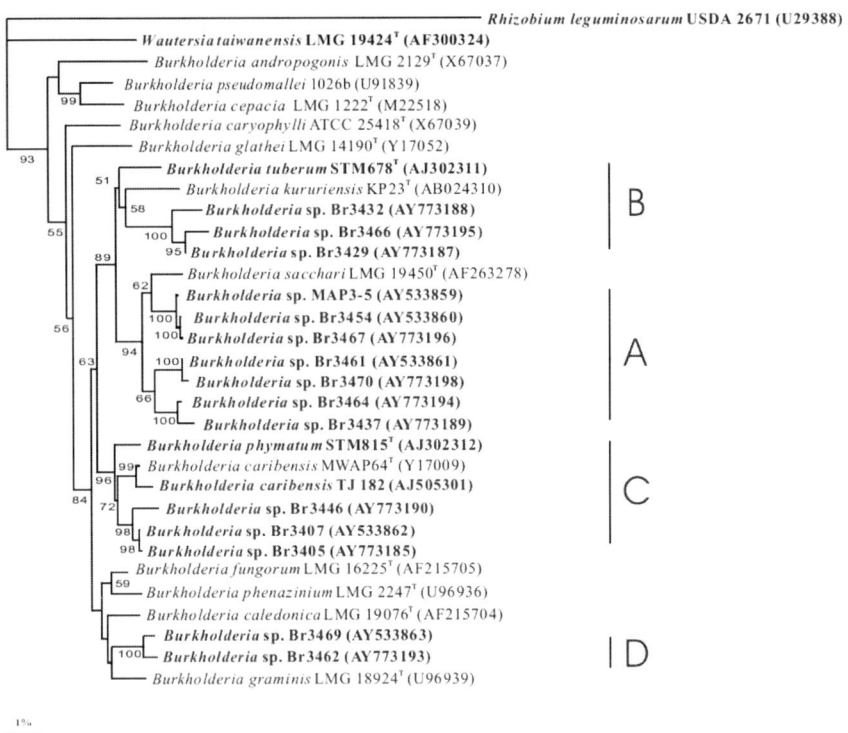

Figure 1. Neighbor-joining showing phylogenetic positions of South American *Mimosa* nodulating strains and *Burkholderia* species within the β-proteobacteria based on 16S rDNA sequence comparisons.

Five of the strains were selected for further studies of the symbiosis-related genes, *nifH*, *nodA*, and the NodD-dependent regulatory consensus sequences (*nod box*), and these were all shown to be present in MAP3-5, Br3407, Br3454, Br3461 and Br3469. In the case of the *nifH* sequences, all five strains were very close to each other as well as to other β-rhizobia, such as *B. phymatum* STM815, *B. tuberum* STM678, *B. caribensis* TJ182, and *Wautersia taiwanensis* LMG19424, and also to free-living diazotrophic *Burkholderia* strains. The *nodA* sequences, however, showed that, although the five South American strains were very close to each other and to *B. phymatum* STM815, *B. caribensis* TJ182 and *W. taiwanensis* LMG19424, they were relatively distant from *B. tuberum* STM678. Similarly, the *nifH* sequences of strains MAP3-5, Br3407, Br3454, Br3461 and Br3469 clustered together as well as with all the known β-rhizobia, with the exception of *B. tuberum*. In addition to nodulating their original hosts, all five strains could also nodulate *M. pudica*, *M. diplotricha*, *M. acutistipula* and *M. pellita*. Furthermore, strains MAP3-5, Br3407, Br3454, Br3461 and Br3469 produced nodules on *M. pudica* that had high nitrogenase (acetylene-reduction) activities and structures typical of effective N-fixing symbioses. Finally, both wild type and green fluorescent protein (gfp) transconjugant strains of Br3461 and MAP3-5 produced N_2-fixing nodules on *M. pudica* as well as on their original hosts, *M. bimucronata* (Br3461) and *M. pellita* (MAP3-5), and hence this is the first confirmation that *Burkholderia* strains can form effective symbioses with legumes.

References
Chen et al. (2003) Mol. Plant-Microbe Interact. 16, 1051-1061
Moulin et al. (2001) Nature 411, 948-950.

INVESTIGATION APPROACH ON ASSOCIATIVE DIAZOTROPHS IN PLANT MICROECOSYSTEM WITH A MULTI-FIELD OF VIEW

Song Wei
College of Life Sciences, Capital Normal University Beijing 100037 PR China

Since 1970s, the nitrogen fixation with non-legumes has been paying extensive attentions and has become an important aspect within the research field of nitrogen fixation. Nine international symposiums have been organized since 1979: (1979 Brazil; 1982 Canada; 1984 Finland; 1987 Brazil; 1990 Italy; 1993 Egypt; 1996 Pakistan; 2000 Australia; 2002 Belgium).It is expected that the associative diazotrophs could play more important part for increasing plant productivity, decreasing the use of chemical fertilizer and improving environment even that the self-nitrogen-fixing could be realized in non-legumes with *nif* gene transfer.

In early days, it was accounted that the yield increases of non-legumes as the associative diazotrophs inoculants which supplied the fixed nitrogen to their host plants. However researches in recent years indicated that the benefits should be the complex interactive synergistic effects of their nitrogen-fixing, phytohormone-producing and plant disease controlling etc..

In 1930s, it was found that some plants production were increased with inoculation of some free-living diazotrophs such as *Azotobacter spp.* etc.. Afterwards it was indicated that the plant growth-promotion was caused mainly by some phytohormones produced by inoculated diazotrophic bacteria.

In early1970s-1980s,the concept: "associative nitrogen-fixation" were proposed from the agro-microbiologist's field of view. And some associated diazotroph strains such as *Azospirillum spp.* were inoculated on wheat, corn, rice, sorghum etc.. There were some reports about their binding site on plant roots, the inoculation benefits and the possible specificity to host plants. However the results of field inoculation experiment were usually inconsistent. Over the same years, it was reported that some plant rhizosphere bacteria such as some strains of *Pseudomonas fluorescens* inoculation on the roots of sweet beet, potato and carrot etc. resulting the bacterial propagating rapidly and the remarkable plant production increase. The term: "plant growth-promoting rhizobacteria, PGPR" was proposed from the plant pathologist's field of view.

Since 1987, six international PGPR workshops: (1987 Canada; 1990 Switzerland; 1994 Australia; 1997 Japan; 2000 Argentina; 2003 India) have been organized. The investigations on PGPR have indicated that some PGPR strains are also associative diazotrophs and some associative diazotroph are also nitrogen-fixing PGPR.

In early 1990s, the discovery of the endophytic diazotrophs in sugarcane led the deep-going way on the investigation on nitrogen fixation with non-legumes.Actually fifty years ago there had been reported that the endophytes occurred but not caused the symptom of diseases in plants. Up to 1990s, at least sixty genera of endophytic bacteria had been isolated from more than thirty species of plants. It was aware that the endophytes occurrence in plants is a general fact. The publications have been increased remarkably since then. It is indicated that the functions of the endophytes have been taking intensive interest in. And it has been found that some of the isolated endophytic bacterial were also diazotrophs.

Summing up the situation mentioned above, it could be considered that the beneficial endophyte and PGPR in plant microecosystem should be classified to the same categories: plant growth-promoting bacteria (PGPB). And the associative diazotrophs should be the special PGPB with nitrogen-fixing function. However, their complex interactive synergistic effect still should be made further investigation.

Since 1990s, our research works have been focusing on the endophytic diazotroph and PGPR with rice, rice PGPB. Some interesting results have been gotten:
1. The selecting pressure for colonized rhizobacteria population on rhizoplane of two rice varieties were different (Song W et al., 1998)
2. The associative diazotrophs belong to some genera and species could enter into rice axial root through lateral root emergence site. (Yang HL et al., 1997)
3. The diversity and dominant species of endophytic diazotrophs were occurred in rice plant (Oryza sativa L.) Yuefu. (Yang HL et al., 1999; Shen. DL et al,. 2000;   Feng YJ et al.,2003)
4. The rice protoplast could incorporate the associative diazotrophic bacteria and

regenerate new rice plant with $^{15}N_2$-fixing activity and increased endophyto hormones producing. (Song w et al., 1991)
5. We have got some valuable diazotrophic PGPB strains *Paenibacillus polymyxa* WY110 (Wang Y.S. 1998; Yao W.L. et al., 2004); *Delftia tsuruhatensis* HR$_4$ (Han JG. et al., 2004); *Pantoea agglomerans* YS19 (Feng YJ et al., 2003) etc.

On these results some questions also would be proposed:
1. Do these phenomena mentioned above are not occasional but general facts?
2. How are the compatible association between plant, PGPR and beneficial endophytes (including endophytic diazotrophs) in plant microecosystem.
3. What is the real pathway of the diazotrophic endophytes enter into the plant? (information on cell biology and molecular biology)
4. How do the nutrients and metabolites exchange on the interface between bacteria and host plant? Do they go in one-way or two-ways? It is very important to distinguish the beneficial and compatible association or saprophytism.
5. Do the associative diazotrophs occur inside plants have the special structure?
6. How is the nif gene expression of endophytic diazotroph effected by enzymes in host plant?

To answer these questions, we have to face a integrate plant-microbe ecosystem but not only the individual associative diazotroph cases with a multi-field of view. With the advanced theories and technologies in biophysics, biochemistry and molecular biology, as well as the full play given to the disciplines crossing of microbial ecology, plant pathology and cell biology, the questions mentioned above would be answered more efficiently and the application of the associative diazotrophs in agriculture will be more reasonable.

Acknowledgment: This work was supported by NSFC Projects. (No:39270066, 39570025, 39770023,39970025,30170035,30370032) and NSFB Project(No:5012004)

references:
Feng YJ et al., (2003) Plant and Soil 255,435-444
Han JG et al., (2004) Systematic and Applied Microbiology (in press)
Song W et al., (1991) High Technology Letter 9,6-9
Song W et al., (1998) In: Nitrogen Fixation with Non-Legumes. Malik KA et al.,(eds.) Kluwer Acdemic Publishers 41-48
Yang HL et al., (1999) Acta Botanica Sinica 41(9), 927-931
Yao WL et al., (2004) Acta Genetica Sinica 31(9), 878-887

nifH-PHYLOCHIP FOR THE FUNCTIONAL DIAGNOSTICS OF NITROGEN-FIXING MICROORGANISMS

Lei Zhang, Thomas Hurek, Barbara Reinhold-Hurek
University of Bremen, Faculty of Biology and Chemistry, Laboratory for General Microbiology, Postfach 33 04 40, D-28334 Bremen, Germany

Biological nitrogen fixation, the enzymatic reduction of N_2 to ammonium, is an exclusively prokaryotic process which is crucial to balance the global nitrogen cycle. Rapid, simple, reliable, quantitative, and cost-effective tools are required for analyzing diazotrophic communities, especially their key functions, in real-time and in heterogeneous field-scale environments. To address these questions, the following tools were established: (i) development of protocols to coextract DNA and RNA from environmental samples for PCR and RT-PCR amplification of the iron-protein gene (*nifH*) of the key enzyme nitrogenase followed by construction of clone libraries and phylogenetic analyses of the obtained sequences; (ii) development of a *nifH* gene based oligonucleotide microarray (*nifH*-phylochip) for assessment of the functional diversity of diazotrophic prokaryotes within the context of complex environmental samples. The developed *nifH*-phylochip is a pioneer application of microarray techniques on the functional diagnostics of diazotrophic communities, having great potential for mapping the spatial and temporal variability of diazotrophic diversity in the environment, allowing rapid comparisons of the relative abundance and activity of diazotrophic prokaryotes.

FLUORESCENCE *IN SITU* HYBRIDIZATION FOR COTTON AND WHEAT ASSOCIATED PLANT GROWTH PROMOTING BACTERIA

Sumera Yasmin[1], Anton Hartmann[2], Michael Schmid[2], Kauser A. Malik[1] and Fauzia Y. Hafeez[1, a].
[1] Nat'l. Inst. Biotechnology & Genetic Engineering (NIBGE), Box 577, Jhang Road, Faisalabad 38000, Pakistan. [2] GSF-Nat'l Res. Ctr for Environment & Health Inst. Soil Ecology, Dept.Rhizosphere Biology, Ingolstaedter landstr. 1 D-85764 Neuherberg/ Munich, Germany.

FISH is a rapid molecular technique whereby microorganisms are directly visualized and identified using fluorochrome-labeled oligonucleotide probes that specifically target the 16S or 23S ribosomal RNA.

Twenty-two plant growth-promoting bacteria (PGPB) were isolated from cotton grown in Pakistani soils. Potential twelve PGPB were selected for further studies. All isolates were found to be nitrogen-fixing strains with substantial variation ranging from 18-1625 nmol acetylene reduced h^{-1} vial^{-1}. Seven bacterial strains produced indole acetic acid ranging from 0.2 – 41.2 µg ml^{-1}. Quantitative estimation of phosphate solubilization by isolate 8N-4 was found to be 188.7µg ml^{-1} using Phospho-molybdate blue color method. Siderophore producing *Bacillus* spp. Z5 and Z11 are potent strains for inoculum production regarding their biocontrol activity. With cotton variety *NIBGE-1*, in particular, plant growth and N uptake were significantly increased by these bacterial inoculations in green house study, suggesting that field selection of these PGPB could increase cotton yield. Most of the isolates were classified in the genus *Bacillus* while the other genotypes were assigned to the genus *Pseudomonas*. FISH using domain, division and subdivision-level probes have been employed in combination with confocal laser scanning microscopy for identification of cotton associated PGPB and *in situ* localization of *Bacillus pumilus* 8N-4 from wheat. Colonization studies showed the incompatibility of *Bacillus pumilus* 8N-4 with the wheat variety *Nexose* grown in a monoxenic system. The results of FISH were found to be in accordance with 16S rDNA sequence analysis. The present study indicated that FISH provides a promising opportunity to define efficient and ecologically competent bacteria and to improve inoculation strategies.

Reference
Yasmin S et al. (2004) J. Basic Microbiol. 44 (3), 241-252.

RHIZOBIA PRESENT IN WHOLE PLANTS OF LEGUMES

Yuxiang Jing[1], Feng Chi[1,2], Shihua Shen[1], Yu Liang[1], Mingjuan Tang[1]
[1]Key Laboratory of Photosynthesis and Environmental Molecular Biology, Institute of Botany, The Chinese Academy of Ssciences, Beijing 100093, China; [2]Graduate School, The Chinese Academy of Sciences, Beijing 100039, China.

During symbiosis, rhizobia differentiated into bacteroids gain photosynthates from the host legume(s) and simultaneously provide the host(s) with nitrogen nutrient fixed from the atmosphere. Recently, it has been found that the rhizobia are also capable of colonizing the roots of non-legumes such as rice, wheat and maize, and are beneficial to their growth (Yanni et al. 2001, Aust. J. Plant Physiol. 28, 1-26). Here we show that rhizobia tagged with the *gfp* gene can endophytically colonize non-nodule root tissue of the host legume in addition to their traditional endosymbiotic existence in root nodules. We also present evidence that rhizobia colonize whole plants of the host and non-host legumes after they are inoculated in the corresponding rhizospheres. *Sinorhizobium meliloti* 1021 was tagged with the *gfp* gene, inoculated into the seedling rhizosphere of the host legume alfalfa (*Medicago sativa* L.), grown in gnotobiotic culture, and examined within roots by laser scanning confocal microscopy (LCSM). Examination of root cross sections clearly showed that the fluorescent rhizobia had disseminated in intercellular spaces and cell interiors of the root cortex, and in vessels and pith of the vascular system. This result was also obtained using three other strains of *gfp*-tagged rhizobia (*Azorhizobium caulinodans* ORS571, *Rhizobium leguminosarum* USDA 2370 and *Mesorhizobium huakui* 93) inoculated on its host legume, sesbania (*Sesbania rostrata* Brem), pea (*Pisum sativus* L.) and milkvetch (*Astragalus sativum* L.), respectively. These endophytic rhizobia colonized not only within the roots of their host legumes, but also migrated from the roots up into the stems and leaves. On alfalfa the rhizobial population density in roots not surface sterilized was similar (10^6 CFU/g fresh weight) to the inoculum size of rhizobia. In contrast, the endophytic population density inside surface-sterilized roots contained only about $10^{3.5}$ CFU/g fresh weight after inoculation with 10^7 rhizobia. When the inoculum size was increased to 10^{10} cells/ml, the endophyte populations increased to more than 10^5 CFU/g/fresh weight. The rhizobia in the stems and leaves of alfalfa seedlings attained similar population densities ($10^{4.7-5.3}$ CFU/g fresh weight) despite varying the inoculum size between 10^7-10^{10}.

STUDY ON A NOVEL DIAZOTROPHIC PLANT GROWTH-PROMOTING BACTERIA STRAIN *DELFTIA TSURUHATENSIS* HR4

Jigang Han [1], Lei Sun 1, Xiaolu Sun [1], Zhengqiu Cai [2], Baocheng Zhu [1,3], Wei Song [2]

[1]College of Life Sciences, Hebei University, Baoding, 071002, China;
[2]College of Life Sciences, Capital Normal University, Beijing, 100037, China; [3]College of Life Sciences, Hebei Agricultural University, Baoding, 071002, China

A novel PGPB *Delftia tsuruhatensis* strain HR4, isolated from the rhizoplane of rice in China, was characterized with phenotypic, physiological, biochemical and phylogenetic analysis. In *vitro* antagonistic assay showed this strain could suppress the growth of various plant pathogens strongly, especially to *Xanthomanas oryzae* pv. oryzae, *Rhizoctonia solani* and *Piricularia oryzae* Cavara by 64%, 78% and 92% respectively. After the treatment especially by foliar spraying on rice Yuefu and Nonghu 6 with strain HR4 culture, the biocontrol activities to rice blast, rice bacterial blight and rice sheath blight were evident in greenhouse. In a modified N-free Döbereiner culture medium strain HR4 showed the ARA and N_2FA were 13.06 C_2H_4 nmol ml^{-1} h^{-1} and 2.052 15Na.e.% respectively. The *nifHDK* blotting showed the *nif* gene of HR4 strain was located in the chromosome. The complete 16S rDNA (1 498 bp) and partial *nifHDK* gene (3 715 bp) of strain HR4 were sequenced, they are available at GenBank under accession number AY302438 and AY544164 respectively.

Strian HR4 was classified as a member of *Delftia tsuruhatensis* based on the characterisation results. Most of reported *Delftia sp.* undertook organic chemical degradations. To our knowledge, this is the first report on *Delftia tsuruhatensis* discribed both as a PGPB strain and a diazotrophic strian associated with rice plant.

GENETIC DIVERSITY AMONG PLANT GROWTH PROMOTING RHIZOBACTERIA AND THEIR ULTRASTRUCTURAL LOCALIZATION WITHIN MAIZE ROOTS

Zakira Naureen, Sohail Hameed, Sumera Yasmin, Kauser A. Malik and
[a] Fauzia Y. Hafeez.
National Institute for Biotechnology and Genetic Engineering (NIBGE),
P.O. Box 577, Jhang Road, Faisalabad 38000-Pakistan.

A variety of bacteria, colonize rhizosphere of most of the cereals including maize and have been reported to enhance plant development either directly by fixing nitrogen and producing phytohormones, or indirectly by improving the uptake efficiency of nutrients and water and by inhibiting pathogens. Based on these criteria, thirty plant growth promoting rhizobacteria (PGPR) isolated from maize grown in Pakistani and Indonesian soils were evaluated for their morphological characteristics, nitrogen fixation, P-solubilization, indole acetic acid (IAA) and siderophores production. Nitrogenase activity was detected in ninteen isolates ranging from 21.8-3624 n moles C_2H_4 produced h^{-1} mg^{-1} protein. Most of the isolates produced IAA, while only four were P-solubilizers. The successful application of PGPR as bio-inoculants depends to a great extent on their ability to colonize roots of the host plants. Ultrastructural studies of *Pseudomonas* sp. F14 was carried out using transmission electron microscopy (TEM). Studies indicated characteristic rhizospheric colonization within 48 h that was observed to change considerably with the passage of time from few bacteria to micro colonies, which shows potential of *Pseudomonas* sp. F14 to reside the host rhizosphere. Understanding the genetic diversity among microbial colonizers is also of great utility in strategic approaches of integrated nutrient management. Random amplified polymorphic DNA (RAPD) analysis of 30 PGPR strains using 30 oligonucleotide primers resulted in considerable level of genetic diversity, with genetic distance ranging from 2-16%. Indonesian isolates were found to be more diverse as compared to Pakistani isolates. The characterization and screening of PGPR of maize rhizosphere has helped in selection of isolates F7, LS-1, 3.1.1.C, F2, F3 and F13 as superior strains for use as bioinoculant to enhance the yield of maize.

Reference

Hafeez FY (2002) Biofertilizer (*Bio Power*): Development, use and economic importance in leguminous and cereal crops. In *Proceedings of Orientation Training/ Workshop on Saline Agriculture*, NIAB, Faisalabad, Pakistan.

ENDOPHYTIC RHIZOBIA FROM COLONIZED ROOTS UPWARD INTO STEMS, LEAVES AND OVULES WITHIN TOBACCO PLANTS

Feng Chi[1,2], Shihua Shen[1], Yu Liang[1], Yuxiang Jing[1]
[1]Key Laboratory of Photosynthesis and Environmental Molecular Physiology, Chinese Academy of Sciences, Beijing 100093, China.
[2]Graduate School, Chinese Academy of Sciences, Beijing 100039, China.

One of diazotrophs is *Rhizobium* which forms nitrogen-fixing nodules on roots of legume, providing leguminous plants with nitrogen nutrient and promoting plant growth. In this aspect, a large amount of research on molecular interactions between *Rhizobium* and legume have been focused on and extensively investigated. However, *Rhizobium* could also colonize the roots of non-legumes such as rice, maize, lettuce, wheat, barley, and canola etc., and make them growth promotion and increase their yield rather than N_2- fixation in gnotobiotic and field experiments, has been found recently.

Sinorhizobium meliloti 1021 strain tagged with *gfp* gene was used to infect non-legume tobacco for confirming whether this *Rhizobium* also can endophytically colonize roots of tobacco plants grown in gnotobiotic condition. It is demonstrated that the rhizobial bacteria entered tobacco roots by crack entry invasion at the lateral root junctions at 6-21 days after inoculation (DAI). They spread and located in root hairs, intercellular spaces and inside cells at the regions of epidermis, cortex and vascular system. After the roots were checked, the sections of stems and leaves were observed to see whether there are any rhizobial bacteria existing in them? Surprisingly, the results showed that the rhizobial bacteria up-migrated themselves into stem and leaf tissues within plants. After we found this phenomenon, we took the ovaries of tobacco plants which were grown in pots in greenhouse for 128-136 ADI. It is amazed that the bacteria actually existed in tissues (ovules) of ovaries which are developed into seeds after fertilization, meaning that the rhizobial bacteria are able to climb up even in reproductive stage and suggesting that the rhizobia have longevity to stay in whole body of plant interior as their 'niche', and make a circulation through plants in nature.

A STUDY ON THE POTENTIAL NITROGEN-FIXING MICROBES IN MALAYSIA

Liew, P. Woan-Ying, Jong Bor-Chyan and Khairuddin A. Rahim
Agrotechnology and Biosciences Division, Malaysian Institute for Nuclear Technology Research (MINT), Bangi, 43000 Kajang, Selangor, Malaysia.

A number of free-living bacterial isolates obtained from Malaysia soil samples were selected on Nitrogen-free Davies' medium for their abilities to fix atmospheric nitrogen. These isolates were later examined by targeting the respective nitrogen fixation (nif) gene sequences using specific primer sets. Bacterial isolates of interest were identified based on the first 500 base pair sequences of the 16S rRNA genes. The 16S rDNA sequences obtained were then cross-matched to the GenBank database. The results demonstrated a variety of Bacillus species. Preliminary characterizations of these positive isolates using specific restriction enzymes successfully distinguished between most of the Bacillus species except for B. subtilis and B. pumilus, which showed closer evolutionary distance based on phylogenetic analysis.

PRODUCTION OF IAA BY *AZOSPIRILLUM BRASILENSE* SP 245 AND ASSOCIATIVE SYMBIOTIC N_2-FIXATION OF RICE SEEDING INOCULATED WITH SOME N_2-FIXING BACTERIA

Hong-Gon Ryang, Sung-Bok Choi and Pil-Gum Li
Research Center for Compound Microorganisms, Academy of Sciences of D. P. R. Korea

As a well known biofertilizer, *Azospirillum* has been reported to show wide adaptability around the world. The action of *Azospirillum* to synthesize indole-3-acetic acid (IAA) is thought to be one of the major factors influencing crop growth and yield.

Cultures of *A. brasilences* sp 245 produce IAA only in the late stationary phase. In order to enhance the production of IAA, D, L-tryptophan was added to medium. The growth of *A. brasilences* sp 245 under aerobic conditions was very good (10^9/ml), but was bad under 1% aerobic or anaerobic conditions (10^{4-6}/ml) in 24 hrs. The production of IAA of this N_2 fixing bacteria was 5.6 ng/ml under aerobic conditions, 0.1 ng/ml under 1% air aerobic conditions and not detectable under anaerobic conditions in 24 hrs.

Another experiment was carried out. In this case, the young seedlings of rice inoculated with *A. lipoferum* A-37 was grown in the test tube containing free-carbon and nitrogen, semi-solid medium. After 7 days, the inoculated bacteria colonized around rice root surface. The distributions of bacteria on the root surface and the cross section of root were observed with scanning electron microscope. The results showed that most of bacteria adhered to the root surface, mainly around the root hairs, and some of them invaded into the cortex. This suggested that the inoculated bacteria to the rice seeding propagated using the nutrients from host embryo and invaded into the root cortex.

When *A. lipoferum* A-37 was inoculated to the rhizosphere of rice seedling in test tube, the nitrogen fixation activity was increased with the incubation time and reached the maximum in 5~7 days, and then decreased gradually after 7 days.

Rice seedlings in the test tubes were inoculated with *A. lipoferum* A-37, *A. chroococcum* A-89 and *K. pneumoniae* K-89, respectively. The length of plant shoot, root, the numbers of root hairs and N_2 fixation activities were compared among the inoculants together with null control. As a result, *A. lipoferum* A-37 was the best.

CULTURE-INDEPENDENT ANALYSIS OF ENDOPHYTIC BACTERIAL DIVERSITY OF RICE

Sun Lei[1,2], Han Jigang[2], Song Wei[1]
[1]College of Life Sciences, Capital Normal University, No.105, Xisanhuan Beilu, Beijing 100037, PR China
[2]College of Life Sciences, Hebei University, Baoding 071002, PR China

Studies on the species diversity of endophytic bacteria have been mainly approached by cultivation-based methods, however some bacteria are not accessible to cultivation methods because of their unknown growth requirements or their entrance into a viable but not culturable state. Some culture-independent methods which utilize polymerase chain reaction (PCR) to amplify 16S rDNA from DNA extract from samples have been applied to study plant endophytic bacteria (Dave Seghers et al.,2004).

A preliminary study on rice endophytic bacteria using culture-independent method was done. A bacterial 16S rDNA primer was designed to amplify bacterial sequence directly from rice roots and leaves by PCR to exclusion of chloroplast DNA and mitochondrial DNA. Total DNA was extracted from the surface-sterilized roots and leaves of rice seedling grown in a greenhouse. Endophytic bacterial diversity was examined by PCR-DGGE (polymerase chain reaction-denaturing gradient gel electrophoresis) of 16S rDNA. DGGE profiles reflected differences of endophytic bacterial diversity in roots and leaves. The diversity of endophytic bacteria from rice roots was more complex than from rice leaves. We also found that the DGGE profiles were influenced by different DNA extraction methods. Our results indicated that the method which was based on lysis with extraction buffer in hexadecyltrimethylammonium bromide (CTAB) was fit to study rice endophytic bacteria (This work was supported by NSFC project No:30170035, 30370032).

Reference:
Dave Seghers et al. (2004) Applied and Environmental Microbiology 70(3), 1475-1482.

AGRICULTURAL MICROBIAL COLLECTION STATUS IN CHINA

Xiaoxia Zhang, Ruibo Jiang, Jingang Gu, Bingquan Fan, Xiaotong Ma, Shigui Li and Zhiyong Ruan
Agricultural Culture Collection of China, Chinese Agricultural Academy of Science, Beijing, China, 100081)

Agricultural Culture Collection of China (ACCC) is an professional agricultural microorganism resource and research centre in China. ACCC holds more than 4000 agricultural microorganism strains covering more than 160 genera and 510 species in total. One of the distinguishing features of ACCC is that it conserves more than 1000 strains of *Rhizobium*. In the past 20 years, more than 10000 *Rhizobium* strains has been isolated from some special geographies and climatic environments around China, such as the dry and semiarid areas in Xinjiang province, low-temperature and dry areas in the Inner Mongolia, high-temperature and rainy areas in Hainan Province. By now, more than 1000 strains with high efficient nodule-forming and stress resistant characteristics were screened and studied on different legume crops and herbages in different areas. Lots of *Rhizobia* strains have been applied to the West and Northeast China widely. ACCC acts on an important part in the strategy of developing the western region and Grain for Green Project. Some rare species such as acid-tolerant *Rhizobium fredii* (pH4-4.5) and salt-tolerant *Sinorhizobium meliloti* (3.48-5%NaCl) and some non-legume Nitrogen Fixation Bacteria, such as *Azotobacter, Aztospirillum* are preserved in ACCC.

With the government financial support increasing and the startup of National Natural Resource Platform, ACCC will play more and more important roles in microbial collection, application, evaluation, as well as the studies of Biological Nitrogen Fixation.

ECOLOGY OF COMMON BEAN RHIZOBIA IN CENTERS OF HOST DIVERSIFICATION

O. Mario Aguilar, O. Riva, E. Peltzer and G. Favelukes
IBBM-Facultad de Ciencias Exactas, Universidad Nacional de La Plata.
Calle 47 y 115, La Plata, Argentina.

In the Southern Andes, wild common beans share the habitat with other legumes both herbaceous and leguminous trees. We have found for that region that species *R. etli* is the predominant rhizobia found symbiotically associated with beans. By using soil samples from North West of Argentina and Bolivia, and common beans and leucaena as trapping hosts, respectively, we found that other genotypes which are also able to nodulate common beans, are present in that environment. Thus, analysis of 16S rDNA from the bean nodulating rhizobial isolates has shown that alleles of *Sinorhizobium* sp., *R. tropici* and a group of isolates were found to be closely related to *Agrobacterium rhizogenes*.

We have investigated the intraspecies diversity by examining symbiotic genes *nodA*, *nodC*, and *nifH* and have demonstrated *nodC* gene to be polymorphic among isolates of *R. etli* from different geographical origins in the Americas. The distribution of these alleles follows that of the centers of bean diversification. Allele δ was found predominant in the Southern Andes whereas allele α was prevalent in Middle America. Furthermore, wild common bean accessions from Meso America were preferentially nodulated by *R. etli* strains from the same geographical origin, and the reverse was also true for accessions from the Andean center. These results suggested that common beans and rhizobia coevolved into species *R. etli* followed by local coevolution in centers of host diversification. This host x rhizobia interaction could be used for the selection of a better inoculant of common beans.

References
Aguilar *et al*. (2004) PNAS 101, 13548-13553

GENETIC DIVERSITY AND PHYLOGENY OF RHIZOBIA ISOLATED FROM SIX GENERA OF LEGUMINOUS PLANTS, IN SICHUAN PROVINCE, P. R. CHINA

Chen Qiang[1,2], Chen Wen-xin[1*], Zhang Xiao-ping[2], Li Deng-yu[2], Kristina Lindstrom[3]
[1]Key Laboratory Agro-Microbial Resource & Application, Ministry of Agriculture, China Agricultural Univ. 100094 Beijing; [2]Dept. Microbiology, Sichuan Agricultural Univ., 625000 Yaan Sichuan, P. R. China; [3]Dept. Applied Chemistry & Microbiology, Univ. Helsinki, Fin-00014, Finland.

The genetic diversity of 103 rhizobial strains from six genera of legume plants of *Kummerowia* spp., *Vigna* spp., *Pueraria* spp., *Albizia* spp., *Campylotropis* spp., and *Acacia* spp., in different regions of Sichuan Province, together with 17 reference strains, were analyzed by AFLP, BOXAIR-PCR, 16S and 23S ARDRA, and the phylogeny of the representative strains were performed by 16S rRNA full sequences. The results showed that, there was high genetic diversity among these strains. Analysis of AFLP, BOXAIR-PCR divided the 103 strains into 26 AFLP genotypes and 18 BOX-AIR PCR genotypes, respectively, and most of the strains grouped with their host plants. 16S and 23S ARDRA analysis revealed that the 56 representative strains formed 13 genotypes, and most of them belonged to the genus of *Bradyrhizobium*, but distinguished from the type strains of *B. japonicum* USDA 6^T and *B. elkanii* USDA 76^T. The phylogenic relationship constructed by the 16S rRNA gene full sequences suggested that these rhizobial strains distributed into three branches of *Rhizobium-Agrobacterium-Allorhizobium*, *Mesorhizobium*, and *Bradyrhizobium*. Of them, *Rhizobium* sp. (*Kummerowia*) CCBAU 61054 located in the phylogenic sub-branch of *Rhizobium-Agrobacterium*; *Bradyrhizobium* sp. (*Kummerowia*) CCBAU 61030, *Bradyrhizobium* sp. (*Vigna*) CCBAU 61073, *Bradyrhizobium* sp. (*Pueraria*) CCBAU 61106, and CCBAU 61116 were in the branch of *Bradyrhizobium*. *Mesorhizobium* sp. (*Pueraria*) CCBAU 61095 was within the range of *Mesorhizobium* phylogenic branch.

Acknowledgment: This paper was supported by the "973" project of China (No. 2001CB108905)

CHARACTERIZATION OF NODULE ISOLATES FROM THE TREE LEGUMES *DALBERGIA* SPP. GROWN IN CHINA

Wenfeng Chen[1], Entao Wang[1,2], and Wenxin Chen[1]
[1]Key Laboratory of Agro-Microbial Resource and Application, Ministry of Agriculture/College of Biological Sciences, China Agricultural University, Beijing, 100094. [2]Departamento de Microbiología, Escuela Nacional de Ciencias Biológicas, Instituto Politécnico Nacional, Prol. De Carpioy Plan de Ayala s/n, Santo Tomás, México D. F. 11340, México.

This is the first report about the diversity of rhizobia associated with Dalbergia trees grown in temperate zone. Twenty-three bacterial strains isolated from nodules of Dalbergia trees grown in China were characterized. The majority of these strains were Bradyrhizobium and Rhizobium, while three of them were Agrobacterium based upon the analysis of numerical taxonomy, AFLP, PCR-RFLP and sequencing of 16S rDNA. The grouping results seemed relating to their geographic origins. The three Agrobacterium-related isolates were from Shandong province. The nine isolates within cluster 4 (Rhizobium sp.) were from Hunan province, where the climate is warmer and more humid than Shandong. The six Bradyrhizobium isolates were from Anhui, Hubei and Jiangxi provinces, where the climate is similar to Hunan, but the soils are acidic with high amount of aluminum. The fact that the Dalbergia spp. nodulated with different rhizobial populations in different environments indicates that the association of rhizobia-legumes is a consequence of interaction among the symbiotic partners and the environmental factors.

Acknowledgment: National Science Foundation of China No. 30270001

DIVERSITY AND PHYLOGENY OF *BRADYRHIZOBIUM* SP. ARACHIS ISOLATED FROM NODULES OF PEANUT IN CHINA

Jiangke Yang and Junchu Zhou
State Key laboratory of Agricultural Microbiology, Huazhong Agricultural University, Wuhan ,430070, P. R. China.

Peanut (*Arachis hypogaea L.*), an important oilseed and subsistence food crop, plays a critical role in agriculture and economy of China. Studies on *Bradyrhizobium* isolated from different legume plants revealed that the bradyrhizobia were very heterogenous so that it was necessary to conduct further polyphasic studies on peanut bradyrhizobia from various regions on a large framework. This study is conducted to clarify the diversity and phylogeny of bradyrhizobia from different geographical regions of China by both phenotypic and genetic methods including phenotypic testes, RFLP analysis of PCR-amplified 16S and 16S-23S rRNA internally transcribed spacer (ITS), 16S rRNA sequencing, ERIC-PCR and REP-PCR. All stains were clustered by phenotypic and genotypic tests into group I - 16S rRNA RFLP genotype 3 and group II - 16S rRNA RFLP genotype 1. The group II strains included the reference strain USDA110, USDA122 and USDA127 of *Bradyrhizobium japonicum* and also strains of genotype 2. Results of 16S rRNA sequencing revealed that peanut bradyrhizobia were phylogenetically related to *B. japonicum* and their sequence divergence was less than 1.1%. Strains were also classified based upon the size of the intergenic spacer region (ITS) and clustered into ITS-I, ITS-II and ITS-III genotypes. Strains could be further divided into five sub-groups or clusters (clusters IA, IB, IIa, IIb and IIc). Host specificity test revealed that all strains tested nodulated *Phaseolus vulgaris* and strains of cluster IIb and IIc nodulated *Glycine soja* efficiently. Our results indicated that bradyrhizobia isolated from peanut were related, but still exhibited phylogenetical divergence with *B. japonicum*.

THE DIVERSITY OF RHIZOBIA THAT NODULATED *DESMODIUM* SPP. IN THE DIFFERENT GEOGRAPHY REGIONS OF CHINA

Jun Gu, Wen-Xin Chen
Department of Microbiology, China Agricultural University, Beijing, China, 100094

Forty- four strains were isolated from nodules of eleven *Desmodium* species in different physical geography regions in China. With the anlalysis of 16S rDNA RFLP, IGS RFLP, 16S rDNA sequences and BOX-PCR fingerprinting , the results showed that the rhizobia nodulated with *Desmodium* spp. were diverse. The microsymbionts of *Desmodium* were mainly different *Bradyrhizobium* species including *B.elkanii*, *B.japonicum* and *B.yuanmingense*. Moreover, there were also other species of *Sinorhizobium* and *Mesorhizobium* associated with *Desmodium*, which indicated that the rhizobia associated with *Desmodium* spp. were great diverse. There were no distinct relationships between rhizobia and different geographic regions.

DIVERSITY OF A BRAZILIAN COLLECTION OF RHIZOBIAL STRAINS

Mariangela Hungria[1,2], Pamela Menna[1,3,4], Mariana G. Germano[1,3], Ligia Maria O. Chueire[1,2], Eliane V. Bangel[5], Rubens J. Campo[1].
[1]Embrapa Soja, Londrina, PR, Brazil. [2]CNPq, Brazil. [3]M.Sc. Univ. Estadual Londrina. [4]CAPES, Brazil. [5]FEPAGRO, Porto Alegre, RS, Brazil.

The Leguminosae is one of the largest plant families with over 18,000 species classified into around 650 genera; most are capable of effective symbioses with rhizobial strains. However, despite reports of high strain diversity, especially in the tropics, there are less than 40 described rhizobial species. Ribosomal sequences, emphasizing 16S rRNA genes ($\approx$1.5 kb), are the method of choice for tracing bacterial phylogenies. Speciation of genera can be enhanced by analysis of other ribosomal genes because both the 23S rRNA ($\approx$2.3 kb) and the 16S-23S rRNA intergenic space (ITS) show higher variability than 16S rRNA. Analyses of rhizobial ribosomal genes have used RFLP-PCR, as well as partial or complete sequencing of bases. In Brazil, there are 200 rhizobial strains recommended for 95 legume species, but little is known of their genetic diversity. A Brazilian culture collection of 119 strains of *Bradyrhizobium*, isolated from thirty-three legume species, representing nine tribes and all three subfamilies, were analyzed by RFLP-PCR of the 16S rRNA, 23S rRNA and ITS region, each with three restriction enzymes. 43 of those strains are recommended in Brazil as the most effective for 31 host legumes. For the 16S rRNA gene, reference strains of *B. japonicum* fit into two major clusters at 50% similarity, whereas two other clusters (at 53% similarity) were composed of strains of *B. elkanii*; all strains were joined at a final level of similarity of 28%. The higher variability in the ITS and 23S rRNA resulted in final groupings at very low level of similarities, 27% and 16%, respectively. Considering the three ribosomal regions, two great groups were visualized, related to *B. japonicum* (with most strains from soybean) and *B. elkanii* (most strains from indigenous legumes), at similarity levels of 54 and 46%, respectively, and at least two new clusters that might represent new species. The complete 16S rRNA sequence for another set of 80 rhizobial strains from 30 different legume species (23 recommended as most effective) were clustered within the genera *Rhizobium, Sinorhizobium, Mesorhizobium* and *Bradyrhizobium*. Three strains, recommended for *Clitoria fairchildian, Piptadenia gonoacantha* and *Ormosia nitida*, were clustered with *Burkholderia* sp. strain TJ182. This collection of strains indicates an extremely high level of genetic diversity in the tropics.
Partially supported by CNPq (PRONEX-41.96.0884.00 and 35216/1992-3).

NUMERICAL TAXONOMY AND 16S rDNA-PCR RFLP OF BACTERIAL ISOLATED FROM ROOT NODULES OF *ALBIZIA* SPP., *ACACIA* SPP. AND *LEUCAENA LEUCOCEPHOLA* IN CHINA

Fengqin Wang, Yongfa Zhang, Jie Liu and Wenxin Chen
Key laboratory of Agro-Microbial Resource and Application, Ministry of Agriculture/College of Biological Science, China Agricultural University, Beijing, 100094, China

Fifty-one isolated strains from root nodules of *Albizia* spp. *Acacia* spp and *Leucaena leucocephala* growing in different regions of China were studied by performing numerical taxonomy analysis of 127 phenotypic characteristics and 16S rDNA PCR-RFLP. Based on the results of numerical taxonomy analysis, 25 of the isolated strains were grouped into six clusters at the similarity of 83.5%, which were related to *A.tumefaciense* (cluster 1), *Agrobacterium* sp. (cluster 2), *Mesorhizobium* sp. (cluster 3), *M.plurifarium*(cluster 4), *Mesorhizobium* sp. (cluster 5) and *S.terange* (cluster 6). 15 strains formed the *Bradyrhizobium* main cluster that was separated into 3 sub clusters related to *B.japonicum* (cluster7), *B.elknii* (cluster 8) and *Bradyrhizobium* sp. (cluster 9) at the similarity of 87.5%. Others formed single branches or small groups (two strains). More reference strains were selected in 16S rDNA PCR-RFLP, in which 51 isolates produced 22 genotypes and formed 9 clusters at the similarity of 95%. The results were quite agreed with that of numerical taxonomy analysis at the level of genus. Strains of cluster 5 and cluster 9 in numerical taxonomy were grouped together with *M.plurifarium* LMG11892^T and *B.yuanmingensis* CCBAU10071^T respectively, while Cluster 2 and cluster 3 did not group together with reference strains either in this method. In addition, the phenotypic characteristics showed that some isolates could grow in stress conditions such as high concentration of some antibiotics (300ng ml$^-$), intense hot shock (60°C, 20min), acid (pH4.0) or alkali (pH12). *Albizia* spp. were promiscuous hosts for *Rhizobium*, *Sinorhizobium*, *Mesorhizobium* and *Bradyrhizobium*. Isolates from *Acacia spp.* were most related to *Mesorhizobium* and *Bradyrhizobium*, while the microsymbionts of *Leucaena leucocephala* were grouped together with *Rhizobium* and *Sinorhizobium*. From these results we inferred that these isolated strains had a characteristic of high phenotypic and genotypic diversities. The nodA genes of the strains in cluster 2 in numerical taxonomy analysis were gotten by PCR although they were grouped closely to *Agrobactium*.

DIVERSE RHIZOBIA ASSOCIATED WITH WOODY LEGUMES *WISTERIA SINENSIS, CERCIS RACEMOSA* AND *AMORPHA FRUTICOSA* GROWN IN CHINA*

Jie Liu[1], En-To Wang[1,2] and Wen-Xin Chen[1]
[1]Dept. Microbiology, College Biological Sciences, China Agricultural Univ. 100094, Beijing, China. [2]Dept. de Microbiología, Escuela Nacional de Ciencias Biológicas, Instituto Politécnico Nacional, México D. F., 11340, México.

Fifty-nine bacterial isolates from root nodules of the woody legumes *Wisteria sinensis, Cercis racemosa* and *Amorpha fruticosa* grown in the central and eastern regions of China were characterized with phenotypic analysis, PCR-based 16S and 23S rRNA gene RFLP, Box PCR and 16S rRNA gene sequencing. Seven main phena were defined in numerical taxonomy, which corresponded to distinct groups within the genera *Agrobacterium, Bradyrhizobium, Mesorhizobium* and *Rhizobium* in 16S and 23S rRNA gene PCR-RFLP. The phylogenetic relationships of the 16S rRNA genes supported the grouping results of PCR-RFLP. Most of the isolates from *Amorpha fruticosa* were classified into two groups closely related to *Mesorhizobium amorphae*. Seventeen of the twenty-one isolates from *Wisteria sinensis* were identified as two groups related to *Rhizobium* and *Agrobacterium* respectively. Six out of ten isolates from *Cercis racemosa* were identified as a group related to *Bradyrhizobium*. Four strains in phenon 8 of numerical taxonomy (inclusive of SH28301, SH283012, CCBAU23151 and CCBAU43142) were further studied by DNA-DNA hybridization and cross-nodulation, and the results showed that it was a new group different from all the defined species of *Bradyrhizobium*. Our results also indicated that each of the investigated legumes nodulated mainly with one or two rhizobial groups, although isolates from different plants intermingled in some small bacterial groups. In addition, correlation between geographic origin and grouping results was found in the isolates from *Amorpha fruticosa*. These results revealed that the symbiotic bacteria might have selected by both the legume hosts and the geographic factors.

* This work was supported by the National Natural Science Foundation of China (30270001).

DIVERSITY OF RHIZOBIA ISOLATED FROM PASTURE AND WOODY LEGUMES GROWN IN PANXI REGION, SICHUAN, CHINA*

Leena A. Räsänen[1], Qinghua Hu[2,3] Qiang Chen[2], Dengyu Li[2], Kristina Lindström[1] and Xiaoping Zhang[2]
[1]Department of Applied Chemistry and Microbiology, University of Helsinki, Finland. [2]Department of Resource and Environment, Sichuan Agricultural University, Sichuan China. [3]Insitute of Natural Resources, Chengdu, Sichuan China. E-mail: Leena A.Räsänen@Helsinki.Fi

Panxi region is located in the southwestern part of Sichuan Province in China, consisting of the area between Changjiang River (Yangtse) and the western mountain range. Due to its geographical location Panxi region has characteristics both of a subtropical river-valley climate and a mountain climate. The area is also well known for its rich flora and fauna. Several legume species grow in this region. Forty-three rhizobial strains were isolated from two different vegetation zones. Fifteen strains were isolated from pioneer plants capable of growing in poor soil, tolerating dry and hot growth conditions (acacias, *Campylotropis delavayi*). Another 28 rhizobial strains were isolated from herbaceous legumes that are typical of Panxi pastures of high quality. The diversity of rhizobia was analysed and compared with 20 reference strains by physiological tests (numerical taxonomy) and genetic methods (BOXAIR-PCR, 16S rDNA PCR-RFLP and 16S sequencing.

According to the physiological tests, the diversity of rhizobia was high. A majority of strains (29) belonged to slow-growers and 14 strains belonged to fast-growers. *Acacia kalkora*, *Aeschymene indica* L., and *Indigofera bungeana* were nodulated both by slow-growing and fast-growing strains but otherwise plants were nodulated either by slow-growers or fast-growers, and according to the numerical taxonomy, strains isolated from the same plant species generally grouped together. The BOX-AIR analysis indicated that only few strains, even if they originated from the same plant species, were closely related. 16S rDNA RFLP-PCR confirmed the high diversity of rhizobia of the Panxi region. The analysis of partial 16S rRNA sequences of some strains revealed that strains isolated from *Acacia kalkora* belonged to or were related to *Mesorhizobium* sp., *Bradyrhizobium japonicum* and *B elkanii*. A strain isolated from *Phaseolus angularis* was closely related to *B. elkanii*. The strains isolated from *Trifolium repens* belonged to *R. leguminosarum* bv. *trifolii*.

*Supported by the "973" project of China (No. 2001CB108905) and Univ. of Helsinki.

GENOMIC FINGERPRINTING OF INDIGENOUS SOYBEAN RHIZOBIA

Sanja Sikora[1], Sulejman Redžepović[1], Andrea Skelin Vujić[1] and Dubravko Maćešić[2]
[1]Departments of Microbiology and [2]Field Crops, Forage and Grassland, Faculty of Agriculture, University of Zagreb, Croatia

The presence of adapted and competitive indigenous soybean rhizobia in soil can reduce the inoculation response even with highly efficient commercial strains. On the other hand, the study of rhizobial biodiversity opens up the possibility of preserving and maybe exploiting some indigenous strains with unknown symbiotic or ecological potential, particularly under unfavourable conditions. In this respect, it is important to characterize the indigenous strains and to obtain information about the actual composition of rhizobial field population. The main aim of the present study was to isolate and to identify indigenous *Bradyrhizobium japonicum* strains from different soybean growing areas in Croatia and to detect genetic diversity within rhizobial natural populations. All rhizobial isolates were characterized by using different PCR fingerprinting methods, such as 16S rDNA PCR-RFLP and RAPD analysis. PCR-RFLP of 16S rDNA clearly showed that all isolates can be identified as *Bradyrhizobium japonicum*. Cluster analysis of combined RFLP patterns, obtained with three restriction endonucleases, revealed that all soybean isolates were identical to *B. japonicum* type and reference strains as well as with *B. lioaningense* type strain, while they significantly differed from *B. elkanii* and *S. fredi* type strains. After the identification at the species level, all isolates and reference strains were further characterized by RAPD analysis that enabled differentiation at the strain level. Total genomic DNAs from 17 field isolates and four reference strains were amplified using four different 10-mer primers. Amplification reactions with four randomly chosen primers generated highly specific and reproducible patterns which enabled strain differentiation. RAPD profiles revealed that all field isolates differed considerably from the reference strains. Cluster analysis showed a high degree of diversity among field isolates. None of the strains were identical or nearly identical to each other. Dendrograms derived from RAPD profiles showed that all indigenous strains could be divided into three main groups. Although dependency upon grouping and origin of isolates was not determined, the higher level of diversity was among the several strains obtained from region where soybean has been recently introduced.

GENETIC VARIATION AMONG NATURAL STRAINS OF *SINORHIZOBIUM MELILOTI*

Sheng Sun, Hong Guo and Jianping Xu
Department of Biology, McMaster University, Hamilton, On., Canada

Sinorhizobium meliloti is an economically important bacterium as it forms nodules and fixes nitrogen for alfalfa, an important agricultural crop. The complete genome of a laboratory strain Sm1021, was published in 2001 and this strain was found to have three replicons: a chromosome with 3.65 million base pairs (MB) and two megaplasmids called pSymA (1.35 MB) and pSymB (1.68 MB). In the present study, we analyzed the genetic variation among a collection of 59 natural strains of *S. meliloti* on both genome size level and gene sequence level by using pulse field gel electrophoresis and multilocus sequence typing (9 genes, 3 from each replicon). These 59 strains belong to 33 MLEE types, 27 strains for ET 1 and one strain for each of the other 32 ETs. The main results are: 1) Genome sizes among the 59 strains vary between 6.6-7.2 MB; sizes of chromosome are stable while sizes of pSymA and pSymB vary between strains; size variation in pSymA is more prominent than that in pSymB and no correlation between sizes of pSymA and pSymB were found. 2) The most frequent alleles in each of the 9 genes matched those in the published genome sequence of strain *Sm*1021; the mean sequence divergence between strains ranged from 0.0011 (aqpz1 on chromosome) to 0.0502 (exoF3 on pSymB). The between strain divergence for chromosomal genes is lower than that for genes on pSymA and pSymB.

Our results indicate that natural *S. meliloti* strains vary in both genome and replicon sizes and DNA sequences, and these two kinds of variation are independent from each other. Preliminary analysis indicated that genes on the chromosome are in linkage disequilibrium and those on the two megaplasmids are in linkage equilibrium (data not shown here), indicating the three replicons may have different patterns of evolution.

References
Eardly, B. D. et al., 1990. Applied and Environmental Microbiology 56, 187-194.
Galibert, F. et al., 2001. Science 293:668-672.

PROGRESS OF STUDIES ON RHIZOBIA ISOLATED FROM LEGUMES GROWING IN ARID-SEMIARID REGIONS OF NORTH-WESTERN CHINA

Gehong Wei[1], Minge Zhu[1], Wenxin Chen[2]
[1]Northwest A & F University, Yangling Shaanxi 712100, P.R. China; [2]China Agricultural University, Beijing 100094, P. R. China.

During the last 12 years, we have investigated rhizobia resources collected from legumes growing in Northwestern China. These regions exist droughty, alkali-saline and barren soils. Nearly 2600 strains of rhizobia were isolated from 150 legumes species belonging to 70 genera under different conditions. The rhizobial strains were catalogued in a database named CCNWAU. More than about 100 rhizobial isolates from root nodules of *Kummerowia*, *Indigofera* spp. and *Astragalus* spp. growing in the Loess Plateau of China were characterized by numerical taxonomy, RFLP and sequencing of 16S rRNA genes, DNA-DNA relatedness, and cross-nodulation with selected legume species. The studies indicated that three novel species, named *R. indigoferae*, *S. kummerowiae*, and *R. loessense*, were discovered. These results were published in the 《IJSEM》 (Wei et al., 2002; 2003). Physiological and biochemical tests, which included 130 phenotypic characteristics, were conducted on more than 700 rhizobial strains. Based on above experiments we found some strains with high nitrogen fixation activity and high resistance to extreme environmental conditions. The most beneficial strains were those that form symbiotic relationships with *Medicago sativa*, *Glycyrrhiza uralensis*, *Sophora alopecuroides*, *Astragalus membranaceus*. In addition, tests were carried out to match rhizobial strains with plant varieties and soils. These results will provide a scientific method for selecting appropriate plants and rhizobial strains in order to establish high quality pasture and rehabilitation of vegetation in Northwestern of China.

Acknowledgments: National Science Foundation of China (30470040, 30000005, 39730010) and Foundation for the Author of National Excellent Doctoral Dissertation of P.R. China (200254)

DIVERSE RHIZOBIA ISOLATED FROM LOESS PLATEAU IN CHINA AND DESCRIPTION OF *BRADYRHIZOBIUM GANSUAE* SP. NOV.

Zhi-Yuan Tan[1, 2], Jie Liu[1], En-Tao Wang[3], Gui-Xiang Peng[4], Wen-Xin Chen[1]*

[1] Dept. Microbiology, College of Biological Sci., China Agri. Uni., Beijing 100094, P. R. China, [2] Lab of Molecular Genetics, College of Agriculture, South China Agri. Uni., GuangZhou, 510642, P. R. China, [3] Dept. de Microbiologia, Instituto Politecnico Nacional, Escuela Nacional de Ciencias Biologicas, 11340 Carpio y Plan de Ayala S/N, Mexico D. F., Mexico, [4] College of Resource and Environment, South China Agri. Uni., GuangZhou, 510642, P. R. China, *Corresponding author. E-mail: wenxin_chen@263.net

Diverse rhizobia isolated from Loess Plateau in China were described as *Rhizobium yanglingense*, *R. loessense*, *R. indigoferae*, *Mesorhizobium amorphae* and *Sinorhizobium kummerowiae* before. More rhizobial isolates from different species of wild legumes at the Loess Plateau were further studied by numerical taxonomy, PCR amplified 16S rRNA gene restriction fragment length polymorphism, DNA hybridization, 16S rRNA gene sequencing, profile of sodium dodecyl sulfate-polyacrylamide gel electrophoresis (SDS-PAGE) and cross-nodulation. The results showed that a new group was formed and different from all the defined species. We described it as *Bradyrhizobium gansuae* sp. nov., CCBAU7128301 is the type strain.

Acknowledgment
This research was a part of projects no. 39730010 and 30300001supported by the National Science Foundation of China.

DIVERSE BACTERIA ISOLATED FROM ROOT NODULES OF *PHASEOLUS VULGARIS* AND SPECIES WITHIN THE GENERA *CAMPYLOTROPIS* AND *CASSIA* GROWN IN CHINA

Su Zhen Han[1,2], En Tao Wang[1,3], Wen Xin Chen[1*]
[1]Key Lab Agro-Microbial Resource & Application, Ministry Agriculture, Dept.Microbiology, College Biological Sci., China Agricultural Univ., Beijing 100094 China. [2]Dept. Micro.,College Biological Sci.,Capital Normal Univ. Beijing 100037 China. [3]Dept. Micro., Escuela Nacional Ciencias Biológicas, Instituto Politécnico Nacional,México DF. 11340 México.

Eighty bacterial isolates from root nodules of the leguminous plants *Phaseolus vulgaris*, *Campylotropis* spp. and *Cassia* spp. grown in China were classified into five groups by phenotypic analyses, SDS-PAGE of whole-cell proteins, PCR-based 16S rRNA gene RFLP and sequencing. Thirty-three isolates from the three plant genera were identified as *Agrobacterium tumefaciens* because they are closely related to the type strain of *A. tumefaciens*. Fourteen isolates from *Phaseolus vulgaris* grown in Yunnan and Inner Mongolia were classified as *R. leguminosarum* bv. phaseoli based on their close relationship with the strain in numerical taxonomy and in 16S rDNA phylogeny. Twenty-seven isolates from *Campylotropis delavayi*, *P. vulgaris* and 4 species of *Cassia* grown in the central zones of China were classified into three groups within the genus *Bradyrhizobium*. One of these, three groups could be defined as *Bradyrhizobium japonicum*. Our results demonstrated that *P. vulgaris* and the species of *Campylotropis* and *Cassia* could form nodules with diverse rhizobia in Chinese soils, including novel lineages associated with *P. vulgaris*. These results also offered information about the convergent evolution between rhizobia and legumes since the rhizobial populations associated with *P. vulgaris* in Chinese soils were completely different from those in México, the original cite of this plant. Some rhizobial species could be found in all of the three leguminous genera.

DIVERSE BACTERIA ISOLATED FROM ROOT NODULES OF LEGUME SPECIES WITHIN THE GENERA *TRIFOLIUM*, *CROTALARIA* AND *MIMOSA* GROWN IN CHINA

Xiao-Yun Liu[1,2], En-Tao Wang[1,3], Ying Li[1] and Wen-Xin Chen[1]
[1]Key laboratory of Agro-Microbial Resource & Application, Ministry of Agriculture/College of Biological Sciences, China Agricultural Univ., Beijing 100094, China, [2]Dept.Bio-conservation, Southwest Forestry College, Kunming Yunnan 650224, China, [3]Dept.Microbiología, Escuela Nacional de Ciencias Biológicas, Instituto Politécnico Nacional, México D. F., México.

Sixteen putative species or lineages related to *Bradyrhizobia*, *Rhizobium*, *Sinorhizobium*, *Mesorhizobium, Agrobacterium,* and *Burkholderia* were defined among sixty-seven bacterial isolates from the root nodules of *Trifolium*, *Crotalaria* and *Mimosa* species grown in the subtropical fields of China by analyses of restriction fragment length polymorphism (RFLP) of PCR-amplified 16S ribosomal DNA, numerical taxonomy, SDS-PAGE of whole cell proteins, sequencing of 16S rDNA, and nodulation tests. Among these bacteria, three groups including *R. leguminosarum*, *Bradyrhizobium* sp. II, and *Agrobacterium* sp. were found in all the three legume genera, while the group *Bradyrhizobium* sp. I was isolated only from *Crotalaria*. These might be a result of interaction among the bacteria, legumes and environments. The two *Burkholderia* strains from *Mimosa* and *Crotalaria* may represent novel lineages of nodule bacteria in the genus. Very diverse rhizobial populations were obtained from *Trifolium* and from *Crotalaria* species, implying that the rhizobia associated with these plants have diversified in Chinese soils. The association of *Mesorhizobium* and *Sinorhizobium* with *Trifolium* species was not reported in previous studies.

BETA-RHIZOBIAL ISOLATES FROM *MIMOSA PIGRA*, A NEWLY-DESCRIBED INVASIVE PLANT

Wen-Ming Chen[1], Euan K. James[2], Jui-Hsing Chou[1], Shih-Yi Sheu[1], Sheng-Zehn Yang[3], and Janet I. Sprent[2]
[1]Department of Seafood Science, Kaohsiung Marine University, Chinese Taiwan. [2]School of Life Sciences, University of Dundee, Dundee DD1 5EH, UK. [3]Dept. of Forestry, Pingtung University of Technology, Chinese Taiwan.

Beta-proteobacteria, particularly strains of *Burkholderia* and *Ralstonia* (*Wautersia*), can be isolated from nodules, however, only *W. taiwanensis* forms genuinely symbiotic N_2-fixing nodules with *Mimosa pudica* and *M. diplotricha* in Chinese Taiwan. This island is interesting with regard to β-rhizobia with isolates throughout the island being mostly *W. taiwanensis*, leading Chen et al. (2003a,b) to suggest that *W. taiwanensis* may actually be the "specific symbiont" of these species. However, *Mimosa* is not native to Chinese Taiwan; both *M. pudica* and *M. diplotricha* were introduced in 1645 as ornamentals, which subsequently escaped and invasively colonized all parts of the island. *M. pigra* is a more recent invasive plant and is now established in at least three colonies. 191 isolates from root nodules of *M. pigra* were taken from the three sites, and ARDRA, whole-cell protein profiles, and 16S rDNA sequences showed that 96.3% were *Burkholderia* sp. and only 3.7% were *W. taiwanensis*. Selected strains of both bacteria were tested on *M. pigra* and all formed N_2-fixing nodules, with *Burkholderia* strains being more effective. An antibody specific to *B. phymatum* was used to detect *Burkholderia* strain PAS44 in nodules on *M. pigra*, thus confirming its nodulating ability. Taken together, these data suggest that *M. pigra*, has a different "specific symbiont" from the previous, much older invasive plants, *M. pudica* and *M. diplotricha*, and that this specific symbiont (or symbionts) is/are most likely to be in the genus *Burkholderia*.

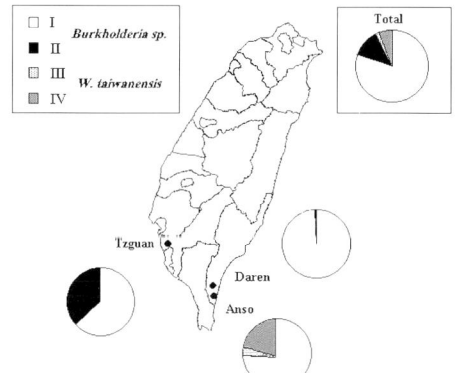

Chen et al. (2003a) Mol. Plant-Microbe Interact. 16, 1051-1061.
Chen et al. (2003b) J. Bacteriol. 185, 7266-7272.

INVESTIGATING *FRANKIA* DIVERSITY BY 16S rDNA-RFLP AND SEQUENCING

Yi Li[1], Li Zhi-zhen[2], Zhou Jun-chu[1], Zhou Qi[1] and Hu Chuan-jiong[1]
[1] State Key Lab. of Agri. Microbiol., Huazhong Agri. Univ., Wuhan, China
[2] Fujian Forest Research Institute, Fuzhou, China

Members of the genus Frankia have the capacity to form root nodules within a wide range of host actinorhizal plants. Although Frankia has been well defined as a group of quite homogeneous actinomycetes, their diversity has been increasingly demonstrated at both phenotypic and molecular level. To day there lacks a clear line of methods to discern Frankia species. The aim of this study is to describe the genetic diversity of Frankia. More attention is paid on Myrica isolates since this species is thought to be more promiscuous than others. By PCR-RFLP analysis, thirteen strains are divided into four distinct groups, namely, Coriaria, Myrica, Myrica-Casuarina-Alnus, and Casuarina-Myrica group. The Coriaria group displays a significant difference with others, whereas the Myrica shows the greatest genetic diversity. Sequence analysis of nearly complete 16S rRNA gene confirms the RFLP grouping (Fig. 1). By phylogenetic analysis, the isolates and nodular samples are clustered with four separate clusters: (I) Myrica-Casuarina-Alnus infective strains/uncultured microsymbionts; (II) Elaeagnaceae-infective strains; (III) other uncultured microsymbionts; and (IV) atypical Frankia isolates/endophytes. Members of Myrica infective strains are placed into more than two clusters showing the greatest genetic diversity. It is evident there is no connection between clusters and the natural geographic range of host species.

Lechevalier MP (1994) Int J Syst Bacteriol 44, 1–8.
Normand P et al. (1996) Int J Syst Bacteriol 46, 1–9.
Clawson ML, and Benson DR (1999) Appl Env Microbiol 65, 4521–4527.

INVESTIGATION ON MOLECULAR DIVERSITY OF 16-23S RDNA ITS IN *FRANKIA* STRAINS

Zhang Liping, Zhang Xiao, Shi Nan, Lu Zhitang, Yang Runlei, Tang Xingmei
College of Life Sciences, Hebei University, Baoding 071002, China

Because of the difficulties of isolation and cultivation, more and more researchers focus on the diversity of uncultured *Frankia* strains of the nodules. But this does little use for the protection and further application of this kind of important resource. In this study, the 16-23S rDNA ITS of five *Frankia* strains of different host plants were amplified by PCR. TGGE profile (Fig.1) of the PCR products showed the tested strains have different spacer copies. Thus the diversity of these strains was analyzed using UPMGA clustering method by Nei & Li's Coefficient. We can see that the five strains are quite heterogeneous and independent each other from the dendrogram (Fig.2). The conclusion that symbiotic *Frankia* are different in different plant hosts, and may different in the same host too. So it is presumed that the symbiotic *Frankia* may be influenced by the reasons of two sides, the host and the environments.

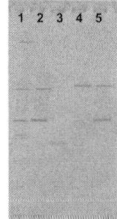 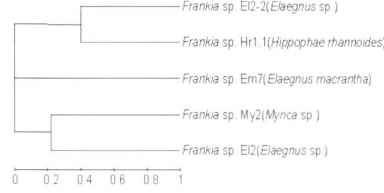

Fig.1 TGGE profile of PCR amplified 16-23S rDNA ITS

Fig.2 UPMGA dendrogram of *Frankia* strains based on 16-23S rDNA ITS PCR-TGGE

Huget V et al.(2001) Appl Environ Microbiol.67, 2116-2122.Clawson ML et al(1999) Appl Environ Microbiol. 65, 4521-4527.
*This work was supported by National Sciences Foundation of China (No.30270002) and Key Discipline Establishment Founds of Microbiology of Hebei University

PHYLOGENETIC RELATIONSHIPS BETWEEN TWO NEW *FRANKIA* ISOLATES WITH KNOWN STRAINS

Zhang Wei, Lu Zhitang, Yang Runlei, Wang Hongbin, Zhang Liping
College of Life Sciences, Hebei University, Baoding 071002, China

Actinomycetes from the genus *Frankia* are a very important group of nitrogen-fixing microorganisms. Because it's very difficult to isolate and grows quite slow, the taxonomy studies are quite lagging than other bacteria. In this study, two *Frankia* strains, strain Em4 and strain Em7, were isolated from the nodules of the same host *Elaeagnus macrantha* located at two far away sampling districts. Genomic DNAs were extract and 16S rDNAs were amplified and sequenced. From the blast and phylogeny analysis results(Fig.1), the two new isolates are identical but quite different from known strains, which have a phylogenetic distance of 1.94% to their nearest neighbor *Frankia* sp. strain M16386 (isolated from *Myrica californica*). So the two new isolates may indicate a new genospecies. The results also indicated that the same hosts might have the coincident symbiotic *Frankia* though they are grown in different places.

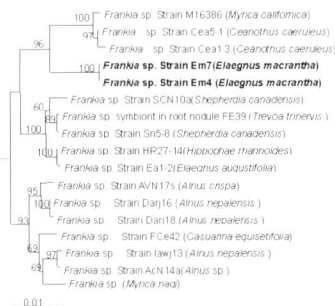

Fig.1 Neighbour-joining phylogenetic tree constructed based 16S rDNA sequences showing relationships between two *Frankia* isolates from *Elaegnus macrantha* and known strains from other actinorhizal plants. *Blastococcus aggregatus* DSM4725^T was using as outgroup.

Clawson ML et al.(2004) Mol Phylogenet. Evol. 31, 131-138. Jeong SC et al.(1999) Mol. Phylogenet. Evol. 13:493-503.
This work was supported by Chinese National Sciences Foundation (No.30270002) and Key Discipline Establishment Founds of Microbiology of Hebei University

DIVERSITY OF 16S RDNA SEQUENCES OF *RHIZOBIUM* SPP. IMPLICATIONS FOR SPECIES DETERMINATIONS

J. M. Young, D.-C. Park, B. S. Weir
Landcare Research, Private Bag 92170, Auckland, New Zealand

The genus *Rhizobium* has been considered to comprise a sub-group comprising mainly nodulating species (but including the tumorigenic *R. rhizogenes*) and another sub-group comprising mainly tumorigenic species [Farrand et al., 2003, Young et al., 2001]. *Allorhizobium* was proposed as a separate genus. Sequences for analysis are usually from type strains of species and the extent of the variability within and between species was unknown. If, for instance, the single sequence of *Rhizobium* (*Agrobacterium*) *rhizogenes* was not representative of the species because of mishandling of the strain or in sequencing, then an authentic sequence might produce a clade representing the genus *Agrobacterium*, distinct from *Rhizobium*. Such a result would invite support *Agrobacterium* as a separate genus. Clade structure is dependent on the choice of sequences and choice of analytic method. The comparison of several sequences representing each species and analysis using more than one analytic method would clarify matters [Young et al. 2004]. In this case, seventy 16S rDNA sequences representing *Rhizobium* spp. were aligned using ClustalX, and a tree obtained using Maximum Likelihood and Neighbour-joining algorithms. The results suggest that 1) The genus forms a single clade. There is no support for the differentiation of *Agrobacterium* or *Allorhizobium* as distinct genera. Sequences representing particular nodulating and tumorigenic species usually form discrete and intermingled clusters. 2) *R. galegae* and *R. huautlense* sequences have been suggested to form a separate clade [Wei et al. 2003], with the single sequence representing *R. loessense* as a further outlier. There is no support for differentiation of *R. galegae* as a novel genus [Wei et al. 2003]. 3) Sequences were received as *R. etli* AY509210 and *R. mongolense* AY509212 (from the rhizosphere of *Allysum* spp.), and *R. gallicum* AF417558, AF417561 (origins unknown) and AY509211 (from the rhizosphere of *Allysum* spp.). These appear to be unrepresentative of their species identifications and may represent a novel species. 4) Anomalously placed sequences (*R. vitis* AY626396; *R. rubi* AY626395) need further investigation. 5) Relationships indicated for some strains with sequences deposited in

GenBank cannot be investigated further because the strains have not been recorded as being deposited in publicly accessible collections [Young et al. 2004].

Acknowledgements The New Zealand Foundation for Research, Science and Technology funded this paper under contract C09X0201.

References
Amarger N et al. (1997) Int. J. Syst. Evol. Microbiol. 53, 1681–168
Wei GH et al. (2003) Int. J. Syst. Evol. Microbiol. 53, 1575–1583.
Young JM et al. (2001) Int. J. Syst. Evol. Microbiol. 51, 89–103.
Young JM et al. (2004) FEMS Microbiology Letters 238, 125-131.

CONTROLLING THE EXPRESSION OF BACTERIAL GENES DURING ROOT NODULE DEVELOPMENT

Jodie M. Box and K. Dale Noel
Department of Biology, Marquette University, Milwaukee, WI 53233 USA

We have been testing a simple strategy for triggering or repressing the expression of bacterial genes conceivably at any point in nodule development. It involves the use of IPTG (isopropyl thiogalactoside) as inducer of genes controlled by *lac* promoter/operators and the upregulated repressor gene *lacI^Q*. Most of our experiments have involved *Rhizobium etli* CE3 and its host *Phaseolus vulgaris*, but we have also demonstrated the efficacy of this induction system in the *Sinorhizobium meliloti* - alfalfa symbiosis. Apparently, IPTG is taken up by the roots and nodules of these plants in sufficiently undegraded concentrations that induction even in the most interior portions of the nodule is possible.

RESULTS

Initial tests of this method utilized transposon mTn5SSgusA10 (Wilson et al 1995) in which *gusA* is under control of *lacI^Q* and Ptac so that it serves as a reporter of gene induction by IPTG. Nod$^+$ Fix$^+$ strains carrying this transposon in *R. etli* CE3 (strain CE424) and *S. meliloti* 1021 (strain MM1) were constructed. If IPTG was added only during the first 4 days after inoculation on bean, GUS staining at 8 days was observed in the infection thread of the root hair and 10 cell layers beneath. If IPTG were present only the next four days on bean, the converse of this staining was revealed at 8 days: staining began in a strand of infection thread about 5 cell layers below the root hair and continued into the central nodule region, which was intensely stained. Alfalfa nodules stained for GUS 21 days after inoculation showed staining only in the most mature third of the nodule if IPTG had been removed after the first seven days and only in the younger portions of the nodule if IPTG were present only during the last seven days.

We used this method to ask whether *R. etli* LPS O antigen is required in nodulation at the stage of bacterial release from infection threads. This is a question that cannot be addressed with typical mutants lacking O antigen, because their infections become blocked within root hairs. First we constructed a plasmid based on pFAJ1700 (Dombrecht et al 2001) by adding an expression cassette (P^Q lacI 03 P_{lac} 01) in front of a multicloning site into which was inserted the ORF and RBS of *R. etli* CE3 *lpsG*. This plasmid (pJB4) was introduced into mutant strain CE358, which is mutated in *lpsG*. This transconjugant showed induction of O-antigen synthesis by IPTG. However, appreciable synthesis of O antigen occurred in the absence of IPTG. Therefore, additional copies of *lacI^Q* were introduced on an IncW plasmid (pJB12). This construct showed greatly reduced O-antigen synthesis in the absence of IPTG and wild-type levels in the presence of IPTG. Figure 1 shows that IPTG addition clearly controlled the extent of infection and nodule development when bean plants were inoculated with the construct carrying both plasmids.

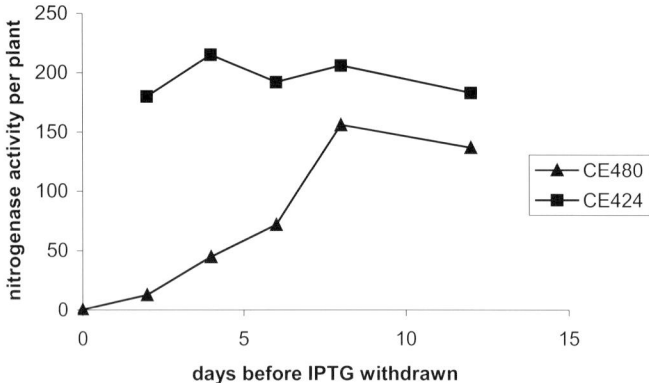

Fig. 1. Nodulated *Phaseolus vulgaris* roots were assayed for acetylene reduction 12 days after inoculation. IPTG was present in the root medium from the time of planting and adding the bacteria until the day noted on the horizontal axis. *Rhizobium etli* inoculant strains were CE424 (Nod+ Fix+ Lps+ ::mTn5SS*gusA10*) [upper curve] and CE480 (*lpsG* /pJB4 /pJB12) [lower curve].

METHODS

Plants were grown in plastic growth pouches. Where present, IPTG was added at 0.1 mM to the root nutrient and watering solutions. IPTG was removed by pouring off the root nutrient solution, two rinses of 100 ml of water (left in contact with the roots for 10 minutes each time), and adding root nutrient without IPTG.

REFERENCES

Wilson KJ et al. (1995) Microbiology 141, 1691–1705.
Dombrecht B et al. (2001) Mol. Plant-Microbe Interact. 14, 426-430.

AUTHOR INDEX

Index

A

Abe, M. 226, 237, 244, 250, 330
Abe, T. 339
Adham, S. 129
Aguilar, O.M. 199, 408
Akao, S. 321
Aktas, M. 95
Akune, M. 226
Alkama, N. 277
Alloing, G. 297
Alloisio, N. 137
Amor, B.B. 165
An, C.S. 207
Ané, J.M. 165
Aoki, T. 226
Aono, T. 383
Araíza, G. 243
Arima, Y. 187, 236
Arja, N. 317
Arrighi, J.F. 165
Asamizu, E. 123
Ashida, K. 238
Aslam, A. 319
Auguy, F. 205
Autran, D. 205
Auvinen, P. 373
Azeb, M. 326

B

Bai, X.L. 335
Banba, M. 233, 237
Bangel, E.V. 413
Barea, J.M. 13
Baron, C. 115
Barrios, E. 391
Battistoni, F. 345
Becker, A. 119, 133, 215, 248, 373
Bédu, S. 77
Belimov, A.A. 279
Berry, A.M. 137
Bi, R.C. 49
Bian, S.M. 47, 48, 49

Bilal, R. 257
Bisseling, T. 153, 165
Biswas, B. 173
Blair, M. 277
Bogusz, D. 205
Böhm, M. 345
Boiangiu, C.D. 106
Boison, G. 349
Boonkerd, N. 301, 322
Bordes, P. 59
Borisov, A.Y. 261, 279
Borthakur, D. 301
Boscari, A. 297, 328
Bothe, H. 349
Boukli, N.M. 217
Bourdes, A. 189
Bovitz, J.J. 127
Box, J.M. 429
Bres, C. 165
Brewin, N.J. 193, 225
Broughton, W.J. 201, 213, 217
Brügel, D. 106
Brusch, M. 95
Buck, M. 59, 100
Buckel, W. 106
Buhrmester, J. 119
Burrows, P.C. 59
Buzas, D.M. 173

C

Cai, Z.Q. 401
Campo, R.J. 315, 316, 413
Cannon, W.V. 59
Cao, R. 35
Cao, Z.X. 43
Capoen, W. 161
Capstick, D. 129
Carlson, R.W. 213
Carroll, B.J. 173
Chain, P. 115, 242
Chan, P.K. 173
Chang, J.S. 334
Chang, S.C. 334

Chang, Y.Q. 107, 323
Chao, T.C. 119
Charles, T. 115, 129, 140, 143, 311
Chavéz-Zamora, J. 141
Checchetka, S.A. 237
Chen, C.R. 35
Chen, D.M. 314
Chen, M. 91, 105
Chen, Q. 409, 416
Chen, S.F. 87, 323
Chen, W.F. 367, 410
Chen, W.M. 334, 391, 423
Chen, W.X. 314, 367, 409, 410, 412,
 414, 415, 419, 420, 421, 422
Chen, Y.C. 101
Cheng, H.P. 215, 248, 381
Cheng, J. 115, 242
Cheng, Q. 387
Cheng, X.G. 228
Chi, F. 379, 381, 400, 403
Choi, S.B. 405
Choisne, N. 137
Chou, J.H. 334, 391, 423
Chou, M.X. 227
Chou, Y.J. 391
Choudhury, A.T.M.A. 271
Chueire, L.M.O. 413
Ci, E. 329
Clark, S. 127
Collavino, M. 199
Consortium, G. 111
Contreras, S. 142
Cook, D. 165
Coulloux, A. 137
Cowie, A. 115
Croonenborghs, A. 265
Cruveiller, S. 137

D

Dakora, F.D. 253
Day, A. 387
Dazzo, F.B. 379, 381
de Faria, S.M. 391

Deakin, W.J. 201, 217
Debellé, F. 165
Deguchi, Y. 237, 238
Demange, N. 137
Demchenko, K. 157
den Herder, J. 161
Dénarié, J. 165, 169
Deng, Y.F. 43
Denison, R.F. 221
D'Haeze, W. 213
Dietz, K.J. 279
Dixon, R. 53, 387
Djordjevic, M. 173
Dobbelaere, S. 265
Dodsworth, J.A. 89
Dong, Z.M. 273
Dowson-Day, M. 387
Drepper, T. 95
Drevon, J.J. 277
Drew, E.A. 269, 333
Du, B.H. 241
Du, S.W. 35
Duc, G. 165
Dunn, M.F. 243
Dupont, L. 297
Durrant, M. 225

E

Edgren, T. 79
Elmerich, C. 87, 91, 105
Encarnación, S. 141, 142, 331
Equi, R. 140
Ezura, H. 332

F

Fan, B.Q. 407
Fani, R. 21
Favelukes, G. 408
Felix, S. 137
Feussner, I. 157
Finan, T.M. 115, 242, 249
Finn, R.D. 59

Flores, H. 142
Forchhammer, K. 73
Foster, R. 361
Fowler, J. 115
Franche, C. 205
Franken, C. 153
Friedman, Y. 303
Friedrich, F. 345
Fritz-Steuber, J. 106
Fujihara, S. 231
Fujikake, H. 287
Fukunaga, A. 203
Furuya, F. 330

G

Gabbarini, L.A. 209
Galera, C. 169
Gao, M. 329
Gao, M.Y. 213
Garcia, A. 277
García, J. 142
Gemmer, S. 345
Germano, M.G. 413
Geurts, R. 153, 165
Girard, M.L. 331
Gleason, C. 169
Göbe, C. 157
Golding, G.B. 115, 242
Gomaa, A.B.M. 379
Goormachtig, S. 161
Gough, C. 165, 169
Granot, G. 326
Grasso, D.H. 199
Gresshoff, P.M. 173
Gu, J. 412
Gu, J.G. 407
Gu, M.H. 309
Gualtieri, G. 173
Guan, F. 46
Gucciardo, S. 225
Guillén-Navarro, K. 243
Guo, H. 418
Guo, Y.J. 327

Gupta, V.V.S.R. 269, 333
Gyllenberg, H. 373

H

Hafeez, F.Y. 257, 295, 319,
                320, 325, 399, 402
Hakoyama, T. 236
Hamaguchi, R. 131
Hameed, S. 257, 295, 320, 402
Han, J.G. 401, 406
Han, S.Z. 421
Hans, M. 106
Har, K.J. 248
Hartmann, A. 399
Hartog, M. 153
Haselkorn, R. 65
Hata, S. 233, 237, 238
Hause, B. 157
Hayashi, M. 139, 195, 233
He, Z.S. 232
Heimer, Y.M. 326
Heinrich, A. 73
Hernandez, G. 277
Hernández, M. 142
Hérouart, D. 297
Hien, N.T. 271
Higashi, S. 237, 244, 250, 330
Hinsinger, P. 277
Hiroki, M. 239
Hiroshi, O. 239, 383
Hiyama, T. 321
Hoa, L.T.P. 131
Hocher, V. 205
Hoffmann, D. 173
Holsters, M. 161
Hori, K. 203
Hosie, A. 189
Hou, S.Y. 43
Hou, W. 324
House, B.L. 127
Hu, C.J. 424
Hu, Q.H. 416
Hu, Y.G. 314

Huang, J.F. 47, 48, 49
Huang, S. 335
Huang, X. 69
Hungria, M. 315, 316, 413
Huo, Y.X. 100, 104
Hurek, T. 345, 398
Hwang, C.H. 173
Hyun, J.J. 207

I

Iguchi, K. 238
Ikeda, J.I. 203
Iki, T. 383
Imagama, Y. 226
Imaizumi-Anraku, H. 195
Indrasumunar, A. 173
Ishizaki, N. 321
Ito, S. 287
Izui, K. 233, 237, 238

J

Jacob, A. 129
Jaillard, B. 277
Jaiswal, H.K. 289
James, E.K. 391, 423
Jayamani, E. 106
Jiang, J.Q. 241, 307
Jiang, Q. 173
Jiang, R.B. 407
Jiang, W. 97
Jiang, Y.G. 313
Jin, H. 127
Jing, Y.X. 379, 381, 400, 403
Johnson, P. 53
Jong, B.C. 404
Jung, G.H. 283

K

Kahn, M.L. 127
Kaló, P. 169
Kambara, K. 217

Kanbe, K. 321
Kaneko, T. 123, 139
Kang, B.S. 35
Kang, L.H. 313
Kanu, S. 253
Kaplan, D. 326
Kato, T. 123, 330
Kaushal, T. 287
Kawaguchi, M. 179, 233
Kawahara, M. 339
Kawasaki, S. 195
Keeskés, M.L. 271
Kennedy, I.R. 271
Khairuddin, A.R. 404
Khodorenko, A.V. 279
Kiers, E.T. 221
Kim, H.B. 207
Kim, J.H. 106
Kim, S.G. 285
Kim, W. 285
Kim, Y.C. 318
Kinghorn, S. 189
Kinkema, M. 173
Kisutin, P.Y. 279
Kloft, N. 73
Kobayashi, H. 217
Kokubun, M. 283
Kolb, A. 100, 102
Kondorosi, A. 147
Kondorosi, E. 147
Kouchi, H. 228, 236, 237
Krause, A. 345

L

Landry, T. 242
Laniya, T. 173
Laplaze, L. 205
Laurent, S. 77
Lavire, C. 137
Layzell, D.B. 240
le Quéeré, A. 217
le Rudulier, D. 297, 328
Lee, H.S. 207

Lee, H.W. 285
Lee, S.H. 207
Lehman, S. 242
Leigh, J.A. 89
Lévy, J. 165
Li, D.M. 41
Li, D.X. 173
Li, D.Y. 409, 416
Li, F.D. 7
Li, H.Q. 105
Li, J. 133
Li, J.L. 46, 87, 96, 97, 98, 107, 323
Li, P.G. 405
Li, S.G. 407
Li, S.X. 49
Li, X.F. 309
Li, Y. 97, 422
Li, Z.T. 102, 103, 104
Li, Z.Z. 424
Liang, Y. 400, 403
Liew, P.W.Y. 404
Lin, M. 91, 105
Lin, Y.J. 41
Lindblad, P. 279
Lindström, K. 373, 409, 416
Little, R. 53
Liu, H.L. 49
Liu, L. 87
Liu, J. 414, 415, 420
Liu, Q.T. 35
Liu, X.Y. 422
Liu, Z.F. 101
Liu, Z.Y. 101
Long, R.J. 325
Long, S.R. 165, 169
Lopez, A. 277
Lu, Z.T. 425, 426
Luo, L. 215, 248
Luyten, E. 265

M

Ma, H.B. 313
Ma, Q.S. 335
Ma, X.T. 407
Maćešić, D. 417
Maeda, D. 238
Maheswaran, M. 73
Mahna, S.K. 13
Mahobia, V. 13
Maillet, F. 165
Maitru, V.N. 253
Mak, L. 225
Malik, K.A. 257, 295, 319,
                                           320, 399, 402
Maloney, S.C. 127
Mandon, K. 297, 328
Mao, C.H. 140, 143
Mao, X.J. 104
Marechal, J. 137
Marie, C. 217
Martínez, G. 142
Martinez-Argudo, I. 53
Maruya, J. 233
Masepohl, B. 95
Matsunami, T. 283
McCarry, B. 115
Médigue, C. 137
Men, A. 173
Mendes, I.C. 315
Menna, P. 413
Mergaret, P. 147
Methe, B. 361
Miao, L.H. 245
Miché, L. 345
Michiels, J. 265
Minamisawa, K. 139, 332, 339
Mirabella, R. 153
Mirza, M.S. 257
Mishra, R.P.N. 289
Mitra, R. 169
Miyahara, A. 173
Momose, A. 321
Montoya, J.P. 355
Mora, J. 142, 331
Mora, Y. 142, 331
Moradas-Ferreira, P. 211
Mortimer, M.W. 127

Morton, R.A. 249
Mubeen, F. 295
Müller, P. 13
Mullin, B.C. 137
Mushtaq, N. 320

N

Nagata, M. 330
Nakamura, H. 93
Nakamura, Y. 123
Nakanishi, Y. 321
Nakatsukasa, H. 244
Nan, B.Y. 104, 246, 247
Naseem, S. 319
Naumkina, T.S. 279
Naureen, Z. 402
Neera, G. 293
Nejidat, A. 326
Newton, W.E. 25
Nicole, M. 205
Nishibayashi, Y. 39
Nishimura, K. 321
Nole, K.D. 429
Nomura, M. 131, 235
Nonaka, S. 332
Nontachaiyapoom, S. 173
Nordlund, S. 79
Normand, P. 137

O

O'Brian, M.R. 303
O'Gara, F. 246
Obertello, M. 205
Oh, C.J. 207
Ohtake, N. 287, 321
Ohwada, T. 139
Ohyama, T. 287, 321
Okazaki, S. 332
Oki, Y. 283
Oldroyd, G. 165, 169

Onda, M. 287
Ooki, Y. 233, 237
Osuki, K.I. 226
Ovtsyna, A.O. 261

P

Pak, Y.N. 318
Pan, J. 91
Pan, M. 46
Park, D.C. 427
Park, H.S. 285
Parniske, M. 183, 195
Patten, C. 115
Paulin, L. 373
Pawlowski, K. 157
Payakapong, W. 301
Pedrosa, F.O. 111
Pellock, B.J. 213
Peltzer, E. 408
Peng, G.X. 324, 420
Penmetsa, R. 165
Peret, B. 203
Perret, X. 217
Peters, J.W. 29
Ping, S.Z. 91, 105
Pitard, R.M. 391
Poggi, M.C. 328
Pohlmann, E.L. 83
Poole, P. 189
Prasad, B.N. 13
Pregelj, L. 173
Prell, J. 377
Prescott, A.R. 391
Priefer, U.B. 377
Provorov, N.A. 261
Pühler, A. 119

Q

Qi, S.W. 241
Qiu, J. 143

R

Ramakumar, A. 345
Rappas, M. 59
Räsänen, L.A. 416
Rasul, G. 257, 325
Rathbun, E.A. 193, 225
Redžepović, S. 417
Reinhold-Hurek, B. 345, 398
Remans, R. 265
Ren, F. 47, 48
Reuhs, B. 217
Reyes, A. 142, 331
Riccillo, P.M. 199
Riva, O. 408
Rizk, R. 379
Roberts, G.P. 83
Rodino, P. 277
Roget, D.K. 269, 333
Rosenberg, C. 165
Roughley, R.J. 271
Ruan, Z.Y. 407
Rüberg, S. 119, 133, 215, 373
Ruppert, U. 73
Ryang, H.G. 405
Rybak, A. 249

S

Saad, M. 217
Saeki, K. 139, 233
Safronova, V.I. 279
Saito, A. 339
Salazar, E. 141, 142
Saldanha, M. 311
Sarita, S. 377
Sarkar, A. 345
Sasakura, F. 250
Sato, A. 321
Sato, S. 123, 233, 330
Sato, T. 321, 339
Schmid, M. 399
Schneider, K. 95
Schroeder, B.K. 127
Schubert, B. 95
Schumacher, J. 59
Schumpp, O. 201, 217
Scott, P. 173
Sen, S. 29
Shang, J.H. 313
Sharma, P.K. 377
Shaw, S. 165
Shearer, N. 53
Shelswell, K. 311
Shen, S.H. 379, 381, 400, 403
Sheng, X.Y. 140, 311
Sheu, S.Y. 423
Shi, N. 425
Shi, Y.M. 69
Shimoda, Y. 237, 330
Sibley, C. 115, 242
Sicking, C. 95
Sikora, S. 417
Simões-Araujo, J.L. 391
Singh, R.K. 289
Singla, R. 293
Skorpil, P. 217
Snocck, C. 265
Sobral, B.W. 140, 143, 311
Somers, E. 265
Song, W. 395, 401, 406
Sooksa-nguan, T. 322
Sprent, J.I. 391, 423
Squartini, A. 379
Staehelin, C. 201, 213, 217
Straliotto, R. 391
Streit, W.R. 213, 217
Su, X.L. 401
Sueyoshi, K. 287, 321
Suga, Y. 203
Sugawara, M. 332
Sui, X.H. 314
Summers, P. 115
Sun, L. 401, 406
Sun, S. 418
Sun, Y.C. 102
Suzuki, A. 226, 237, 244, 250, 330
Svistoonoff, S. 205
Sy, M.O. 205

T

Tabata, S. 123, 233, 330
Tajima, S. 131, 139, 228, 235
Takahashi, Y. 287
Takeda, N. 195
Takekuma, S.I. 39
Takenouchi, K. 250
Tan, Z.Y. 324, 420
Tang, M.J. 400
Tang, X.L. 335
Tang, X.M. 425
Tavares, F. 211
Taylor, C.A. 127
Teaumroong, N. 301, 322
Terakado, J. 231
Terefework, Z. 373
Tewari, K. 321
Thierfelder, H. 13
Thies, E.J. 322
Thu, M.H. 235
Tian, Z.X. 99, 100, 102, 103, 133
Tikhonovich, I.A. 261, 279
Tittabutr, P. 301
Tomkins, J. 137
Toshihiro, A. 239
Trinchant, J.C. 297
Truong Cong, Y.C. 137
Tsai, K.R. 43
Tsyganov, V.E. 261, 279

U

Uchiumi, T. 139, 147, 226,
237, 244, 250, 330
Uemura, S. 39

V

Vallenet, D. 137
Vanderleyden, J. 265
Vargas, M.C. 142, 331
Vermöhlen, S. 95

Vgenopoulou, I. 106
Vieira, J. 211
Vinardell, J.M. 147
Vinuesa, P. 13
Vujić, A.S. 417

W

Walker, G.C. 213
Wall, L.G. 209
Wan, H.L. 43
Wang, C.X. 140, 143, 311
Wang, E.T. ..367, 410, 415, 420, 421, 422
Wang, F.Q. 414
Wang, G.Y. 91
Wang, H.B. 426
Wang, H.P. 47, 48
Wang, H.R. 324
Wang, L. 241, 307
Wang, L.Y. 105
Wang, Y.Z. 229, 230
Wang, Y.P. 99, 100, 101, 102,
103, 104, 133, 246, 247
Wang, Z.P. 49
Ward, K.L. 127
Wei, G.H. 419
Wei, H. 240
Wei, W. 307
Weidner, S. 119, 373
Weir, B.S. 427
Wells, T. 173
Wen, B.Q. 247
Wen, J. 100, 246, 247
Weretilnyk, E. 115
Werner, D. 13
White, J. 189
Wiethaus, J. 95
Wigneshweraraj, S.R. 59
Wisniewski, J.P. 225
Wu, B. 335
Wu, C.H. 334
Wu, X.T. 3

X

Xie, B. 234
Xie, F.L. 245
Xie, L.S. 307
Xie, R. 232
Xie, Z.H. 91
Xing, Y.H. 41
Xu, J. 115
Xu, J.P. 418
Xu, J.Q. 41

Y

Yamaya, H. 187
Yamazaki, A. 287
Yang, E.C. 101
Yang, G.H. 69
Yang, J. 387
Yang, J.K. 411
Yang, J.H. 303
Yang, M. 309
Yang, R.L. 425, 426
Yang, S.S. 241, 307
Yang, S.Z. 423
Yang, Y. 91
Yanni, Y.G. 379
Yano, K. 233
Yao, S.Y. 215, 248
Yao, T. 325
Yao, T.Y. 323
Yasmin, S. 320, 325, 399, 402
Ye, B. 339
Ye, Z.G. 327
Yi, I. 424
Yokoyama, T. 139, 183, 236
Yoneyama, T. 231
Yoshida, S. 183
Yoshida, Z.I. 39
You, M. 339
Young, J.M. 427
Young, J.P.W. 391
Yu, G.Q. 133, 215, 219,
                                   229, 230, 232, 248
Yu, J.X. 323
Yu, Y.X. 309, 329

Yuan, Z.C. 115, 249
Yue, W. 247
Yukihiro, M. 239
Yurgel, S. 127

Z

Zaheer, R. 115, 249
Zdyb, A. 157
Zehr, J.P. 361
Zeng, Z.H. 314
Zhang, C.C. 77
Zhang, D.G. 325
Zhang, G.X. 324
Zhang, L. 345, 398
Zhang, L.P. 425, 426
Zhang, W. 426
Zhang, W.H. 283
Zhang, X. 425
Zhang, X.D. 59
Zhang, X.P. 409, 416
Zhang, X.X. 407
Zhang, Y. 69
Zhang, Y.F. 414
Zhang, Y.P. 83, 96, 98
Zhang, Y.T. 102, 103, 133
Zhang, Z.H. 102
Zhao, D.H. 46
Zhao, J.D. 69
Zhao, J.F. 47, 48, 49
Zhao, Y. 47, 48, 49
Zhao, Y.H. 69
Zhernakov, A.I. 279
Zhou, H.N. 47, 48, 49
Zhou, J.C. 227, 234, 245, 411, 424
Zhou, K. 245
Zhou, Q. 424
Zhou, X.H. 41
Zhou, X.Y. 87, 98, 107
Zhou, Z.H. 43
Zhu, A.N. 327
Zhu, B.C. 401
Zhu, J.B. 215, 219, 229, 230, 232, 248
Zhu, M. 419
Zhu, R.Y. 96

Zidan, M. 379
Ziemkiewicz, H.T. 127
Zou, H.S. 133, 219

Current Plant Science and Biotechnology in Agriculture

1. H.J. Evans, P.J. Bottomley and W.E. Newton (eds.): *Nitrogen Fixation Research Progress.* Proceedings of the 6th International Symposium on Nitrogen Fixation (Corvallis, Oregon, 1985). 1985　　　　　　　　　　ISBN 90-247-3255-7
2. R.H. Zimmerman, R.J. Griesbach, F.A. Hammerschlag and R.H. Lawson (eds.): *Tissue Culture as a Plant Production System for Horticultural Crops.* Proceedings of a Conference (Beltsville, Maryland, 1985). 1986　　　　ISBN 90-247-3378-2
3. D.P.S. Verma and N. Brisson (eds.): *Molecular Genetics of Plant-microbe Interactions.* Proceedings of the 3rd International Symposium on this subject (Montréal, Québec, 1986). 1987　　　　　　　　　　　　　　　　ISBN 90-247-3426-6
4. E.L. Civerolo, A. Collmer, R.E. Davis and A.G. Gillaspie (eds.): *Plant Pathogenic Bacteria.* Proceedings of the 6th International Conference on this subject (College Park, Maryland, 1985). 1987　　　　　　　　　　　　ISBN 90-247-3476-2
5. R.J. Summerfield (ed.): *World Crops: Cool Season Food Legumes.* A Global Perspective of the Problems and Prospects for Crop Improvement in Pea, Lentil, Faba Bean and Chickpea. Proceedings of the International Food Legume Research Conference (Spokane, Washington, 1986). 1988　　　　　　　　ISBN 90-247-3641-2
6. P. Gepts (ed.): *Genetic Resources of* Phaseolus *Beans.* Their Maintenance, Domestication, Evolution, and Utilization. 1988　　　　　　　　ISBN 90-247-3685-4
7. K.J. Puite, J.J.M. Dons, H.J. Huizing, A.J. Kool, M. Koorneef and F.A. Krens (eds.): *Progress in Plant Protoplast Research.* Proceedings of the 7th International Protoplast Symposium (Wageningen, The Netherlands, 1987). 1988
　　　　　　　　　　　　　　　　　　　　　　　　　　ISBN 90-247-3688-9
8. R.S. Sangwan and B.S. Sangwan-Norreel (eds.): *The Impact of Biotechnology in Agriculture.* Proceedings of the International Conference The Meeting Point between Fundamental and Applied in vitro Culture Research (Amiens, France, 1989). 1990.
　　　　　　　　　　　　　　　　　　　　　　　　　　ISBN 0-7923-0741-0
9. H.J.J. Nijkamp, L.H.W. van der Plas and J. van Aartrijk (eds.): *Progress in Plant Cellular and Molecular Biology.* Proceedings of the 8th International Congress on Plant Tissue and Cell Culture (Amsterdam, The Netherlands, 1990). 1990
　　　　　　　　　　　　　　　　　　　　　　　　　　ISBN 0-7923-0873-5
10. H. Hennecke and D.P.S. Verma (eds.): *Advances in Molecular Genetics of Plant–Microbe Interactions.* Volume 1. 1991　　　　　　　ISBN 0-7923-1082-9
11. J. Harding, F. Singh and J.N.M. Mol (eds.): *Genetics and Breeding of Ornamental Species.* 1991　　　　　　　　　　　　　　　　ISBN 0-7923-1094-2
12. J. Prakash and R.L.M. Pierik (eds.): *Horticulture – New Technologies and Applications.* Proceedings of the International Seminar on New Frontiers in Horticulture (Bangalore, India, 1990). 1991　　　　　　　　　　ISBN 0-7923-1279-1
13. C.M. Karssen, L.C. van Loon and D. Vreugdenhil (eds.): *Progress in Plant Growth Regulation.* Proceedings of the 14th International Conference on Plant Growth Substances (Amsterdam, The Netherlands, 1991). 1992　　ISBN 0-7923-1617-7
14. E.W. Nester and D.P.S. Verma (eds.): *Advances in Molecular Genetics of Plant–Microbe Interactions.* Volume 2. 1993　　　　　　　ISBN 0-7923-2045-X

Current Plant Science and Biotechnology in Agriculture

15. C.B. You, Z.L. Chen and Y. Ding (eds.): *Biotechnology in Agriculture*. Proceedings of the First Asia-Pacific Conference on Agricultural Biotechnology (Beijing, China, 1992). 1993   ISBN 0-7923-2168-5
16. J.C. Pech, A. Latché and C. Balagué (eds.): *Cellular and Molecular Aspects of the Plant Hormone Ethylene*. 1993   ISBN 0-7923-2169-3
17. R. Palacios, J. Mora and W.E. Newton (eds.): *New Horizons in Nitrogen Fixation*. Proceedings of the 9th International Congress on Nitrogen Fixation (Cancún, Mexico, 1992). 1993   ISBN 0-7923-2207-X
18. Th. Jacobs and J.E. Parlevliet (eds.): *Durability of Disease Resistance*. 1993
    ISBN 0-7923-2314-9
19. F.J. Muehlbauer and W.J. Kaiser (eds.): *Expanding the Production and Use of Cool Season Food Legumes*. A Global Perspective of Persistent Constraints and of Opportunities and Strategies for Further Increasing the Productivity and Use of Pea, Lentil, Faba Bean, Chickpea, and Grasspea in Different Farming Systems. Proceedings of the Second International Food Legume Research Conference (Cairo, Egypt, 1992). 1994   ISBN 0-7923-2535-4
20. T.A. Thorpe (ed.): In Vitro *Embryogenesis in Plants*. 1995   ISBN 0-7923-3149-4
21. M.J. Daniels, J.A. Downie and A.E. Osbourn (eds.): *Advances in Molecular Genetics of Plant-Microbe Interactions*. Volume 3. 1994   ISBN 0-7923-3207-5
22. M. Terzi, R. Cella and A. Falavigna (eds.): *Current Issues in Plant Molecular and Cellular Biology*. Proceedings of the VIIIth International Congress on Plant Tissue and Cell Culture (Florence, Italy, 1994). 1995   ISBN 0-7923-3322-5
23. S.M. Jain, S.K. Sopory and R.E. Veilleux (eds.): In Vitro *Haploid Production in Higher Plants*. Volume 1: Fundamental Aspects and Methods. 1996
    ISBN 0-7923-3577-5
24. S.M. Jain, S.K. Sopory and R.E. Veilleux (eds.): In Vitro *Haploid Production in Higher Plants*. Volume 2: Applications. 1996   ISBN 0-7923-3578-3
25. S.M. Jain, S.K. Sopory and R.E. Veilleux (eds.): In Vitro *Haploid Production in Higher Plants*. Volume 3: Important Selected Plants. 1996   ISBN 0-7923-3579-1
26. S.M. Jain, S.K. Sopory and R.E. Veilleux (eds.): In Vitro *Haploid Production in Higher Plants*. Volume 4: Cereals. 1996   ISBN 0-7923-3978-9
27. I.A. Tikhonovich, N.A. Provorov, V.I. Romanov and W.E. Newton (eds.): *Nitrogen Fixation: Fundamentals and Applications*. 1995   ISBN 0-7923-3707-7
28. N.L. Taylor and K.H. Quesenberry: *Red Clover Science*. 1996 ISBN 0-7923-3887-1
29. S.M. Jain, S.K. Sopory and R.E. Veilleux (eds.): In Vitro *Haploid Production in Higher Plants*. Volume 5: Oil, Ornamental and Miscellaneous Plants. 1996
    ISBN 0-7923-3979-7
30. R.H. Ellis, M. Black, A.J. Murdoch and T.D. Hong (eds.): *Basic and Applied Aspects of Seed Biology*. 1997   ISBN 0-7923-4363-8
31. C. Elmerich, A. Kondorosi and W.E. Newton (eds.): *Biological Nitrogen Fixation for the 21st Century*. 1998   ISBN 0-7923-4834-6

Current Plant Science and Biotechnology in Agriculture

32. S.M. Jain, D.S. Brar and B.S. Ahloowlia (eds.): *Somaclonal Variation and Induced Mutations in Crop Improvement.* 1998 ISBN 0-7923-4862-1
33. S.J. Bennett and P.S. Cocks (eds.): *Genetic Resources of Mediterranean Pasture and Forage Legumes.* 1999 ISBN 0-7923-5522-9
34. R. Knight (ed.): *Linking Research and Marketing Opportunities for Pulses in the 21st Century.* Proceedings of the Third International Food Legumes Research Conference. 1999 ISBN 0-7923-5565-2
35. C. Oropeza, J.L. Verdeil, G.R. Ashburner, R. Cardeña and J.M. Santamaria (eds.): *Current Advances in Coconut Biotechnology.* 1999 ISBN 0-7923-5823-6
36. A. Altman, M. Ziv and S. Izhar (eds.): *Plant Biotechnology and In Vitro Biology in the 21st Century.* 1999 ISBN 0-7923-5826-0
37. E.C. Lefroy, R.J. Hobbs, M.H. O'Conner and J.S. Pate (eds.): *Agriculture as a Mimic of Natural Ecosystems.* 1999 ISBN 0-7923-5965-8
38. F.O. Pedrosa, M. Hungria, M.G. Yates and W.E. Newton (eds.): *Nitrogen Fixation: From Molecules to Crop Productivity.* Proceedings of the 12th International Congress on Nitrogen Fixation. 2000 ISBN 0-7923-6233-0
39. N. Maxted and S.J. Bennett (eds.): *Plant Genetic Resources of Legumes in the Mediterranean.* 2000 ISBN 0-7923-6707-3
40. M. Unkovich, J. Pate, A. McNeill, D.J. Gibbs (eds.): *Stable Isotope Techniques in the Study of Biological Processes and Functioning of Ecosystems.* 2001
    ISBN 0-7923-7078-3
41. Y.-P. Wang, M. Lin, Z.-X. Tian, C. Elmerich and W.E. Newton (eds.): *Biological Nitrogen Fixation, Sustainable Agriculture and the Environment.* Proceedings of the 14th International Nitrogen Fixation Congress. 2005 ISBN 1-4020-3569-1